A City on Mars

Also by Kelly and Zach Weinersmith

Soonish: Ten Emerging Technologies That'll Improve and/or Ruin Everything

A City on Mars

Can We Settle Space, Should We Settle Space, and Have We Really Thought This Through?

KELLY AND ZACH WEINERSMITH

PENGUIN PRESS
New York
2023

PENGUIN PRESS
An imprint of Penguin Random House LLC
penguinrandomhouse.com

LIBRARY OF CONGRESS CONTROL NUMBER: 2022951665

ISBN 9781984881724 (hardcover)
ISBN 9781984881731 (ebook)

Printed in the United States of America
1st Printing

Set in Adobe Caslon Pro
Designed by Cassandra Garruzzo Mueller

To the space-settlement community.
You welcomed us and you shared
your wisdom. Also, your data. We worry
that many of you will be disappointed by
some of our conclusions, but where we
have diverged from your views, we haven't
diverged from your vision of a
glorious human future.

Contents

PART III • POCKET EDENS: HOW TO CREATE A HUMAN TERRARIUM THAT ISN'T ALL THAT TERRIBLE

PART IV • SPACE LAW FOR SPACE SETTLEMENTS: WEIRD, VAGUE, AND HARD TO CHANGE

PART V • THE PATHS FORWARD: BOUND FOR MOONSYLVANIA?

Introduction

A Homesteader's Guide to the Red Planet?

> It is no longer a question of *if* we will colonise the Moon and Mars, but *when*.
>
> —Tim Peake, astronaut

Wherever you are on this planet, you've recently given some thought to leaving it. Space is looking more promising every day. There's no political corruption on Mars, no war on the Moon, no juvenile jokes on Uranus. Surely space settlement presents the best chance since about 50,000 BC to try out something completely new and leave all the bad stuff behind. After five decades of stagnation in human spacefaring, we now have the technology, the capital, and the desire to go beyond the age of quick forays to the Moon and seize our destiny as a multiplanetary species.

Well . . . maybe not. If you're like most of the nonexperts we've talked to as we researched this book, you might have some ideas about space settlement that aren't quite right. We don't blame you—the public discourse around space settlement is full of myths, fantasies, and outright misunderstanding of basic facts.

In 2020, for example, SpaceX's internet service provider, Starlink, released a Terms of Service agreement that declared that "no Earth-based government has authority or sovereignty over Martian activities." This clause is like many statements about outer space settlement: it was promoted by a powerful advocate, widely shared and commented upon, and profoundly misleading. Earth-based governments *do* have authority over Mars activities—Mars is regulated by long-standing treaties and is an international commons. Admittedly, the treaties are weird and vague, but they do exist and can't be de-existed via a Terms of Service agreement.

Not all the bad space-settlement discourse comes from rocket billionaires. Consider the 2015 *Newsweek* article "'Star Wars' Class Wars: Is Mars the Escape Hatch for the 1 Percent?" which claims "the red planet will likely only be for the rich, leaving the poor to suffer as earth's environment collapses and conflict breaks out." The only way you could believe this would be if you had no idea how thoroughly, incredibly, impossibly horrible Mars is. The average surface temperature is about -60°C. There's no breathable air, but there *are* planetwide dust storms and a layer of toxic dust on the ground. Leaving a 2°C warmer Earth for Mars would be like leaving a messy room so you can live in a toxic waste dump.

The truth is that settling other worlds, in the sense of creating self-sustaining societies somewhere away from Earth, is not only quite unlikely anytime soon, it won't deliver on the benefits touted by advocates. No vast riches, no new independent nations, no second home for humanity, not even a safety bunker for ultra elites.

Yet we find ourselves in a world where space agencies, huge corporations, and media-savvy billionaires are promising something else. According to them, settlements are coming, perhaps as soon as 2050 or so. When they are built, they will fix just about everything. They will save Earth's biosphere or enable a wildly creative frontier civilization or provide huge economic advantages for the United States or China or India or whoever else makes the first big move.

While we believe all these claims are false, they are buoyed by genuinely game-changing technological developments that have made accessing space much cheaper. In the next decade, it will almost certainly be easier to build outposts in space than ever before. The problem for any would-be settler is that most of the problems, especially those pertaining to things like biology and economics, are far more complex than making bigger rockets or cheaper spacecraft. As we'll see, ignoring these problems while trying to force a near-term settlement is a recipe for social calamity and potential danger to the home planet.

Meanwhile, the international legal structures that govern space have barely been updated since the 1970s. Space law is often vague, ambiguous, and if you accept the interpretation favored by the United States, highly permissive. In the modern world of fast-growing space capitalism and an ever-increasing number of countries with launch capability, we have the makings of a new Moon Race. But racing in the 2020s or 2030s will be very different from racing in the 1960s, in that it will likely involve attempts to gain priority access to the highly limited best portions of the Moon. In terms of the risk of conflict, it's much less like two kids seeing who can run the fastest and much more like a growing group of kids scrapping over a small pile of candy.

That's dangerous. If we convince you that there's no clear return on investment here, then it's needlessly dangerous. Oh, and actually let's ruin the metaphor here a little and make it so the kids also have nuclear weapons.

So. Space settlements. Have we really thought this through?

If humanity survives the next few centuries, it's probable we'll expand into space. People, nations, and the international community have options about how to proceed. The choices we make now—about the pace of expansion and the rules underpinning it—will shape that future in ways we can't yet imagine. The wrong choices wouldn't merely slow us down, they might create existential risk for humanity.

We can't make these choices properly unless people actually know what the truth is about space settlement. All of it. Not just the size of the rocket or the power needs of a settlement or the available minerals in asteroids, but the big, open questions about things like medicine, reproduction, law, ecology, economics, sociology, and warfare. Detailed treatments that are honest about the severe difficulty of these things are almost invariably left out of books and documentaries about space settlement.

Why is this discourse so often bad? We believe there are two major reasons. First, the general public knows very little about space. Most people can name exactly one astronaut, and with an appropriate mnemonic can say the planets in order. Outside of a few weirdos, most of us don't know things like what lunar soil is made of, or what the Outer Space Treaty says, or the history of nuclear weapon detonation in space.

Given the limited public knowledge of space science in general, knowledge of its weird little cousin—space-*settlement* science—is almost nonexistent. And that's where we arrive at the second problem. If you are ignorant about space settlement and want to become educated, many of the articles you'll read, many of the documentaries you'll watch, and pretty much every single book on the topic have been created by an *advocate for space settlement.*

Now look, there's nothing wrong with advocacy. The space-settlement geeks we've met are smart, thoughtful people. Most of them, anyway. But reading about space settlement today is kind of like reading about what quantity of beer is safe to drink in a world where all the relevant books are written by breweries. Even when they're trying to be evenhanded, they leave things out. One of the most prominent books on space settlement, *The Case for Mars*, is over 400 pages long, including obscure historical information on Mars conferences of the 1980s as well as detailed chemical equations for plastic production at the Martian surface, but never once mentions the existence of international space law. Of the five decades of legal precedent that will dictate the political nature and geopolitical consequences of any Martian future, not a word.

The little book you're reading right now, which admittedly begins with a Uranus joke and contains an explainer on space cannibalism (stay tuned), is nevertheless the only popular science book we're aware of that offers the whole picture without trying to sell you on the idea of near-term space expansion.* Rather, we'll try to clear up a lot of misconceptions and then replace them with a much more realistic view of how feasible space settlements are and what they might mean for humanity.

But first, we should introduce ourselves. Hi. We're Kelly and Zach Weinersmith. Kelly is a biologist and Zach is a cartoonist. We're also a wife-and-husband research team who've spent the last four years trying to understand how humans will become space settlers. We've gone to conferences, conducted endless interviews, and collected, at last count, twenty-seven shelves of books and papers on space settlement and related subjects. We are space geeks. We love rocket launches and zero gravity antics. We love space history's strange corners like red cubes and tampon bandoliers. We love visionary plans for a glorious future. We are also very skeptical people. If you want to visualize us, imagine John F. Kennedy giving a beautiful, uplifting speech on sailing "this new ocean," and then notice in the background two people squinting at the middle distance, thinking "but is it *really* like an ocean?"

After a few years of researching space settlements, we began in secret to refer to ourselves as the "space bastards" because we found we were more pessimistic than almost everyone in the space-settlement field, and especially skeptical about the most grand plans of space geeks. We weren't always this way. The data made us do it. Frankly, we are cowards and would very much like to agree with the consensus. We didn't like being this

*That said, there is a longstanding critical literature with a growing number of recent entries, such as *Space Forces* by Fred Scharmen and *Off-Earth* by Erika Nesvold.

pessimistic, especially about an endeavor that so many people think embodies the best of human nature. It makes one feel like, well, a bastard.

We think space settlement is possible, but the discourse needs more realism—not in order to ruin everyone's fun, but to provide guardrails against genuinely dangerous directions for planet Earth.

How We Became Space Bastards . . . and You Can Too!

If you're new to the study of space, you may not be aware of the scale of the revolution in the cost of access, and in the space business generally, that has been ongoing since the mid 2010s.

Most of us have some sense that the 1950s and 1960s were awash in glorious space promises: Moon bases, orbital vacations, Martian pioneers, and especially if we're talking late-'60s space books, weird low-gravity erotic possibilities. All this gave way to the shag-carpeted misery of the '70s and forty years of moderate and decidedly chaste human presence in space. This failure is sometimes blamed on a loss of imagination or ambition, but a pretty simple explanation is cost. Changes in the price of launch explain both the wild dreams of the early post–Moon landing era and the forty years of disappointment. If you look at just the period from the first orbit in 1957 to the end of the 1960s, the price of putting something in orbit fell by around 90–99 percent. If each subsequent decade did likewise, sending a package to space would now be cheaper than sending international mail. This is why if you want to find truly extravagant space-settlement proposals, the groovy years are when all the best books got published.

Sadly for many a geeky heart, the prices stopped falling around the early 1970s, and the Space Shuttle, which was supposed to make travel routine, cheap, and safe, failed on all three fronts, remaining, by one estimate, the costliest way to put mass in orbit for decades. That was the state of play until the 2010s when, largely as a result of a US policy shift and SpaceX in particular, the cost of putting stuff in space began to fall dramatically again.

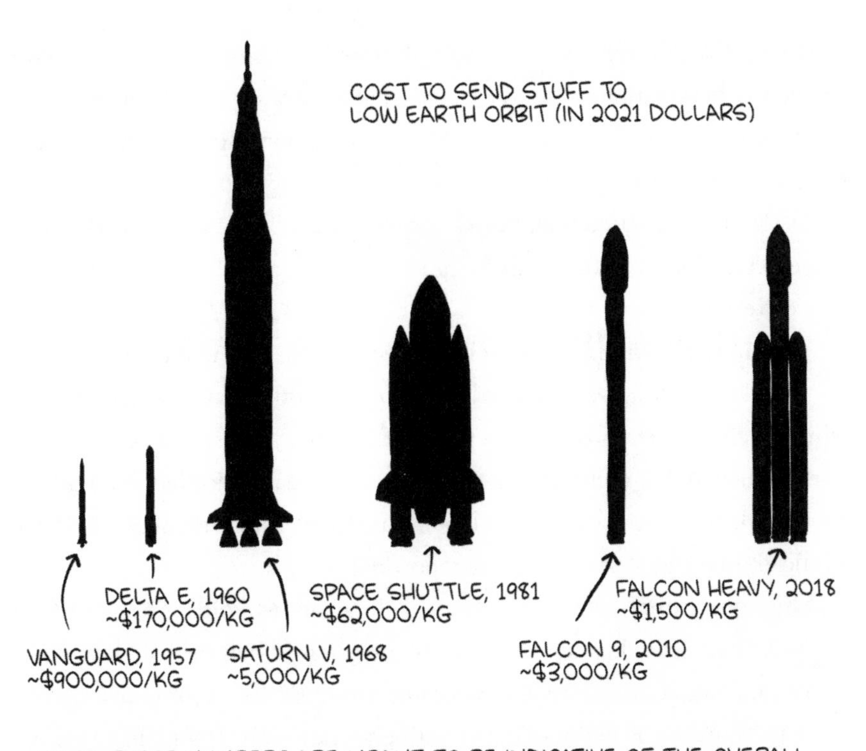

This doesn't just mean more rocket launches, it means more spacecraft. In 2015, there were about fourteen hundred active satellites. As of 2021, there were about five thousand; and as of October 2022, around three thousand working satellites are controlled by SpaceX's satellite internet service, Starlink.

Space tourism, long promised but rarely delivered on, appears to actually be happening. Jeff Bezos's rocket company Blue Origin regularly sends people on 100-kilometer-high hops, and SpaceX has contracted to send tourists around the Moon. Where once there were only a few government agencies doing space launch, there is now a growing number of private entities competing on cost. Meanwhile, humanity's appetite for high-speed data everywhere all the time continues to boom, with one paper finding people in the United States interact with satellites an average of thirty-six times a day. Estimates vary, but investor prospectuses put out by financial organizations tend to agree that the overall space business will be worth at least a trillion dollars by 2040 or so, assuming no huge uptick in the pace of growth.

In short: the hype is real. Being concerned about the laws pertaining to space expansion in 2005 would've been very premature. But 2025 is apt to be a different story.

Watching this trend was a weird experience for us. As it began to ramp up, we were writing a book called *Soonish* about futuristic technology, which included sections on the effects of cheaper access to outer space. In late 2015, reusable rockets, one of the keys to cheap space launch, had become real. By the time the book landed in stores, they were routine. What would humanity do with these new powers?

One clue came from research we'd done on asteroid mining, the attempt to harvest valuable matter from the asteroid belt or near-Earth objects. Our analysis was that harvesting asteroids for commodities to be used on Earth was economically unlikely and, well, try to imagine explaining to a hadrosaur about your plan to hurl heavy space objects toward Earth for processing.

But if you are starting a settlement in space, asteroids are quite interesting indeed. The asteroid belt contains over 2 *sextillion* kilograms worth of stuff: metals, carbon, oxygen, water, all of it already boosted away from Earth and ready for use. With the new rocket technology and huge sums of money pouring into the business, effectively you had a way to get to the space frontier plus homebuilding materials waiting on-site.

Even the legal picture for space settlement appeared to be improving. While there was debate about whether or not the existing international space treaties allowed resource extraction for profit, in 2015, the United States passed a law specifically codifying the idea that Americans *can* exploit space resources without limit. And at least Luxembourg seemed to agree, passing a similar law and dumping a ton of money into two US-based asteroid mining companies. Space access was getting easier, resources in space were plentiful, countries were starting to give the green light for developers to go nuts, and the guy in charge of the biggest rocket company was Elon Musk—a dynamic tech geek whose stated goal was Mars settlement in his lifetime.

Okay, sure the path to space *settlements*, as opposed to space hotels or research bases, was a little hazier, but then again there was *so much* money going into the design of rockets, spacecraft, and even some life-support technology. At the very least, space settlement was coming closer. The dreams of the 1950s seemed like they might finally manifest by the 2050s.

We wanted to contribute. We saw space settlement as a near-term possibility and intended to write a sort of sociological road map—how to scale to one hundred, one thousand, ten thousand people, and beyond. A little guide to explain to the public what comes next. But we also had a few nagging concerns. Things we didn't understand, like how to design the legal regime to make it safe to live in a solar system where dozens of nations, corporations, and possibly single individuals can sling dinosaur-annihilation-size objects at the homeworld. A clear protocol would be nice. What we found was that, with just a few exceptions, concerns of this sort were ignored, sometimes even treated with hostility by space-settlement advocates.

As we dug in, our stack of concerns got bigger and bigger. How does democracy function in a society where air is rationed—and possibly under corporate control? How does sociology change if humans can't reproduce unless they're in Earth-normal gravity? How do we avoid a scramble for territory if some regions of space are better than others? Incidentally, what is the actual space law today, how did it get that way, and is it likely to change? These questions seemed very basic to space settlement, and frankly really interesting, but were typically skipped over as things that would just get worked out as the rockets got bigger. So the book became less about explaining the deal on future settlements and more about getting to the bottom of unexplored questions, the pursuit of which led us to some weird places.

We read about caves on the Moon, uncomfortably detailed orbital mating concepts, space madness, Moon law, plans for Martian company towns, hopes for new ways of life in distant worlds. We read dozens of old space books going back to the 1920s, many of them predicting imminent space settlements. We talked to experts in the economic and political fields who had little interest in space, but also to space advocates and space entrepreneurs. Friends, we are practically bursting with weird space knowledge. Did you know the Colombian constitution asserts a claim to a specific region of space? Did you know the first woman to step foot in a space station was "gifted" an apron and asked if she'd handle cooking and cleaning for the rest of her mission? Did you know an early space life-support concept involved a substance that could double as shelving and as breakfast? Did you know former US Republican Party presidential nominee Barry Goldwater once advocated sending bull semen to orbit to separate sperm for sex-selection purposes?

While we fell in love with space settlement as a field of study, we became more concerned about all the proposals for doing it in the coming decades. It turns out when you just talk about technical things like the size of rockets, or whether Mars has water and carbon, the picture can look pretty solid. When you get into the more squishy details of human existence, things start to look, well, squishy.

Especially squishy, for example, are space babies. Can we make them? Proposals for settlements often just assume you can safely have natural population growth. We don't know if this is true, and there are good reasons to suppose it isn't. A start-up called SpaceLife Origin announced in 2018 their goal of the first human birth in space by 2024. In 2019, their

CEO left, citing "serious ethical, safety, and medical concerns." That's exactly right. Out of all the NASA astronauts, only five have spent nine consecutive months in space, only two of those five have been women, and none of them had to do it while being a fetus. As for the person around the fetus, they might have concerns too. Moms on Earth worry about things like eating sushi or having a beer. Try 1 percent bone loss per month while doing several hours of resistance training every day in a high-radiation, high-carbon-dioxide atmosphere without Earth-normal gravity. It's certainly possible everything will be fine, but we wouldn't want to bet on it. Given that population growth requires babies not just to be born, but to grow up to have their own babies, getting appropriate safety protocols would take decades, even if we unethically began doing experiments on humans starting tomorrow. But we aren't. The current state of the art is short, unsystematic experiments in orbit, like the one where geckos were sent up for some highly documented together time, before the experiment failed and everyone froze to death. *C'est la vie dans l'espace.*

Elon Musk says we'll have boots on Mars in 2029 and a million-person city is possible by twenty or thirty years later. We'll assume he's got space babies worked out for now so we can deal with a bigger problem: space sucks. Our impression talking to nongeeks is that while they realize space sucks, they have underestimated the scale of suckitude. We said above that you'd be crazy to leave Earth for Mars. This is true, but we should add that Mars is *easily* the most inviting place for space settlement. The runner-up is the Moon, which among its many shortcomings is very poor in carbon, the basic building block of life.

The result of the general awfulness of space is that you're likely living underground to keep the environment from touching you. Survival for a million people will require a very good seal-in, enormous amounts of electricity, an insanely large structure, and hardest of all, an artificial ecosystem to sustain everyone inside. Can we do this? The biggest such system ever built was Biosphere 2, created in the 1990s, which sustained a total of eight people for two hungry years. Can we realistically scale from eight people to one million in the next thirty years? Like with space babies, the problem here isn't just that the technology is challenging. Computers were challenging and so were airplanes, but we still built them. The problem is that getting from here to there is going to require understanding an extremely complex biological system that settlers will be reliant on for food, clean water, air, and not-dying in general. We can do it, but at the pace of ecology, not venture capital. Speaking of which, as with space babies, nobody is spending the kind of money necessary to get answers in a hurry, perhaps because there's no obvious profit in things like orbital obstetrics or airtight greenhouses the size of two Singapores.*

We still don't know a lot of first-order stuff, and getting that knowledge is going to be expensive, time-consuming, and without an obvious return on investment. If you're like us, at this point your thought is—okay, the science and tech are hard, but we can still do it, and we *should* do it because it's awesome. This unfortunately leads to a problem bigger than science or technology: law.

Believe it or not, there is space law and there are space lawyers. They are not briefcase-toting people in space suits, but scholars of international law. They have conferences, institutes, moot courts, and as far as we can tell are very annoyed that space-settlement fans often pretend they don't exist. We'll get into the details later, but the overarching problem is that the way space law interacts with modern technology and geopolitics is practically designed to produce crisis if humanity moves toward space settlements. Here's why: space is a commons. It is shared. Nobody is allowed

*Biosphere 2 was about 3.14 acres for eight people. If we scale that to a million people, you're around 1,600 square kilometers of greenhouse.

to appropriate any territory. However, under many modern interpretations, and absolutely under the American interpretation, everybody can use as much of the surface as they like. Let's sit on that a second: you can use the entire lunar surface any way you please, ad libitum, as long as you never say "This is mine in the sense of being my territory." Legally, we could probably write "The Moon Belongs to the Weinersmiths, You Filthy Earth Scum" in giant letters visible from Earth, as long as we didn't claim to actually believe it.

Other players could do likewise: China, India, the European Space Agency, or private launch corporations, for that matter. Add in the fact that only a tiny amount of the lunar surface is especially useful, and that the most likely parties to an argument are nuclear powers, and you have what might be called an interesting situation. Kelly attended the 2019 International Astronautical Congress—think space-nerd prom only with major officials from world government and space agencies—where she sat in on a session on space law. The going opinion among US officials? Space law is too slow and nobody agrees on the path forward, so we should just pass national rules, try to get friendly nations to agree to go along with them, and do our thing. The problem as we see it is that doing our thing may involve quasi-territorial claims that push the interpretation of international law to its limits.

Most worrisome, this decision to rush headlong into crisis might be taken *even if there's no good economic or military reason to do it.* Zach once talked to some international security scholars about why nations do things that make no sense. His specific question was about something called helium-3, which is a substance several governments, companies, and space agencies say they will mine from the Moon for its economic value. For reasons we'll discuss later, we believe this is a plainly silly idea, and we wondered why all these different players were claiming to be interested. The response was along the lines of "Well, the attitude is . . . if China does it . . . we have to do it too." Space officials aren't coy about this either. In a 2022 interview with *The New York Times*, NASA administrator Bill Nelson said regarding the Chinese presence on the lunar surface: "We have to be concerned that they would say: 'This is our exclusive zone. You stay out.'"

If you want to safely settle space, technology is hard, but it isn't enough; we also need at least somewhat harmonious international relations. That's not looking great on Earth right this second, and space may not be all that much different. In a 2022 report put out by the Defense Innovation Unit, written by workshop attendees hailing from organizations like the US Space Force and Air Force, the authors say a new "Space Race" with China has already started. As they write: "The competition represents a major inflection point not just for the 21st Century but for all of human history. The New Space Race seeks to achieve nothing less than the permanent establishment of the first off-planet, human settlement propelled and sustained by a thriving to-, in-, and from-space economy."

But there's some room for optimism. Humanity has peacefully regulated Antarctica and the bottom of the sea—areas that are similar to outer space, in terms of being basically terrible and largely inaccessible until the mid-twentieth century. Whether we can continue to do that in outer space, which since the 1950s has been deeply tied to national prestige, is trickier.

But now suppose we pull all this off. We've got bubble ecologies, China and the United States are getting along great thanks to a brilliant new legal framework, and we're all making top-notch space babies. We still face one last problem: us.

Given the difficulty of settling space, those who favor it generally come to the table with aspirational goals for humanity. One of the most plausible is that a second human civilization is essentially a backup copy in case we accidentally nuke this one. Or cook it. Or it gets hit by an asteroid. In this vision, space settlement is a Plan B for our species, which makes space settlement a worthy goal regardless of risk or short-term return on investment.

But are we certain a Plan B strategy actually delivers *increased* likelihood of species survival? It may not.

Space Bastardry: The Long View

The most detailed treatment of the issue comes from international relations scholar Dr. Daniel Deudney and his book *Dark Skies: Space Expansionism, Planetary Geopolitics, and the Ends of Humanity*. It's an involved argument, but the basic idea is this: humans being what we are, the move into space creates at least two forms of existential peril: the risk of nuclear conflict on Earth due to a scramble for space territory, and the risk of heavy objects being thrown at Earth if humans are allowed to control things like asteroids and massive orbital space stations.

The first point could at least in principle be resolved by a proper legal regime, but the second point is trickier. The more capacity we have to do things in space, the more capacity we have for self-annihilation. That doesn't require anything like interplanetary war either. Terrorism would be enough, and would probably be harder to eliminate.

Deudney is not popular among space-settlement geeks,* but we think he needs to be taken seriously. If he's right, then even if we can make the needed technology and can work out the law, there still remains a strong argument against a massive human presence in space. Note that there are

*Well, we, the space bastards, like him. He seems like a genuinely nice guy.

at least two different ways things could go badly: The first is simply more human presence in space increasing the odds of a bad outcome. The second issue is what you might call the tendency to space bastardocracy. This will be detailed further, but there are reasons to suppose space settlement as generally imagined might be especially likely to produce cruel or autocratic governments.

Making Deudney's arguments especially concerning is the fact that among space-settlement advocates, which let's remember includes two of the richest men on Earth, both of them rocket company owners, there are all sorts of questionable beliefs about how space will *improve* humanity. Space settlement is something people have wanted to do since the Victorian era. There are long-standing societies dedicated to the idea, and over the years they have built up all sorts of arguments for why humans must go to space, must go soon, and how everything will be great when we get there.

Depending on which theory you believe, space is supposed to: lessen the chance of war, improve politics, end scarcity, save us from climate change, reinvigorate a homogenized and rapidly wussifying Earth, and in one widely held notion called the "overview effect," make us all as wise as philosophers. If any of these were true, they might defeat Deudney's arguments. If we're all going to be philosophers up there, why worry about war? Or if we have a shot at eliminating scarcity, maybe the existential gamble is worth the danger. The problem is that, for reasons that will be detailed in the rest of the book, these ideas are almost certainly wrong.

But they remain widespread and influential among powerful technologists in the space-settlement movement and in space agencies. One long-standing thread of space-settlement ideology is broadly libertarian and conservative, seeing the modern Earth as increasingly homogenized and bureaucratized and in need of the influence of a space-frontier civilization to show us a tougher, freer, better way. Elon Musk likely believes some form of this. Consider his recent tweet arguing that "Unless it is stopped, the woke mind virus will destroy civilization and humanity will never reached [*sic*] Mars." A related version of this idea is that space will be like the old American West, which purportedly made the United States its modern, dynamic, and ruggedly individualistic self. This idea goes back to the nineteenth century but hasn't been mainstream among historians since the 1980s. Yet it lives on in government and military documents, political speeches, the National Space Society's Statement of Philosophy, and is promoted by Dr. Robert Zubrin, president of the Mars Society.

Jeff Bezos likely got his theory of space settlement from Dr. Gerard K. O'Neill, a professor at Princeton whose lectures Bezos attended as a young student. O'Neill's philosophy for space oriented around large solar-powered space stations as the way to save Earth's economy and ecology. This argument may have been plausible circa 1970, when it was widely believed that space would keep getting cheaper and that energy and food crises would result in unprecedented worldwide famines by the 1980s. Today you can do a much better job of saving Earth's biosphere with Earth-based solar and wind power. Even if we thought space settlements could take pressure off of Earth's seas and lands, they will absolutely not arrive in time to thwart any environmental calamity.

Whatever else you could say about these ideas, they do appear to be sincerely held. In our experience, people often think that space billionaires are hucksters or liars or even Ponzi schemers. It's never fun being in the position of saying "Guys, wait! These billionaires are misunderstood!" But look, setting aside the hype and showmanship, there is every reason to believe rocket billionaires really care about space settlement. Jeff Bezos gave his valedictory speech as a high school student on the topic of space colonies and today is the most important advocate for large, rotating space-station settlements of the type advocated for by O'Neill. When Elon Musk first got rich off the sale of PayPal, before he created SpaceX, he looked into sending a mouse colony or a small greenhouse to Mars. There is no money to be made doing this sort of thing; Musk wanted people to see his vision for space during a time when space activity was lackluster.

In our experience, a lot of people think SpaceX in particular is some kind of scam, using old government-created space technology for personal enrichment, or somehow hiding the true costs of space launch to fleece public coffers. We've encountered this idea again and again, and all we can say is that it's so contrary to the plain facts as to verge on a conspiracy theory. However you feel about Musk, SpaceX has genuinely revolutionized space launch, and every space agency on Earth, including NASA, has failed to duplicate their technology. In fairness, Musk's SpaceX, Bezos's Blue Origin, and other rocket launch companies have gotten plenty of government contracts, but that's been the standard way space has been done in the United States since the early days of space flight. The revolution in pricing only arrived with SpaceX.

Both Bezos and Musk overhype things, yes, but the evidence is that they actually believe in a space-settlement future. What concerns us is not that they're lying, but that they have weird beliefs about human sociology that may shape the future in undesirable ways.

The Case for Space at a Moderate Pace

So here's the position we're in: space settlement isn't going to eliminate scarcity or make us wise or save the environment. Even if it could, the technological and scientific barriers to doing it safely in the near term are enormous and underappreciated. Even if we had the technology, the legal structures right now would likely produce a conflict as parties scramble for turf. If we're really unlucky, international competition might force pointless geopolitical escalation among nuclear powers. And even if all that stuff were handled, there would still be good reasons to curtail our ambitions for the long term. And with all that said, very powerful people, aided by recent national laws and multilateral agreements, are pushing to make these things happen as soon as possible.

We don't think this *has* to mean that space settlements should never happen. What we do think is that space settlements probably are, and ought to be, a project of centuries, not decades. In particular, we'll argue that if humanity wants space settlements, we should take a "wait-and-go-big" approach. Wait for big developments in science, technology, and international law, then move many settlers at once.

But the waiting isn't just sitting around. In the following pages, we learn about spider bots on the Moon, baby making on Martian roller coasters, the number of humans needed to represent a viable breeding population, and also some weird stuff. Even if our species never settles Mars, deciding how we *might* do it is a project that requires objectively awesome and bizarre research and development in almost every field of human endeavor, from artificial wombs to international law. No amount of science guarantees that we can eliminate long-term existential risk, but if space-settlement plans are operating on the scale of hundreds of years, we've at least got time to work things out.

A quotation used in 99.9999 percent of all books about space settlement comes from rocketry founding father Konstantin Tsiolkovsky, who wrote in a 1911 article, "The earth is the cradle of humanity, but one cannot forever live in the cradle." Perhaps. But we should remember that what emerges from a cradle is not a full-grown adult, but a toddler—lacking in knowledge, very excited, and prone to self-destruction. If we do plan to leave this place, better to do so as an adult. Let's spend the awkward years learning and *then* strike out for new vistas.

Your Introduction to Space Bastardry

Think of this book as the straight-talking homesteader's guide to the rest of the solar system. If you're new to space settlement, most of it will be unfamiliar to you, and we hope quite surprising. If you're already a bit of a space-settlement geek, we expect you will find a vision that's more realistic and holistic than you've seen in any other book on the topic.

The book is divided into six chunks. The first is about what space does to human bodies and minds. The second is about where we might place those bodies and minds in space. The third is about how to have them not all die. The fourth section is about whether any of this is or ought to be legal. The fifth section is about how we might update the law to better accommodate human settlement, while keeping an eye on the humans back home. The final section concerns sociology, growth, whether we can achieve a Plan B for humanity, and whether it's desirable to do so.

Because we're trying to cover so much ground without losing too much nuance, we close each section with a Nota Bene—a weird yarn from our research that doesn't necessarily contribute to the overall vision, but which provides a respite from the firehose of information you're about to imbibe.

We also wish to introduce you to Astrid:

Astrid is a space settler and she's ready to bid farewell to this pale blue dot. With each passing chapter we'll change her to illustrate what we've learned so far. As we go, we'll scale from what she's wearing to where she's living to her new space nation, which hopefully won't get nuked by Earth. By the end, you'll be able to decide for yourself whether her decision to settle the solar system was a smart one, either for herself or the world she left behind.

1.

A Preamble on Space Myths

> Idyllic views of the future always seem to come with the hidden assumption that human nature will change. That somehow, the flaws of mankind will just melt away amongst the awesomeness of living among the stars. People will abandon mundane flaws like booze and drugs, and also everyone will be super-efficient like some kind of environmentalist's dream. But that's never been the case as we march forward, so I don't see why it would happen in the future.
>
> —Andy Weir, world famous sci-fi author who also writes really insightful commentary in books about booze in space

Outlandish ideas about space settlement often function as a justification for the whole project, typically promising vast wealth, an improved humanity, or an escape from Earth-awfulness. Because much of this book hinges on the idea that there is no urgent need to settle space, here we'll try to convince you that most of the pro-settlement arguments are wrong. Some of these arguments may be unfamiliar to you, but all of them have at least some powerful advocates in government, military, or business settings.

Bad Arguments for Space Settlement

Argument 1: Space Will Save Humanity from Near-Term Calamity by Providing a New Home

The idea of a multiplanetary humanity as more resilient to extinction is a common one and is plausible over the very long term. However, over the short term, space settlement won't help with any catastrophe you're imagining right this second. Not global warming, not nuclear war, not overpopulation, probably not even a dinosaur-style asteroid event. Why? In short, because space is so terrible that in order to be a better option than Earth, one calamity won't do. An Earth with climate change and nuclear war and, like, zombies and werewolves is still a way better place than Mars. Staying alive on Earth requires fire and a pointy stick. Staying alive in space will require all sorts of high-tech gadgets we can barely manufacture on Earth. We'll elaborate on all of this over the course of the book, but the basic deal is that no off-world settlement anytime remotely soon will be able to survive the loss of Earth. Getting any kind of large settlement going will be hard enough, but economic independence may require millions of people.

We believe there's a decent case for a Plan B reserve of humanity off-world, but there isn't a good case for trying to do it fast. A commonly made argument for urgency is what's sometimes called the "short-window" argument. The idea is that historically, "golden ages" don't last long, so our current age of space travel might come to an end before we get to Mars. We don't know if that's a good analysis of history, but what we can say is that the current age is simply not golden enough to deliver an independent Mars economy. If you want a Mars that can survive the death of Earth, you'd better make sure Earth doesn't die for a very long time.

Weinersmith Verdict:

Nah.

Argument 2: Space Settlement Will Save Earth's Environment by Relocating Industry and Population Off-World

There are various flavors of this argument, many of which are popular with the rotating-space-station settlement community, including Jeff Bezos.

One version of this idea is that the solar system contains more than enough mass to create rotating space stations that can accommodate an almost endless number of humans in space. This is literally possible in the sense that there is lots of stuff in space, and the stuff could be refashioned into space bases, but we need a sense of proportion here. The Earth of 2022 puts on about 80 million people per year. If saving our ecology requires us to reduce Earth's human population, then we need to launch and house 220,000 volunteers *per day* just to tread water.

A related idea is that space should be zoned for heavy industry, while Earth returns to an unpolluted Edenic state. All the nasty mining and manufacturing can be done elsewhere, with by-products cleanly disposed of into the vast landfill that is the solar system. As Jeff Bezos says, "Earth will be zoned residential and light industrial." Again, this is literally possible, and perhaps as long as you're just thinking in terms of big concepts like pollution and mass it sounds doable. But the details are where the difficulty lives. Consider for example cement. It's a major contributor to global warming, so can we make it in space?

Technically, most of the components of cement by mass exist on the Moon, but they won't be easy to dig up. Construction equipment will need to be built to function in an airless environment at low gravity with equatorial temperature swings from -130°C to 120°C. Little things start to loom in this context. Just getting a lubricant that can handle these temperature shifts without degrading is nearly impossible. The same goes for the machines themselves. At extreme cold some metals can undergo a ductile-to-brittle transition; below a certain temperature, metals behave more like stone. However strong they may be, they can't flex and bend. It's speculated that the *Titanic* sank because its steel hull experienced a ductile-to-brittle transition before hitting the infamous iceberg. That's a nontrivial problem when you desire to use construction equipment that regularly slams into hard surfaces.

And that's just one detail of one part of the process, never mind replicating all those factories. How soon can we plausibly get all these problems solved and then scaled to the needs of Earth, which currently requires over 3.5 *billion* metric tons of cement per year? And does it sound economically competitive with Earth-made cement even if we could do it? And, by the way, what are the rules for dropping 3.5 billion tons of rock on Earth annually?

Part of what's supposed to make these ideas work is cheap, plentiful energy thanks to space-based solar power. This is another bad idea. Space-based solar power figures prominently in space-settlement proposals for giant rotating space stations. It's also frequently proposed by governments and private space companies as a way to make money while greening the planet. You may have read an article recently about Chinese universities or the European Space Agency, or some new start-up planning to field this technology in the near future. They probably shouldn't.

It's certainly true that there's a whole Sun's worth of sunlight in space, unobstructed by annoying Earth features like weather and the atmosphere. Exactly how much more energy you might get per panel depends on exactly what assumptions you're prepared to make, but different estimates expect about an order of magnitude improvement. That sounds like a lot until you ask yourself what the cost differential will be between a panel in space and a panel in Australia.

It's conceivable that in a world where solar panels are incredibly expensive and there's an extreme collapse in the cost of launching objects to space, you might want to maximize your energy per panel by putting them above the atmosphere. But panels are cheap, and even if we assume pretty steep drops in the cost of space launch, the numbers don't add up. This becomes especially clear when you start to think about maintenance. Try to imagine acres upon acres of glass panels in space, regularly pelted by intense radiation and bits of space debris while enduring the extreme heat of perpetual sunlight.* They'll have to be repaired and cared for either by

*A major problem here is also that heat has to go somewhere. On Earth, we normally dissipate heat by dumping it into something like air or water. In space you don't have this, which is why the ISS, for example, has gigantic specialized radiator systems. These too will have to be maintained.

astronauts or an army of advanced robots. Solar panels in Australia can be cleaned by a teenager with a squeegee.

When dumping solar power back to Earth, you have another problem. Solar panels on the ground can send their power right into the grid or to batteries. Space-based power has to be beamed to huge receivers on Earth, losing energy en route. But it can't be beamed at too high an intensity, lest it endanger birds and planes.

Space solar *is* valuable if you're already in space, as a way to generate energy without burning fuel. It may also be valuable on Earth in some very narrow cases, such as beaming energy to military bases where fossil fuel delivery would be dangerous. For more practical uses, you're better off with conventional boring renewables. Cover every rooftop with solar panels, followed by the Sahara desert, and then if the planet still needs energy, we can talk about space.

We are skeptical that it will ever be a great financial idea to harvest massive amounts of solar power in space and then use that energy to convert moondust into cement or steel or industrial chemicals. But even if we believe that this'll all happen one day, that one day will not come in time to spare us from any environmental concern of today.

Weinersmith Verdict:

Argument 3: Space Resources Will Make Us All Rich

It's certainly possible, but right now the economics of it aren't looking great. As we'll explore later, no place in space has something like a giant hunk of pure platinum or gold. What space resources do exist are likely to be very expensive to acquire and will remain so even with big improvements in technology.

Also, there's a real difference between access to commodities and universal wealth. Consider aluminum. Discovered in 1825, early on it was so valuable that only the wealthy could afford it. Victorian-era jewelry sometimes includes aluminum as a precious metal. Today, it's a way to cover lasagna. That's because by the late nineteenth century, industrial processes had made aluminum incredibly cheap, effectively flooding the market with a former luxury good. This is a great development, and of course aluminum has uncountable valuable applications from the kitchen to airplanes. But the fact that most of us can buy large quantities of a once-precious metal doesn't mean we're all millionaires.

In our experience, people tend to assume raw minerals are the major factor in human well-being. Although they're necessary inputs into our economies, according to a recent report by the World Bank, nonrenewable resources, in the sense of valuable stuff found in the ground, make up about 2.5 percent of Earth's wealth. And a lot of that is fossil fuels, which are not available in space. The really valuable thing for economies is humans, and our ideas and technology. You can convince yourself by melting down your phone and assessing the value of the resulting glass, metal, and plastic.

Even if space does produce inexpensive access to all sorts of commodities that make *someone* rich, there's also no reason to assume anything like an equal distribution of wealth back on Earth. In fact, if you believe there's big money in space, the United States is uniquely poised to go get it, potentially harming the economies of less-developed countries dependent on commodities. Some readers will care about this more than others, but even if you don't think wealth distribution has much moral significance, it may still have geopolitical significance. As we'll see later, under some conditions, changes in the balance of power among nations can make war more likely. If space really does make some country especially rich, the consequences don't have to be uniformly good.

Weinersmith Verdict:

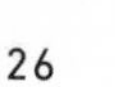

Argument 4: Space Settlement Will End, or at Least Mitigate, War

There are a few versions of this one, but we've found these three pretty common: space settlement will create more territory so we'll fight less about territory; space settlement will make us rich so we won't want to fight anymore; and space settlement will allow unhappy citizens to just leave for other settlements, which will reduce tension here on Earth.

The territory argument is the most silly. Nations don't fight over land, they fight over *particular* land. You can't solve disputes over Jerusalem or Kashmir or Crimea by promising the parties involved equally large stretches of Antarctica. It'd be like going to a nasty divorce proceeding and trying to solve the custody fight by offering to just grab some other kids. Also, if we're defining land as "built structures humans live in," which is the definition you *must* use for space habitats, well then, on Earth we are creating land all the time. Individual buildings create far more square footage than any space settlement likely to be built anytime soon. Meanwhile, if you personally just want any sort of land there's plenty. Google it. Small towns all over the developed world are offering *free land* to people willing to move there instead of big cities.

The argument about riches may sound tempting; if humans are rich, why would we fight? But the "money makes us all friends" argument isn't one that all war scholars buy. Wars start for all sorts of reasons that have nothing to do with a bunch of people looking at their resource base and saying "hey, this is pretty good." A nonexhaustive list of causes of war includes: religious differences, leaders who don't bear the cost of the violence, and misperception about the other party's strengths or intentions. Even if space activity left everyone better off, it wouldn't stop nations from having religious differences, bad leaders, or suspicion about rivals.

As for peace through allowing people to just move between settlements, well, we should consider that most people aren't even allowed to do this between nations on Earth. Space will likely be worse. However you feel about immigrants coming to your country, one thing you probably don't fear is the possibility that they'll breathe too much air. In space, the atmosphere is constructed, as is the ground beneath your feet, and individual settlements will only be rated for certain population sizes. That's not obviously an environment where you'd expect to see open borders. Some advocates note that you can always just create a new place to live in space, but then the argument becomes "you can just pull up stakes by creating a million-ton space station," which, we suspect, will not be a live option for most of us. Even if it were, it's still not clearly desirable. Dr. De Witt Kilgore, one of the few historiographers of ideas about space, called it a form of celestial "white flight." That is, space not as a solution to politics, but as an escape from political realities one group finds uncomfortable.

Weinersmith Verdict:

Argument 5: Space Exploration Is a Natural Human Urge

This is a popular one. The basic idea is that yeah, maybe there's not a good return-on-investment reason for space exploration, but if we don't do it, we'll be thwarting our own nature, resulting in widespread human stagnation. The prettiest version of this argument is of course from Dr. Carl Sagan: "For all its material advantages, the sedentary life has left us edgy, unfulfilled. Even after 400 generations in villages and cities, we haven't forgotten. The open road still softly calls, like a nearly forgotten song of childhood." It's a nice idea, and much better written than any of our Uranus jokes. Also, it can be hard to argue against views like these because it's not always clear what the exact claim is. However, when people do get specific, they tend to point to two things: famous human explorers, and the fact that humans have spread around the world.

The appeal to famous explorers is moving but not very convincing. Most of us are not in fact famous explorers. Most of us prefer to vacation in places that have pastries and air-conditioning, not Mount Everest or the Amazon basin. It's cool that some people are into this stuff, but it's hard to argue that they represent universal human nature. Some people are competitive mayonnaise eaters, but you never hear anyone say they embody deep human truth.

Plus, if you actually look at the stories of explorers, priority claims seem to be at least as important as exploration. When the Peary expedition said they'd reached the North Pole in 1909, they entered into a public priority fight with Dr. Frederick Cook, who said he'd gotten there first. One person who *didn't* lay a priority claim was Roald Amundsen. He'd been planning to be first to the North Pole when he heard about Peary's success. What'd he do? He immediately switched his expedition to the yet unreached *South* Pole even though plenty of the Far North remained unexplored. Does it *sound* like the major goal was always pure curiosity? If exploration is a natural human urge that must be satisfied, why are so many of us happy to sit on our couches, and why are the few actual explorers so concerned specifically with exploration that will make them famous?

The second argument—the appeal to humans spreading around the world—is also questionable. *Homo sapiens* indeed has spread to every continent. But then so have roaches. So have a lot of plants that don't *seem* very concerned with their cosmic destiny. Humans go to new places all the time for reasons that have nothing to do with an exploratory impulse. Modern mass migration often has to do with warfare, persecution, and starvation. It's plausible that it was the same way in the distant past.

Lastly, if failing to explore causes stagnation, well, where's all the stagnation? "Stagnation" is an opinion, of course, but the world's surface has been completely mapped since the mid-twentieth century, and a lot of cool stuff has happened since the 1950s. We have trouble imagining a serious argument that culture has ceased to be creative or that science has ceased to advance. This book you're holding specifically exists because the development of space launch technology has suddenly accelerated in just the last ten years—a feat that would be impossible without the decades of rapid advances in computing that came before.

Weinersmith Verdict:

Vague, but probably not true

Argument 6: **Space Will Unify Us**

Probably not. Or anyway, ask yourself if the last twenty years have been a time of especially great cooperation internationally, and in particular between Russia and the United States. They ought to have been—since 2001, international crews have been working in harmony in the International Space Station (ISS). Yet here in 2022, what we find is, in the aftermath of the Russian invasion of Ukraine, American astronaut Scott Kelly telling the head of the Russian space agency, Dmitry Rogozin, "Maybe you can find a job at McDonald's if McDonald's still exists in Russia." This was in response to Rogozin sharing a video in which flags of countries that had sanctioned Russia were covered up, despite having been painted on a rocket during more harmonious times. Rogozin fired back, "Get off, you moron!" This was all, of course, on Twitter. Rogozin followed up by saying Kelly probably had dementia from his lengthy stints in orbit. Even the space people aren't unified by space.

A more likely guess, and one favored by people who've studied space politics, is that space doesn't unify us. Rather, we do joint space activities when we are already getting along. Other than the International Space Station's construction and maintenance, the major space cooperation events of the twentieth century happened in 1975 during the short Cold War détente, and then in the late 1990s after the USSR had collapsed and Russia was no longer considered a threat.

No doubt, space cooperation confers some amount of fellow feeling and gives some segments of our populations practice with working together, but there are almost certainly cheaper ways to achieve the same goal. Consider that the cost of fielding the ISS to date is around $150 billion, which makes it the most expensive human-made object ever built. That's almost enough to send every single Russian—man, woman, and child—to Disneyland. For the cost of putting ten times as many people in space we could probably get them season tickets and an ice cream. That's a lot of unity.

Even supposing space activity caused nations to get along, it's not clear that this would be desirable. Nations often fail to get along for excellent reasons. Part of why there was no major US cooperation with the USSR in space after 1975 is that the Carter administration had concerns about Soviet human rights abuses. Do we want to live in a world where Carter's human rights position changed because some nice people in space got together for a meal of apple pie and borscht? Some international conflicts may really be about the need to stand together and see each other as members of the same human family, but many are down to actual differences in values and goals. Those disputes can and should be solved by conventional politics.

Weinersmith Verdict:

Argument 7: Space Travel Will Make Us Wise

There are different flavors of this argument, but the most famous is philosopher Frank White's notion of the overview effect. White, and many other people in the space community, believe that the view of Earth from space confers special insights about nature and human oneness. As he says, "People who live in space will take for granted philosophical insights that have taken those on Earth thousands of years to formulate."

If so, they don't seem to have been forthcoming with anything terribly Earth-shaking. After almost seventy years of space flight and over six hundred spacefarers, your local library contains no *A Critique of Pure Reason . . . in Space*, no *A Treatise on Human Space-Nature*. As far as we can tell, most of the philosophizing by spacefarers could fit nicely on a Hallmark card: the standard observations are that the Earth is beautiful and fragile, and that you "don't see borders up there." The latter claim, by the way, isn't even true. We were told by one cosmonaut that you can see the India-Pakistan border as well as the border between North and South Korea. In any case, supposing you really couldn't see those borders from space, would that be wisdom? We believe an insightful person *ought* to see that there's a border between the Koreas. The people trapped on one side certainly do.

Another serious problem for this theory is that there's no good evidence for it. A few papers have been written with incredibly leading surveys, asking astronauts whether time in orbit had increased their interest in things like the environment and human interconnectedness. One paper included a free-form questionnaire in which several astronauts noted that they were not given the option to declare reduced or unchanged interest in these things.

That's not to say going to space is dull. No doubt it's a meaningful, transcendent experience. But other transcendent experiences are available, too, generally at a lower fee. One attempt to measure the overview effect found that if it exists, the effect is about in line with that experienced by new moms. We don't plan to make fun of new moms and alienate our audience this early in the book, but can we just agree that if every new mom got the insights of philosophers and sages, Facebook would be substantially more pleasant. Also, making new moms is cheaper and easier than making new astronauts. Nobody becomes an astronaut by accident.

Most damning to the theory is the fact that, while there have only been about six hundred people in space, there are about six thousand stories of astronauts behaving badly. Alcoholism, adultery, flying planes while on drugs, lying to medical staff, denying climate change, promoting pseudoscience, fighting publicly with other astronauts, and the time an astronaut drove across the country in order to kidnap her ex-boyfriend's new girlfriend. The ex was also an astronaut and had arguably been stringing her along. Least sagely of all perhaps was the time Valentina Tereshkova, beloved first woman in space, proposed a constitutional amendment in the Russian Duma granting Vladimir Putin the option of two additional terms as president. She was later sanctioned by the US government in response to her support for Russia's annexation of Ukraine and the sham referendums used to justify it. Perhaps when viewed from space we are all equal, but some of us remain more equal than others.

Weinersmith Verdict:

Argument 8: Creating Nations in Space Will Reinvigorate Our Homogenized Bureaucratic, and Generally Wussified, Earth Culture

Whether Earth is becoming more homogenized is a subject of intense debate among sociologists. Globalization does affect small local cultures, but it also results in cultural hybridization, which makes new stuff. Whether this counts as a net loss is a matter of perspective, but perhaps the most quantifiable version of the homogenization argument is the loss of rare languages around the world. This is indeed happening, but we talked to linguists about the question, and they were skeptical of the idea that going to space would change anything. To get new languages, you need genuine long-term separation. Complete separation from Earth anytime soon will be neither possible nor desirable. If you want a new language, stick people on a lonely island without internet for a few centuries. Mars is going to have Netflix.

For some, it's not about homogenization, but wussification. Many American space advocates favor a version of what scholars call the Turner Thesis or the Frontier Thesis.* The claim is that the United States became dynamic, democratic, ruggedly individualistic, and generally awesome due to a long-standing frontier culture. Sometimes this is a simple rhetorical flourish about space as a place of newness and adventure, but often the

*We've been told by European friends that this is a US quirk, but we did find at least two European sources using this argument as well.

frontier is seen as something more—as a sort of process of social resurrection. In this vision, space settlers will forge a hard, serious, creative civilization, and that borderland society will show Earth people a tougher and more democratic mode of life, just as the American West purportedly did for the United States. The problem here is that this once-popular theory is now rejected by pretty much all mainstream historians as a misleading oversimplification.*

Even if it were true, if you read the original literature, the Turner Thesis relies on the idea that US settlers had cheap land, isolation from nonfrontier areas, and ominously, the need to organize to seize land from the native population. Space is expensive, will have internet, and thankfully lacks a local population to exploit and murder. As with the issue of homogenization, even if the theory is true, it's not at all clear that a Mars base is the way forward.

A more generalized version of this frontier argument says that the harsh world of space and the need for robotics will result in a vast increase in creativity. Again, this is hard to measure and debated by scholars, but there are reasons to be skeptical that space is the optimal solution. To illustrate why, consider an idea we call the "Necrosphere," in contrast with the Biosphere. The Necrosphere is a built structure on Earth. Inside it, the ground is poison, there is no air, and cascades of radiation are fired at the inhabitants on a perpetual basis.

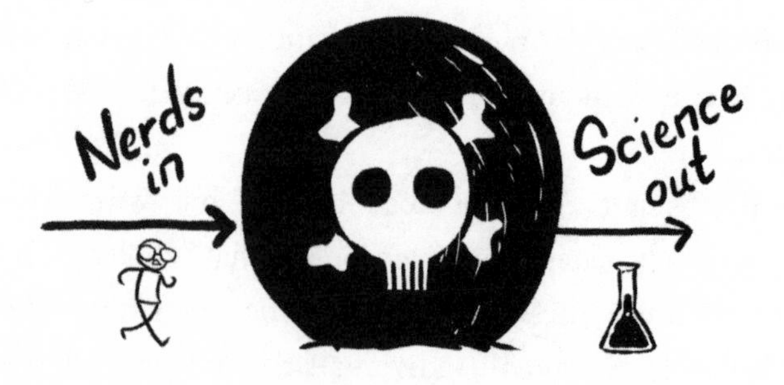

*By academic standards, the overturn was fairly brutal. Here's an excerpt from a 1987 paper by Yale historian Dr. William Cronon, who is notable for his attempt to highlight the *positive* aspects of Turner's work: "After all the articles and books and dissertations, what could possibly justify yet another excursion on the 'blood-drenched field' of the frontier thesis? . . . In the half century since Turner's death, his reputation has been subjected to a series of attacks which have left little of his argument intact."

Why did we build it? In the sure knowledge that we can stick engineers inside who, due to the harsh environment combined with their need not to die, will spew forth valuable ideas like a spigot spews forth pressurized water. If this sort of thing seems implausible to you, you should ask yourself why anyone would expect a Mars base to generate all these supposed benefits. You should also ask yourself why it is that so many innovations on Earth come not from anarchic wastelands but from cities where an engineer's main hardship is eight-dollar espressos.

In any case, if the goal is creating off-world nations, it's worth noting that you're simply not allowed to under international law. Given how dependent on Earth any future space settlement will be, this is a nontrivial impediment.

Weinersmith Verdict:

There are other arguments, but the ones above are those we've most frequently encountered. You should note, however, that we have not made any arguments that people *shouldn't* settle space—just that a lot of the purported benefits are implausible. That leads to our last question here:

OKAY FINE, Is There Any Good Case for Space?

Sorta. We think there are two arguments that at least don't rely on implausible economics or incorrect sociology. They share a problem, but it's a speculative problem about the future.

Argument 1: The Cathedral of Survival

There are very few philosophers willing to argue against human life on net. In their defense, though, they have some of the best book titles, including *Every Cradle Is a Grave* and *Better Never to Have Been*. However, most of us would like this whole human-existence thing to keep muddling along. If you look at long-surviving species on Earth, they tend to have a few factors in common: large populations, genetic diversity, wide geographic spread. Having a Mars settlement with enough people and habitat that it could survive an earthly calamity seems to fit the bill. It will not save us from climate change or from any of the other likely ways to destroy ourselves in the near term, but it may still be a worthwhile long-term endeavor. As with the cathedrals of Earth, those of us who cast the first few bricks may not be around to see the spire placed on top, but we might nevertheless want to start building.

Exactly what that might enjoin us to do right this second is trickier, but if we agree we ought to build this thing for our grandkids' grandkids, taking first steps right now seems wise.

Weinersmith Verdict:

Argument 2: The Hot Tub Argument

When you want to buy a hot tub, nobody says "It is the destiny of Man to put His butt in bubbly hot water." Nobody tries to convince you that a hot tub–less humanity is bound to stagnate. Nobody tells you the proliferation of toasty warm outdoor bathtubs will end human conflict. It's just that you want a hot tub and someone is selling it, and nobody has any right to stop you. This is not the most noble or uplifting argument for any particular action, but then we suspect appeals to human nobility tend to co-occur with appeals for taxpayer money. If the reason for going to space is not philosophical or even just about return on investment, well, that's okay. "Because it's awesome" is still a perfectly serviceable argument.

Weinersmith Verdict:

The Fly in the Space Ointment

The potential snag to both these arguments is the question of whether going to space might put species survival at *greater* risk. Let's say space increases the risk of species-annihilating war or terrorism. In that case, we've now built a "survival cathedral" that has the potential to topple onto all of us. If our justification for space settlement is long-term species survival, we need to be reasonably confident that settlement actually increases our chances.

In the case of the hot tub argument, you can visualize a spectrum running from hot tubs to nuclear weapons. For most of us, hot tub acquisition does not put anyone else in danger. Maybe someone who accidentally peeks over the fence, but that's their problem. Nuclear weapons acquisition is different. Your possession of a nuclear-tipped intercontinental ballistic missile is not a personal matter, because I might get blown up. That gives me a right to prohibit you from having it. Note that this is true even if you are a very nice person and have no plans to actually use the nuke in your garage. The question then is whether a space settlement is closer to the hot tub or the nuke—is it a free choice for individuals or a case in which strict regulation is needed?

We consider these two arguments to be the strongest arguments for settlement, but we've reluctantly become convinced that the fly in the ointment is more like an elephant. We'll spend the rest of the book on what space is like and the regulations that govern it, after which we'll return to these arguments and assess them in light of all that information.

A Brief Note on Language and the Profound Chauvinism of the Authors

The two words most commonly used to indicate a place in space humans intend to live permanently are "colony" and "settlement." A debate that we can trace back at least to the 1970s concerns which of these terms carries less offensive historical baggage.

A few alternatives have been suggested, but to be honest they're not great. The Beyond Earth Institute, for example, favors the clunkily septisyllabic "communities beyond Earth." A shorter option is "space outposts," but it doesn't capture the idea that one day people will have families, children, and institutions. Meanwhile, "space cities" sounds substantially more grand than anything we expect to exist anytime soon, and "space villages" sounds like it involves peasants in burlap pressure suits with horse-drawn Mars rovers. The prolific if not always poetic Isaac Asimov once suggested our favorite option—"spome."* Try saying it. *Spome!* Look, you're already smiling. Spome is short for "space home," and while it certainly lacks historical baggage, it also sounds like an off-brand bar soap or a buildup a surgeon has to extract from your kidneys. We'll be using "settlement," because frankly it's the preferred word right this second among space geeks. If "spome" creeps in from time to time, we spapologize.

*Impressively, "spome" isn't even the most awkward term. That honor goes to a term proposed by Dr. David Criswell in a 1985 book chapter: the unpronounceable "s'home" as a contraction of space and home. David Criswell, "Solar System Industrialization: Implications for Interstellar Migrations," in *Interstellar Migration and the Human Experience*, ed. Ben R. Finney and Eric M. Jones (Berkeley: University of California Press, 1985), 57.

Finally, and with some trepidation, we wish to note that in some places the book may be America-centric. Your authors are both American—we like drip coffee, we like bad cheese, and when we see overseas tourist attractions, we clap inappropriately. This is regrettable, but can't be changed. We have therefore done our best to talk to non-American readers and scholars and listen to their views, especially when they contradict ours. That said, one thing we want to emphasize is that when it comes to space, for better or worse, some level of America-centrism is justified. While the United States is no longer the only major power on Earth, it remains the overwhelming player in space. The US government spends far more on space than any other, and the revolutionary new space launch companies are all US-based. If from time to time we talk about myths of the American frontier or legal theories the United States is apt to favor, it's because those are the myths of the current hegemonic power in space.

In 1976, as part of the Bogota Declaration, eight equatorial nations claimed territorial rights to geostationary orbit, which by the nature of orbital mechanics is perpetually over their heads. Geostationary orbit is valuable, and the idea was that these less-developed nations ought to get paid rent by anyone using their space. But, although to this day the Colombian constitution asserts special rights to a slice of that orbit, the claim has largely been ignored by the international community. If the same claim had been made by the country with the most powerful military in history and the most advanced space launch technology, we suspect the newly installed "for rent" sign would have been taken a bit more seriously.

In a room full of animals, the gorilla may not be the wisest, but you probably want to know what's on its mind.

PART I
Caring for the Spacefaring

IF YOU ARE A SPACE SETTLER, AND THE GOAL IS TO HAVE A LARGE population in space, one of the best things you can do is not die. Or, if you're planning to die, don't do it right away. If you can swing it, have a few kids, raise them to reproductive age, then go gentle into that good night.

We don't know how to do this. Not really, anyway. Humans have been in space for over sixty years now and we do know plenty about what astronauts experience. But astronauts aren't normal people. To be frank, they're better than us. Although this is changing a little as the era of space tourism permits the schlubby among us to reach orbit, the average astronaut is someone who combines deep specialist skills with the ability to pass a battery of physical and mental tests that most of us would bail on in a few days. Many of the early spacefarers were ultraskilled test pilots, but even when space agencies opened up a little, the résumés remained intimidating. Dr. Sally Ride wrapped up a PhD in physics from Stanford before becoming an astronaut, and during her flight training was said by Navy test pilot Jon McBride to be "the best student I ever had." She was twenty-seven when selected. While Dr. Rhea Seddon was awaiting her turn on the Space Shuttle, she combined astronaut training with continuing her work as a surgeon, and in her spare time raised several children. Feel bad about yourself? These are normal stats for an astronaut. That's great for mission success, but not great if we want to know what space medicine will be like for more average people in large space settlements.

Combine this with the facts that literally nobody has been to space for longer than 437 days in a row, and the total time the Apollo missions spent in the Moon's one-sixth Earth gravity is just under two weeks, and it becomes very clear how little we know about the most pertinent question for space settlement: can regular people flourish for long periods off-world?

Here, we want to get a much firmer grip on what is known and what isn't. Many of the most likely problems might be easily surmounted with near-term technology, or just a boatload of money. Others may pose intimidating long-term barriers to any kind of permanent space settlement.

2.

Suffocation, Bone Loss, and Flying Pigs: The Science of Space Physiology

The Prehuman Space Age

A human is, to a first approximation, a pillar of liquid about two meters high, in which are suspended various moist and jiggly biological systems—digestion, waste storage, sense of balance, the movement of blood. All of these systems evolved in an environment where a 6-billion-trillion-ton sphere called Earth sat at the pillar's foot.

So, let's say it's April 12, 1961. You're Yuri Gagarin, about to do this brand-new thing where you ride the warhead part of a missile into orbit. Why are you confident that your stomach won't turn inside out?* That blood will continue to supply oxygen to your brain? That your lungs and liver and kidneys will all work just fine floating around inside your body cavity?

For one thing, you're not the first voyager to make this trip. Not even the first ape. The US and USSR both studied the effects of space on nonhuman life for over a decade before Gagarin went up. Soviet engineers favored dogs for flight. America, in order to demonstrate the superiority of

*Wally Schirra, the only astronaut to fly on all the first three US human space-flight programs—Mercury, Gemini, and Apollo—claimed in his biography that he thought these sorts of fears were silly, because in his Air Force days he'd once observed a base commander drinking a martini while standing on his head. Or, as he put it, "negative one gravity." Walter M. Schirra, Jr., with Richard Billings, *Schirra's Space* (Boston: Quinlan Press, 1988), 23.

liberal democracy, launched monkeys.* But if nonhuman species were allowed to stake priority claims, humans would be in competition with a cat, dogs, mice, tortoises, chimps, fruit flies, and a variety of monkey species. It's like Noah's Ark, only not everyone made it to the rainbow.

Laika, a gentle mutt from the streets of Moscow, was the first animal to orbit Earth. She died when her spacecraft unexpectedly overheated about six hours after launch on November 3, 1957. This shortened her life, but not much—her ship, Sputnik 2, didn't come with a return capsule. In 1960, Soviet space pups Belka and Strelka became the first dogs to orbit and safely return to Earth, where they remain to this day, in their taxidermied glory, on display at the Cosmonautics Memorial Museum in Moscow. Upon landing, they were apparently none the worse for space wear. The same was not true for all space animals, even those who survived. Ham the

*This isn't entirely a joke. Oral histories report that Dr. Wernher von Braun, the ex-Nazi missile designer who became the architect of the rockets that sent men to the Moon, insisted primates were necessary to one-up the Soviets. By the way, fun fact: the only cat ever to go to space was Félicette, sent by the French.

chimp—an American—went up shortly before Gagarin and had to briefly endure an acceleration of fifteen times Earth gravity. He was pretty pissed during his press conference the next day, screaming and baring his teeth at the reporters and photographers. Not typical behavior for NASA astronauts, but from a purely physiological perspective, the good news was that Ham was very much alive.

And indeed, after his hour and a half in orbit, so was Gagarin. He not only survived, he enjoyed humanity's first space meal: two tubes pureed meat, one tube chocolate sauce.

This single orbit marked the beginning of human space medicine. But, like every voyage from the heady early years of space competition, it tells us little about space settlement. If you know anything about the heroic first decades of space travel, with their split-second decisions, hair's-breadth escapes, their triumphs and their tragedies, well, sorry but that stuff is mostly useless for our purposes. Armstrong and Aldrin spent about three hours walking around the Moon. Prior to the 1970s, the longest space trip lasted about two weeks. That's not enough time for the most serious known space medicine issues to crop up.

Long-term space medicine doesn't start until the Soviet Union, having lost the Moon Race, inaugurated the era of space stations with Salyut-1, first fielded in 1971. Since then, there have been so few space stations that we can document them completely on the following page.*

Over this time period, voyages generally got longer, until the peak 437 consecutive days was reached by cosmonaut Valeri Polyakov during his 1994–1995 stint aboard Mir. Since then, only a handful of astronauts have gone for more than six months at a time. Still, the majority of our space medicine knowledge comes from the International Space Station, which remains the largest ever built, having about six times the space of the Salyut series, and a regular crew of six, whereas earlier space stations had two or three doing long-term missions.†

The good news from space is that it doesn't kill you immediately. As

*Salyut-2 was never crewed.

†If you've seen pictures with larger numbers of cosmonauts together in a space station, it's because there are more people during changeovers or when crews came up for short-term missions.

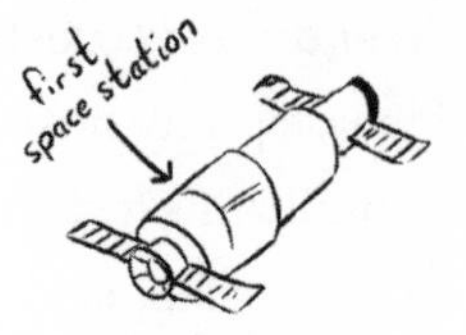

SALYUT-1 (1971)
HABITABLE VOLUME:
90 CUBIC METERS

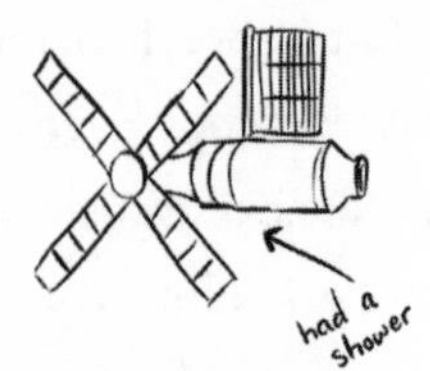

SKYLAB (1973-74)
HABITABLE VOLUME:
361 CUBIC METERS

SALYUT-3 (1974-75)
HABITABLE VOLUME:
90 CUBIC METERS

SALYUT-4 (1974-77)
HABITABLE VOLUME:
90 CUBIC METERS

SALYUT-5 (1976-77)
HABITABLE VOLUME:
100 CUBIC METERS

SALYUT-6 (1977-82)
HABITABLE VOLUME:
90 CUBIC METERS

SALYUT-7 (1982-91)
HABITABLE VOLUME:
90 CUBIC METERS

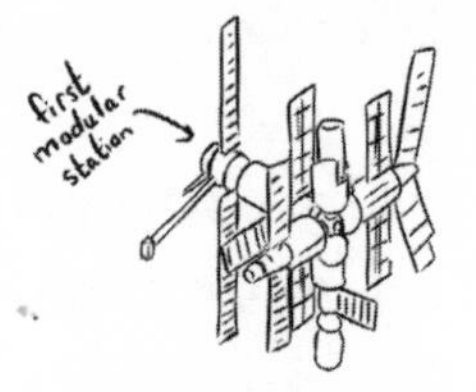

MIR (1986-2001)
HABITABLE VOLUME:
350 CUBIC METERS

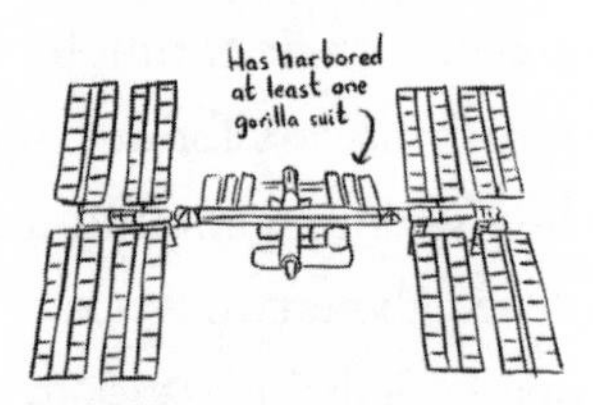

ISS (1998-TODAY)
HABITABLE VOLUME:
388 CUBIC METERS

TIANGONG-1 (2011-18)
HABITABLE VOLUME:
15 CUBIC METERS

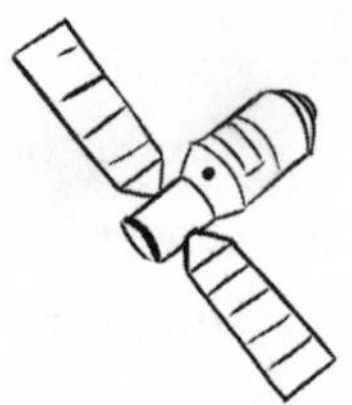

TIANGONG-2 (2016-19)
HABITABLE VOLUME:
15 CUBIC METERS

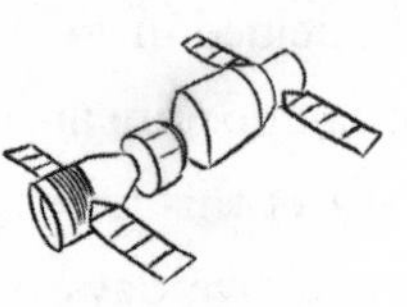

TIANGONG SPACE
STATION (2021-TODAY)
HABITABLE VOLUME:
110 CUBIC METERS
(PLANNED)

DRAWINGS NOT EVEN KIND OF TO SCALE.
HABITABLE VOLUMES ARE MAXIMUMS.

long as the equipment is working, space itself doesn't seem *that* dangerous, at least over moderate periods of time. When astronauts *have* gotten killed, it's always been due to a problem with the vehicle, never due to the slow, detrimental effects of the space environment. But, as one of your authors reminds the other whenever it's his turn to watch the kids, there are many intermediate points between alive and dead. Learning them gives us the best available look into the likely health problems of space settlers.

The Vacuum of Space, or Your Body Is a Soda Can

On Earth, air pushes on your skin from every direction with a consistent pressure of about 14 pounds per square inch, or using ridiculous non-American units, 1 atmosphere. That's about the weight of 1 liter of water on every square centimeter of your skin. You don't notice this for the same reason a seabottom shrimp doesn't notice that the surrounding liquid could implode a submarine—your body is adapted to the pressure near Earth's surface. It counterbalances the typical push of your surroundings, and you only rarely experience sudden pressure changes.

But consider a soda. When you buy a sealed bottle of Diet Pepsi, you know it's full of gas, but you don't see a lot of bubbles. That's because the bottle is held at about four times the surface air pressure of Earth, keeping carbon dioxide suspended sedately inside. When you open the top, you expose its contents to Earth's relatively gentle atmosphere. All that dissolved gas rushes out in the familiar bubbling foam. If you want to avoid the sudden burst of gas, you can always open your bottle forty meters under the sea, where the pressure will keep the gas in place, and the seawater will make the Diet Pepsi taste no worse.

Your body is like the soda, except that the gas suspended in your fluids is nitrogen,* absorbed from the atmosphere. If you were teleported to outer space, where the air pressure level is "none,"† your bodily fluids would react like the Diet Pepsi when opened, only instead of a burst of foam, you'd get nitrogen bubbles blocking your veins and arteries, preventing the normal flow of blood, oxygen, and nutrients. This danger is familiar to

*Technically, this means your body is more like a Guinness.

†We are informed by physicist friends that technically the amount of particles in outer space is nonzero. Kelly thinks this is a good point. Zach encourages them to try running a barometer in high orbit.

divers going from low depths back to the surface. If you switch from high to low pressure too quickly, you get "decompression sickness," colloquially known as "the bends" because it often affects joints, causing the sufferer to bend in agony. If it's in your lungs, that's "the chokes." If it's in your brain, you've got "the staggers."

If you're exposed to space, most likely you'll just have the death. In fact, the only people who've ever died in space* were killed by sudden loss of pressure. It was June 30, 1971, and cosmonauts Georgy Dobrovolsky, Viktor Patsayev, and Vladislav Volkov were returning from Salyut-1. The three cosmonauts spent weeks performing zero gravity acrobatics, televised for the adoring Soviet public. They entered the capsule, and after some brief issues getting the hatch to seal, undocked and began their descent. When the ground crew arrived and the capsule was opened, the men were found, still seated, serene in death. Attempts to revive them proved useless—each had suffered massive brain hemorrhaging. Subsequent investigation determined that when they undocked from their space station, a valve on the return craft had unexpectedly popped open, exposing them to a near-perfect vacuum.

Decompression sickness isn't just a danger during accidents; it's an issue any time you use a pressure suit. You may imagine a space suit as something like bulky clothing, but normal clothing doesn't have to provide a sealed habitat inside itself. It'd be more accurate to imagine a leather balloon that happens to be shaped like a human. And like a balloon, the higher the internal pressure, the harder it is to bend. In a human-shaped balloon, high pressure means difficulty bending at the joints. Like, a lot of difficulty. A phenomenon called "fingernail delamination" is well documented, and we encourage you not to learn what it is. Thus, although the International Space Station is kept at Earth pressure, both American and Russian space suits only have around one third of that.

So, why don't astronauts get bendy, choky, staggery, and deathy when they don space suits? Because they prebreathe pure oxygen before space-

*There have been other spacecraft-related deaths, but they occurred in Earth's atmosphere.

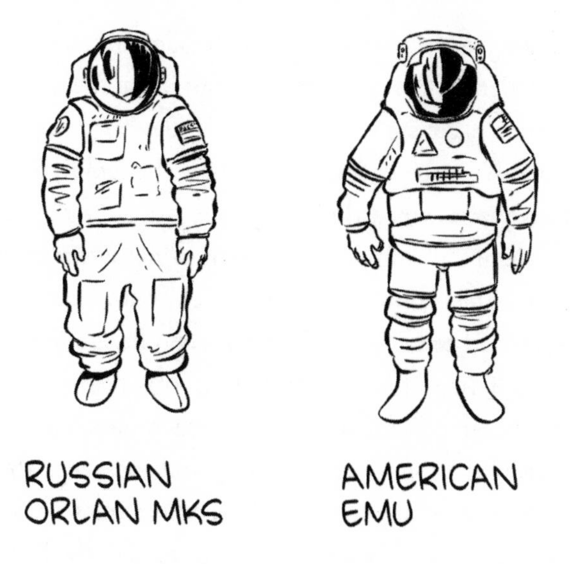

walks, removing most of the nitrogen from their blood. No nitrogen, no nitrogen bubbles.* Movies may have led you to believe heroic astronauts can slip on a space suit and leap to the rescue, but under current designs this would result in Brad Pitt clutching his joints and shambling to a very painful (if handsome) death.

The astute nerd will ask why not just keep the ISS at the same low pressure as the suit. The short answer is that although humans can survive in low pressure as long as there's enough oxygen floating around, engineers would have to design all equipment to operate in a low-pressure, pure-oxygen environment.

But pure oxygen is dangerous. In 1967, during prep for the Apollo 1 flight, a spark went off in the crew's capsule, causing an intense fire in the pure oxygen environment. The three astronauts—Edward White II, Roger Chaffee, and Gus Grissom—could not be rescued, because the sudden increase in temperature and pressure made it impossible to use the inward-opening hatch, while the intense heat prevented rescuers from saving them.

*You may wonder why oxygen doesn't bubble out like nitrogen. The answer is that oxygen gets bound up in various molecules used by the body, so when the pressure drops, you don't get big bubbles of oxygen gas as you do with nitrogen.

Less well known is a similar and earlier incident from the Soviet Union. In early 1961, Valentin Bondarenko was training to be a cosmonaut, and one of the training exercises was to spend ten days in a high-oxygen pressurized chamber. Near the end of confinement, he removed a medical sensor from his body and wiped the sticky glue from the sensor off with an alcohol swab. He absent-mindedly threw it aside, where it landed on an electric hot plate. The resulting fire quickly got out of control, consuming his suit. Oxygen had to be bled out of the chamber before rescuers could reach him, and he died of shock soon after. This happened just a month before Gagarin became the first human to reach outer space. The Soviets preferred to keep their failures a secret, and so when the Apollo 15 astronauts left a plaque on the Moon with the names of astronauts and cosmonauts who lost their lives in the race for the Moon, Bondarenko was not included. His story was only finally shared a quarter century after his passing.

The lack of air in space is well known, which makes it easy to forget just how many aspects of life will have to change if humans expand into space. Vacuum is potentially fatal and perpetually annoying. The risk also changes for the worse as the human presence in space increases because going to space necessarily involves objects at high speed relative to each other. Orbital velocity is about 8 kilometers per second. Around 3 kilometers per second, if an object smacks into your spaceship it delivers roughly the kinetic energy of its own weight in TNT. As long as objects are moving in the same direction at the same speed, things are okay, but in reality they often aren't. This is especially true when nations decide to destroy satellites, sending debris in every direction, as the US, Russia, China, and India have all done.

This is a bigger problem for open space-settlement concepts than for surface settlements, but regardless of where your habitat is positioned, the fact that death is ever present is going to have social and political consequences.

In a space settlement, oxygen will be created by chemical or biological means, but either way it's going to be constructed using systems built top to bottom by human hands. Someone's going to own those. Some space-

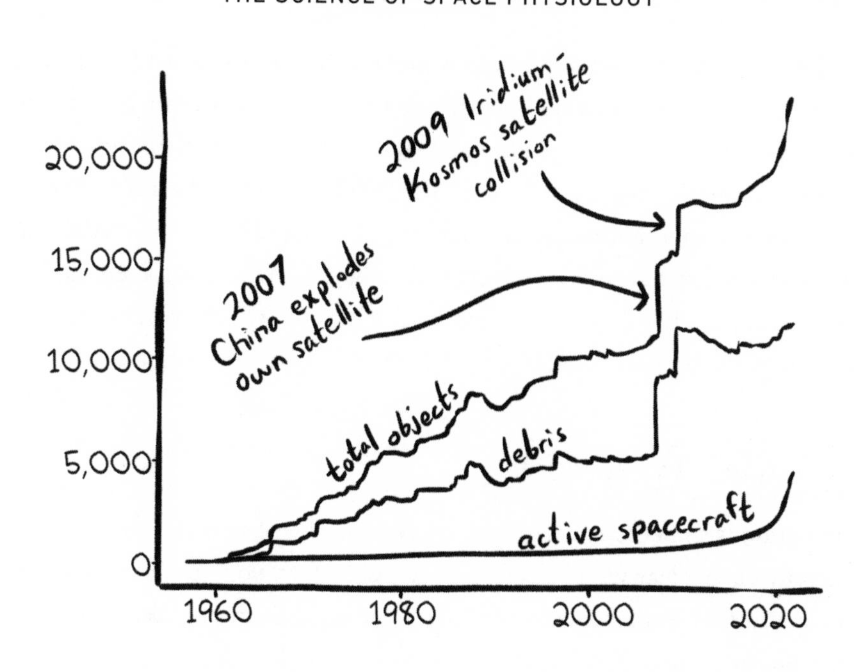

Based on figure in A. Lawrence et al., "The Case for Space Environmentalism," arXiv (April 22, 2022), https://arxiv.org/abs/2204.10025v1.

settlement theorists have argued that the need for artificial atmospherics creates the potential for autocratic control of the stuff of existence. Dr. Charles Cockell, an astrobiologist, has argued on this basis that there should be "engineering for liberty," such as making sure oxygen-creation systems are distributed rather than centralized. We don't know whether this sort of thing will work, but it's likely that the politics of space settlements will be influenced by the physics of space settlements, and not necessarily for the good.

Explosions and Nudity: Lessons on Radiation in Space

Assuming you have found a way to deal with the near emptiness of space, you now should think about an unfriendly presence in the void: radiation. Under special circumstances, radiation can kill fast, but the bigger concern in space is a slow but serious increase in the risk of medical issues, particularly cancer. This is especially worrisome when we imagine going from a world of middle-aged professionals doing a year in orbit to a world where children grow up in space. The need to stop radiation is one of the major factors that will shape human habitation designs off-world, and it will therefore have major effects on quality of life. The problem is that with current knowledge, it's hard to predict the effect of radiation on the body.

Radiation Is Everywhere!

Radiation is sometimes thought to be unnatural—something you encounter only via nuclear waste or atomic bombs. In fact, you are surrounded by materials that emit radiation—radiation comes from the sky, from the ground, from food. The "banana equivalent dose" is a frequently used metric for radiation exposure, since bananas contain potassium-40, a radioactive isotope, yet remain a healthy dietary option that 50 percent of your authors believe is excellent in pudding. Some behaviors apt to increase your relative radiation dose include eating Brazil nuts, going to bed with a human next to you, or taking a trip to Denver, Colorado. Some daredevils may have done all three at once.

As with pressure, the human body evolved in an environment with certain types of radiation at ranges found on Earth. The reason cancer and superpowers remain rare is that most radiation we experience is quietly dealt with by our bodies. The thin shell of dead skin around you provides a natural shield, and your internal machinery is reasonably good at repairing or destroying radiation-damaged cells.

The problem in space is that unless you have sufficient shielding, which we'll explore later, you get higher doses of radiation, and radiation of different types than you'd get on Earth. This radiation comes largely from two places—the Sun and the rest of the universe.

The Sun Wants You Dead

As a radiant ball of plasma, the Sun already spends most of its time blasting out hot ions in every direction. Earth's magnetosphere and atmosphere protect you from most of these. If you're in space, all things being equal, you'd rather avoid solar radiation, but it doesn't cause instant death. However, now and then, the Sun undergoes a "solar flare," when it suddenly increases in brightness.

And then there's something worse: sometimes a solar flare is accompanied by a "solar particle event," which is a particle event in the same sense that a tsunami is a water event. Visualize a relatively small region of the Sun suddenly ejecting a huge stream of protons that move in one direction like a flashlight beam of death.

The good news about these things is that, as science fiction deity Douglas Adams put it, "Space is big." Randomly aimed death beams will likely miss a teeny tiny human ship. But if you happen to be caught in the headlights, the result will be acute radiation sickness, whose symptoms include vomiting, skin burns, heart issues, lung damage, compromised immune system, and—if the dose is large enough—a painful death.

You may wonder what the plan is if you happen to be aboard a spaceship for this sort of thing. For near-term efforts to go back to the Moon, the procedure in the words of NASA scientist Dr. Kerry Lee is to "make use of whatever mass is available." That is, redistribute whatever stuff in the spacecraft or station that you can find because it is now your radiation shield. Why not a dedicated radiation blocker? Because that's a huge amount of mass that costs a lot to send to the Moon and then just sits there. Space settlements will need to do better, and as we'll see the likely solution involves living underground.

Actually, the Whole Universe Wants You Dead

Sometimes a star explodes. This is rare, but happens frequently enough that space is a cross fire of the results. Although they are at relatively low density, very fast charged particles are everywhere. Most of this stuff is low mass—individual protons or helium atoms, but a small percent of the products of these explosions are heavy, fast, charged particles.

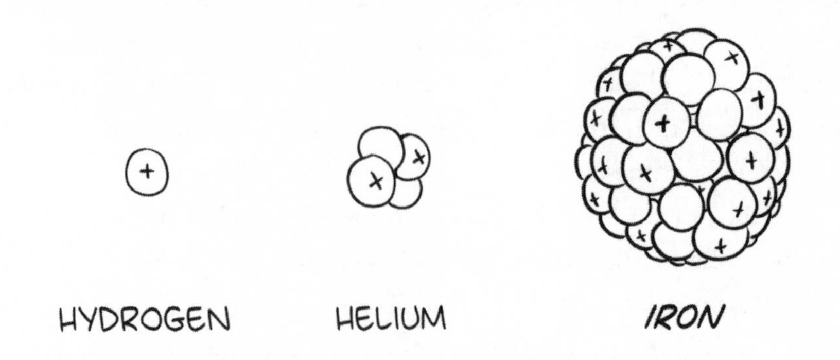

These guys are dangerous. In one experiment, iron nuclei with high energy were fired into a gel-like substance to simulate what space might do to the human body. Individual iron nuclei—*single atoms*—blasted out tunnels as thick as a human hair.

Human hair
~50 microns across
iron nucleus
~ one trillionth of a micron across

NOT EVEN KIND OF SORT OF VAGUELY CLOSE TO SCALE

Exposure to this "galactic cosmic radiation" is a constant part of life in space. Astronauts sometimes report "flashes of light" that only they can see, and this may be the result of their eyes being struck by these far-traveling bits of doomed stars. Current estimates say that once you leave Earth's protective atmosphere and magnetosphere, every single cell nucleus in your body will be struck by a proton every few days, and by a larger charged particle every few months.

Radiation Is Coming for Your Equipment

Radiation can also mess with our technology. In 1859, our planet was hit by a solar particle event that became known as the Carrington Event, after British astronomer Richard Carrington. Records from August 28, 1859, all over America document the effects. In one Boston office at about 6:30 p.m., all the outgoing telegraph lines simply stopped working. In Pittsburgh, sudden currents in their electrical system compelled telegraph operators to disconnect the batteries; when they did, "streams of fire" and sparks jumped out. That night, from California to Britain to Greece to Australia, something like the Northern Lights was observed, with one man describing "a cupola on fire, supported by columns of diverse colors." Disruption continued throughout the day because, as one author wrote, "Earth's atmosphere was still ringing with electrical and magnetic energy." Streams of fire will likely be even less pleasant if you live in a built habitat where electrically powered machines are responsible for ventilation. Nothing like this has happened since, though in 2012, an event of similar power to the one in 1859 missed Earth by what astronomer Dr. Phil Plait told us was "a not-large-enough-to-make-me-happy amount."

Radiation appears to cause problems from time to time on the ISS as well. Here's a story from *How to Astronaut*, Terry Virts's memoir of life on the ISS: It was during a 2014 expedition when he suddenly heard a blaring klaxon. The crew raced to see what was up, and found the "ATM" alarm lit. ATM for "atmosphere." Virts assumed it was a minor leak or false alarm, but Italian astronaut Samantha Cristoforetti remembered exactly what the alarm meant, shouting: "NO—*ammonia leak!*"

Ammonia is not something you want in your space station. Lovely for your cooling system, but toxic for humans and hard to remove. If enough leaked in, it could render the whole station uninhabitable, and in the very worst case, extra gas might overpressurize and rupture a module.

The protocol they were supposed to follow went something like this:

1) Oxygen masks on
2) Float over to the Russian segment, close the first hatch
3) Get naked
4) Close the second hatch to seal off the American segment

You might be wondering about one of those steps, but there's a simple explanation: you seal the American segment because the Russian coolant system uses glycol, not ammonia, so if there's an ammonia leak, it's on the American end.

Oh, step 3. Well, ammonia contaminates clothes, so for safety you should leave them back in the ammonia zone and hope the Russians have spare undies. Following the ammonia alarm, the crew decided not to follow the letter of the law. Mostly because they didn't smell like ammonia, so they figured it was a false alarm. Or maybe they decided sudden zero gravity views of naked coworkers was worse than death.

This entire routine happened a second time, including the now customary elision of step 3. Let this be a lesson for any space-settlement planner who expects to control human behavior or to predict it based on rational self-interest.

Anyway, what's this got to do with radiation? Although there's no way to be 100 percent certain, a suspected cause of the false alarm was radiation thwacking into the equipment.

Radiation has caused problems farther out as well. In 2003, while the Mars Odyssey Orbiter circled the Red Planet, the Sun fired off a huge blast of radiation, which knocked the orbiter out of contact and into safe mode. One of its sensors, designed to detect radiation, was permanently destroyed. In the words of one scientist, it "choked" on the data. Try to imagine being in a Mars-orbiting spacecraft, suddenly losing contact with home, and then being informed that your radiation detector stopped working due to *too much radiation*.

Your Shielding Wants You Dead Too

Oh, and this is fun—even if you provide thick shielding against radiation, you can still get hit by "spallation." Sadly, "spome" and "spallation" are linguistically unrelated—"spallation" comes from a much older word meaning "chip" in the sense of a broken-off bit. When fast, heavy ions smack into your shielding and slow down, they can generate cascades of secondary and biologically dangerous particles, sometimes known as a "nuclear shower."

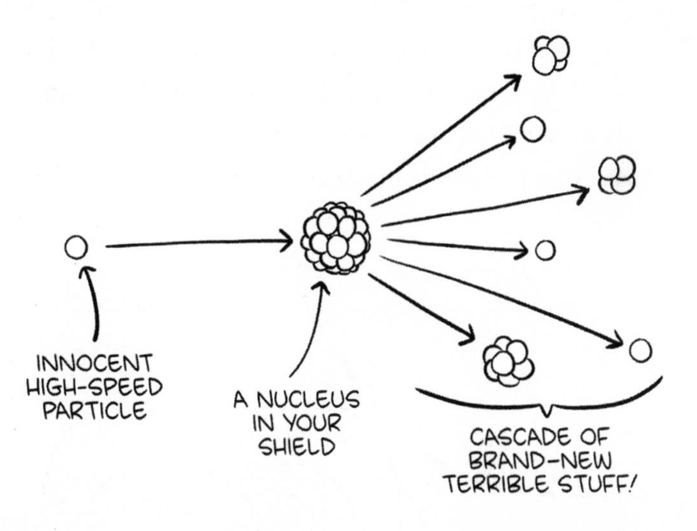

Spallation is particularly illustrative of how mind-bendingly complex designing for space is once you get into the details. For example, if your radiation shield is super thick and is made from aluminum, you can end up being exposed to even more radiation than you would have if you had no shielding at all.

The Decisive Answer on Radiation:

So, space radiation bad.

Right?

Well . . . yes. Probably. We think.

It turns out it's hard to do science on this stuff. The best available data comes from lab animals, from people who've worked with radioactive materials, and from times when something horrific happened, such as the disaster at Chernobyl or when the United States dropped atomic weapons on Hiroshima and Nagasaki. Even these data are imperfect for assessing space radiation. For instance, atomic weapon victims got a sudden massive dose of what was probably mostly neutrons, whereas the typical experience of space involves long-term exposure to charged particles. Lab animal studies are also imperfect because their application to humans is not one to one, and anyhow it's hard to generate space-like radiation in a lab setting.

But wait, you say, we have fifty years of humans on space stations. Doesn't that tell us anything? Sure, it tells us something, but because all space stations orbit under the protection of Earth's magnetosphere, astronauts are exposed to space radiation at doses that are something like two to three times less than what you'd get out in deep space.

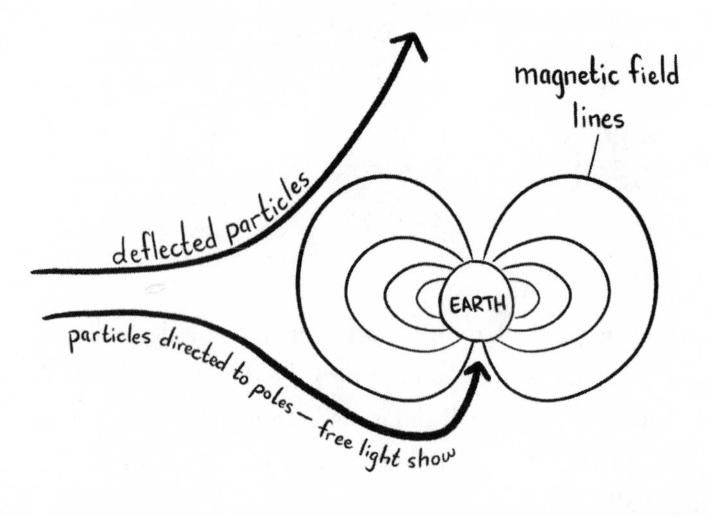

The only direct source we have, then, on the physiological effects of beyond-the-magnetosphere radiation comes from those Apollo missions that went to the Moon and back, the longest of which was the Apollo 17 mission, which only lasted about twelve and a half days.

A typical Mars transit time is around six months. The good news is that although the Apollo guys got a huge dose of extra radiation, they don't appear to have had higher rates of cancer. That's encouraging, but the problem is that it's a tiny sample set—exactly twenty-four men. Also they're not the least bit random—they're ultra elites, most of them test pilots, and all of them got through medical exams that were so thorough they bordered on the sadistic, including intimate encounters with something they referred to as the "steel eel." If these men had less cancer than you'd expect, perhaps it's because they were just a little more robust than most of us.

A more ominous explanation is that the radiation-cancer link is real but not perfectly understood. Doubly so for space. If you've ever wanted to read the scientific equivalent of a plaintive shrug, here's a line from the 2018 paper "Limitations in Predicting the Space Radiation Health Risk for Exploration Astronauts": "there is no definitive evidence that space radiation causes human cancer, but it is reasonable to assume that it can."

The science isn't perfectly clear, but space agencies still need policies. Based on models from the National Radiation Council, NASA has rules for radiation dose levels—nobody gets more than a 3 percent "risk of exposure-induced death."* This apparently reasonable limit creates odd consequences. As it happens, death is not equal opportunity. Some evidence from atomic bomb survivors indicates that ovarian and breast tissue are especially susceptible to radiation damage. Because of this, a few space advocates even call for a no-girls-allowed rule for long-term Mars missions. This might put a damper on plans for population growth in a future settlement. However, under current standards, Mars might be no boys allowed either. The current record for total months is held by Gennady Padalka, who spent about twenty-nine months in space across five flights. That amount of exposure would exceed current NASA limits for all ages. A realistic Mars trip would take longer, with no breaks between trips.

So it's no girls allowed, in that it's no *humans* allowed. It's also certainly no-go for experienced spacefarers who are already near or beyond the radiation limit. The solution? A recent paper by the National Academy of Sciences has argued, in short, that if we want to use veteran astronauts for a Mars trip, we must use Earth's most powerful defense mechanism: a signed waiver.

---

*For the dedicated nerd, here's what that means, according to a recent National Academy of Sciences Report: "3 percent REID implies that within a cohort of 100 astronauts, 3 are likely to die of radiation-induced cancer at some point in their lifetime."

Space's Special Problem: Microgravity

Although astronauts are well within the pull of Earth's gravity, their circular orbit means they are perpetually "falling" toward Earth, sort of like a roller-coaster drop that never ends. Thus, they float, just as they would if untouched by any powerful gravity. Like a roller coaster, the sensation of falling often induces vomiting, but this typically goes away in a few days. Once settled in, astronauts often describe microgravity as one of the singular joys of orbit.

But it's not great for the human body. Microgravity causes predictable physiological problems, some of which are short term, some of which may be permanent, some of which may be still undiscovered because humans haven't spent enough time in space yet.

If this book were about planning a one-year camping trip in orbit, we could offer you some excellent advice. But you're planning to settle space. Here we confess to a major problem with the data. No serious space-settlement plan calls for extended life in microgravity, but this is exactly the gravity regime from which almost all space medicine data comes. The vast majority of proposals for space settlement are about the Moon, which has about one-sixth Earth gravity, or Mars, which has about two-fifths Earth gravity. Most other proposals are for rotating space stations that can artificially create regular old Earth gravity.

We have almost no medical data about life in partial-Earth gravity. The best we have is the twelve guys who collectively spent less than a month on the lunar surface. If there are serious negative effects of life in partial gravity, they likely take longer to show up.

This leaves a lot of scope for wrongness. It may be that the low gravity of the Moon is nevertheless Earth-like enough to spare the human body the worst problems found in microgravity. Or it may be that the effects are broadly similar to those found in microgravity, but take longer to show up. It may also be that there are weird new problems we didn't anticipate.

In terms of understanding the space environment, the best thing we can do in general is tell you what it's been like so far. That's what we'll do here, but understand that our ability to draw conclusions is limited and that getting better data will likely await a large-scale rotating-spacecraft experiment or a first Moon base.

Your Soon-to-Be-Pathetic Physique: The Human Body in Microgravity

From the perspective of bones, walking around is just getting picked up and slammed down over and over again. Your body is prepared for this, but your body is a cheapskate. Bones and muscles are a use-it-or-lose-it affair, and when you're floating in microgravity, you often don't use it.

In space, legs provide thrust off a wall and help anchor your body to the ubiquitous Velcro straps, but most maneuvering is done with the upper body, leaving the rest of you to languish. Four months in space means about 1 percent loss of mass in the spine *per month.*

Incidentally, while your spine is degrading, it's also lengthening in zero gravity. Because of this, lower-back pain is quite common in space and postflight. According to an account by Mike Mullane, on a 1984 flight, the five men aboard had serious back issues. The only person without pain was the lone female flyer, Dr. Judy Resnik. Mullane recalls her saying, "I don't believe it! Here I am going to bed with five men, and they all have backaches."

Muscles suffer a similar fate to bones. One study of ISS astronauts found their calf muscles had shrunk 13 percent after six months in orbit. That might not sound too bad, but note that during those six months the astronauts were doing regular multihour exercise routines. Most spacefarers are back to normal after a month or two on Earth, but in some cases it can take six months to three years.

So after a little time in space, you'll have osteoporosis, weak muscles, and back problems. And, by the way, all that lost bone calcium can contribute to constipation and renal stones. You have left the cradle of Earth for the nursing home of orbit.

After more than fifty years of research in space stations, the best solutions available for these problems are the two worst words in any language—diet and exercise. Vitamin D supplements and osteoporosis medication appear to help with bone loss, and astronauts are still expected to put in 2.5 hours of exercise a day, 6 days a week, to slow muscle and bone deterioration. This is despite the fact that space exercise is the most frequent source of space injury.

For a space settlement with partial-Earth gravity, it's possible that settlers could simulate Earth gravity by wearing weighted clothing. For now, the very long-term effects of exposure to zero or partial Earth gravity are unknown. And as to what the effects would be on a developing human child, we haven't got a clue.

It's in Your Head: The Horrible and Apparently Sexy World of Fluid Shifts

We sometimes think of our circulatory system as a pump with pipes. The heart beats away and blood goes where it needs to go. This is kind of true, but substantially too simple. Blood above your heart just has to fall down to reach it. Blood in your feet needs a good strong push to make its way upward. Well, unless you're doing a handstand, in which case it's reversed. In a way that is unimaginable for the plumbing of a house, your circulatory system stays functional whether you're flat on your back, lying sideways, or hanging upside down.

But when you stay in zero gravity for a while, things get weird. Your legs pump away as if they're still fighting the gravity humans evolved with. Fluid shifts upward, causing your legs' fluid volume to decrease. The result is what one paper referred to as "Puffy-Face Bird-Leg" syndrome. Plus, with your body baffled by various fluids being up so high, you'll make more frequent trips to the bathroom.

Is this bad? Depends on who you ask. As veteran astronaut Lieutenant General Susan Helms said in an interview for the ahead-of-its-time classic book *Sex in Space*, you lose weight and wrinkles while gaining skinnier legs and more height. Some of you may not take heart at the idea of billionaire space settlers also becoming incredibly sexy en route to Mars, but the good news is that during the trip, their bodies will forget how to manage blood flow against gravity, possibly causing dizziness or even fainting as they exit the landing craft.

Our best countermeasure to date is pretty low tech: drink something salty like consommé or sports beverages. Load up on electrolytes and fluid before you return to normal gravity, raising your blood pressure and getting your total fluid load close to normal.

Your Eye in the Sky . . . May Be Permanently Damaged

Microgravity appears to screw up your vision. We don't know why, but the best guess is that the upward fluid shift increases the pressure in your head, altering the shape of your eyeballs and the blood vessels that feed them.

Problems appear to get worse the longer you spend in space. In one survey of three hundred astronauts, 23 percent of those who went on shuttle missions reported difficulty seeing things close up after coming home. That's pretty bad, given that shuttle missions tended to last around two weeks or less. Among those who did longer duration ISS trips, the numbers rose to 50 percent. The problem is more common in astronauts over forty, and they are therefore prescribed spectacles in anticipation of far-sightedness. These are called space anticipation glasses. "SAG." Thanks, NASA acronym team. The over-forty crowd salutes you.

We know most about how this problem impacts eyeballs, though it's possible eyes are just the most detectable form of damage, and more subtle effects are happening to the brain itself. We aren't going to linger on this because at the moment the data are unclear, but there is a possibility that space conditions have cognitive effects. Frankly, between the radiation and the fluid shifts, cumulative brain damage doesn't sound like an especially wild conjecture. If this is a subtle issue that gets worse over time, it'll take on a lot of importance for any space-settlement plan.

And all of the microgravity-related problems we've discussed could get more serious given longer time horizons. Suppose you spend ten years adapting to Mars gravity. Can you come back to Earth? We don't know, and we *really* don't know the answer if your ten years started at age zero.

There aren't any great countermeasures here, but one device tested over the years puts your entire lower half in reduced pressure to convince your fluids to head south. These "pants that suck," as Scott Kelly referred to them, may come with risks. According to his account, a cosmonaut using them once passed out as his heart rate dropped. Scott Kelly nearly did, too, due to an error on the pressure setting.

On Earth, the eyeball problem is studied by making people lie down for days at a time, which produces changes in the eye that we think are similar to what's happening in space. This has allowed for tests of new equipment, and recent work has moved beyond the primitive world of sucky pants. One on-the-ground study trialed out an entire sucky sleeping bag, and it did seem to be helping. Whether sucky pants and sucky sleeping bags will be required on space settlements is a matter for intrepid voyagers of the future.

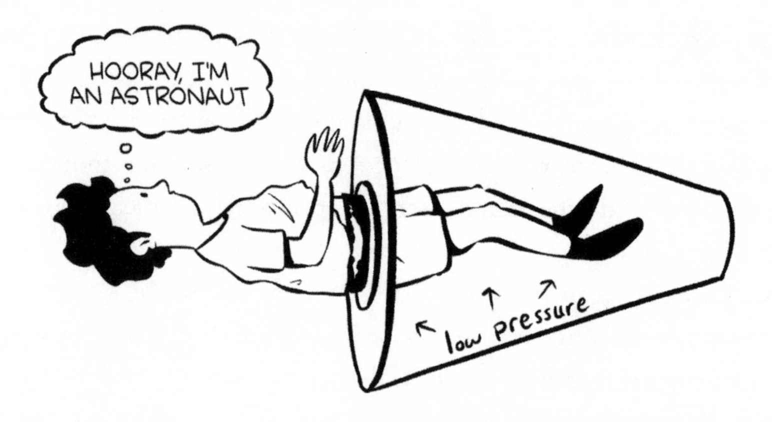

Fluid shifts aren't necessarily the first thing you think of when you imagine space settlement, but given enough time they could be among the most important. Life in *partial* gravity may not produce the same sort of damage, but if it does, space settlements will have to plan for relatively serious vision impairment in most of their citizenry. If there are more serious problems such as cognitive effects, well, let's just hope the sucky pants get a bit more streamlined.

At-Spome Medical Care: Trauma Medicine Off-World

One of our most potent medical strategies is simply to screen out anyone with even a moderate medical issue. But trauma is harder to avoid because you can't screen for "people who won't have an experiment accidentally crush them."

On Earth, common traumatic injury is stuff like airway obstruction, fractures, head injuries, and in general, bleeding where you ought not to bleed. The basic job of a trauma surgeon is to make sure you're breathing and not bleeding too much—things that should be done on-site as quickly as possible, ideally in a matter of minutes. But even in the ISS, which is only a few hundred miles from Earth's surface, evacuation takes six to twenty-four hours.

There are good reasons to worry about this. After a long space journey, spacefarers have all the bodily weakness issues we described above, plus short-term effects pertaining to the change in gravity regime, such as nausea, dizziness, and clumsiness.* They then enter an unfamiliar environment where there is an onus on them to get to work. Accidents will happen.

Unless there's a dedicated sick bay with a proper air-filter system, space-settlement surgeons won't have an ideal operating room. This is especially true in microgravity, where food, microbes, and bits of human waste may be floating around. If surgery is required, the doctor must be trained for the particular gravity regime they're working in. For instance, according to one paper, blood "tends to pool and form domes that can fragment on disruption by instruments." Space trauma surgeons may need specialized imaging training for the gravities they plan to experience.

Although there has never been a dedicated medical center in space, space settlements will need them. Near-term proposals include dedicated

*Clumsiness is a problem because your sense of proprioception—that is, the sense of where your body parts are—gets thrown off because in changed gravity your body doesn't hang in its normal way. Also, lots of astronauts report breaking objects because they've forgotten that when you let go of something in gravity it goes down.

space-station modules called "traumapods" or a "surgical workstation"—a sort of inflatable tent that goes around the patient, protecting them from the environment, and vice versa.

An alternative proposal is to use "minimally invasive surgery." Make a tiny sterile incision and do your work inside, effectively using the body cavity as the surgery workstation. This is theoretically great and works well on Earth, but works only for a subset of medical issues. Also, at least in microgravity, there are problems no Earth doctor has ever experienced. As one paper said, "Bowel floats in the operating field." Another noted, "The tendency of organs to eviscerate has also been reported."

Well, you'll at least get some anesthesia while this is going on, right? Yes, but you can't use inhaled anesthetics, because if there's a leak, you've just released laughing gas into the sealed atmosphere. Another option is spinal anesthesia, but with your fluids shifting upward, the anesthesic might not end up where you want it to go. Your best bet will likely be injection at the site. And by the way, it may not work right. Evidence suggests that human bodies don't absorb nutrients and medication at the same rate in zero gravity. This isn't surprising considering all the fluid shifting and the fact that your stomach contents float, but it does mean that any medication, especially serious stuff like anesthesia, will need to be recertified for each new gravity regime.

Incidentally, you may be wondering how we know so much about zero gravity trauma surgery. How many dramatic trauma surgeries have been performed in space? The answer is zero. At least on humans. We do have, for instance, two Space Shuttle trips where surgical experiments were done on rodents, including one that validated the use of local anesthesia. But that doesn't explain why we know a lot of other weird stuff. Sutures can be done in zero gravity, and apparently it's possible to make "burr holes" in heads while floating. Blood forms domes, viscera wander!

The curious reader will be referred to the copious scientific literature containing the words "porcine" and "parabolic," with titles like "Cardiopulmonary Resuscitation in Microgravity: Efficacy in the Swine During Parabolic Flight." Parabolic flights are a common way to test equipment and protocol in simulated weightlessness. In short, if you fly a plane in a parabola* with just the right arc, you get about thirty seconds of free fall.

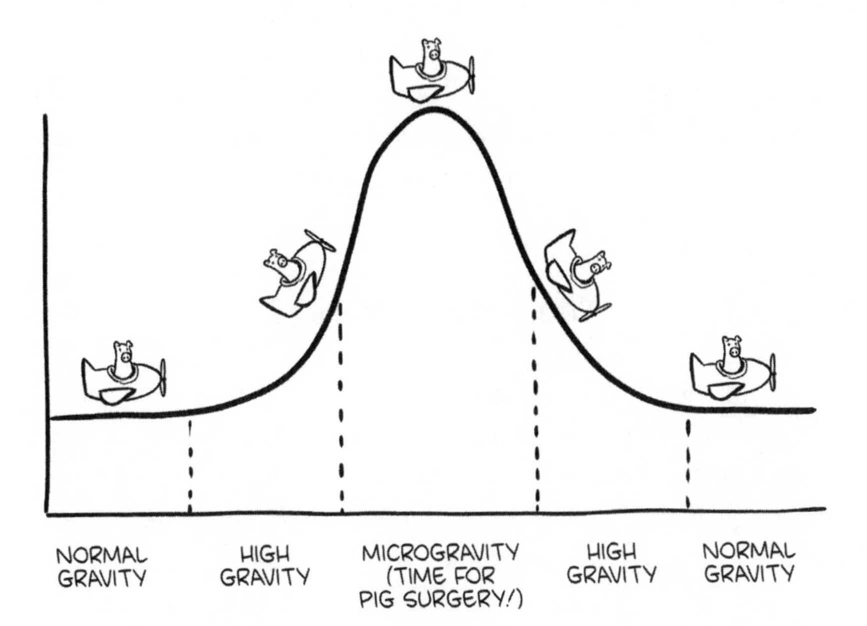

Do this over and over frequently enough and you can get a good hour of simulated weightlessness in a single day. And if you bring along a dead pig and some very very very very very dedicated medical practitioners, you can find out a thing or two about space medical procedures.

We can't go in depth here because this book is sadly not about the brave men and women who endured repeated roller coaster–like plane trips to perform precision surgery on a once-oinky friend. No poet will sing their song. Nor should they. But if you've ever wondered "can we find out how to do a craniotomy in space?" the answer is yes. When pigs fly.

The good news for space-settlement surgeons is that partial gravity probably makes a big difference. If someone spurts blood on the Moon, it'll fall in slow motion, but it will eventually find its way to the ground and ideally a drain. The bad news is that you've got an unfamiliar environment with

*NASA's parabolic flight plane goes by the highly descriptive name "Vomit Comet."

tight working conditions, limited supplies, and no ability to helicopter your patient out to the trauma ward. Accidents should be planned for. Even the badass space heroes of the Apollo program reported frequent falls just walking around the lunar surface and setting up experiments. Any near-term space settlement will be an active construction site employing workers who grew up in a totally different environment. Trauma medicine is a must, even if it calls for a costly amount of mass.

The Future: Worse?

Part of why there has never been space surgery is that we don't generally make astronauts of people likely to have medical problems. But things are changing as space becomes more commercialized. As space tourism grows, the major qualifications for an orbital trip won't be ability, training, and good health, but rather a substantial bank account and the willingness to sign a waiver. The future of space medicine will therefore be less about selecting healthy individuals and more about managing preexisting conditions—a much harder problem about which we know much less. But one we must master if we want to have cities on other worlds.

For now, take a moment to appreciate the scale of our ignorance. Our longest-term space voyage is about 1.3 years. Very few astronauts have gotten anywhere near that total. We have no data on long-term effects of radiation on human bodies outside Earth's magnetosphere. We have almost no data for partial-gravity regimes. We have almost no data on people with chronic health issues.

We suspect that none of these problems preclude space settlement. Assuming a budget of infinity dollars and a lot of technological development, a giant rotating space base with thick shielding might fix pretty much everything. We'll explore whether that's a good idea later, but the important thing is that the above issues may not be insurmountable obstacles. You shouldn't think of them as necessarily deciding whether or not we go to space, but as shaping the constraints whenever we do go to space. The more science and technology we have, the less annoying those constraints will be.

What does this mean for our intrepid settler, Astrid? A pressure suit, some sucky pants, and SAG.

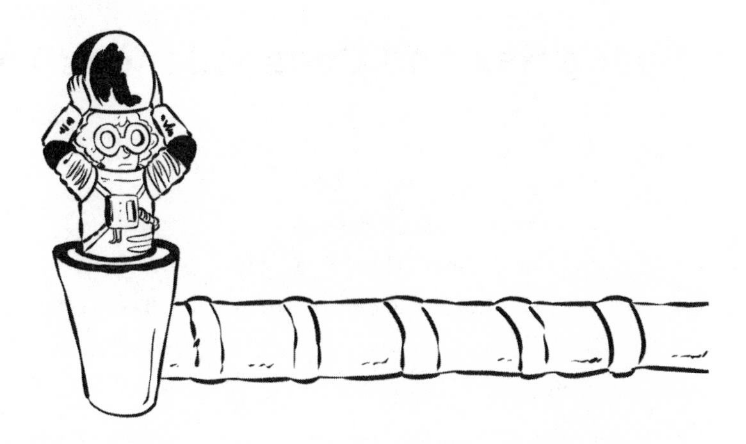

The good news is right now she only feels physically awkward. That will change as we consider a more intimate topic.

3.

Space Sex and Consequences Thereof

> Earth's the right place for love:
> I don't know where it's likely to go better.
>
> —Robert Frost, from the poem "Birches"

Getting Strange in the Lagrange, or, Can You Do It in Space?

Sex probably "works" in zero gravity. In the mechanical sense. Mind you the physics will be a little tricky because every action has an equal and opposite reaction. Also, there isn't a meaningful top or bottom in zero gravity, at least in the physical sense. Here on Earth you can construct a novel sex euphemism just by using the word "horizontal" and the name of a dance, but in zero gravity there is no horizontal. All this may make the "orientationless mambo" awkward. As space popularizers James and Alcestis Oberg once wrote, those who attempt the act "may thrash around helplessly like beached flounders until they meet up with a wall they can smash into."*

On the assumption that this is undesirable, you'll want something to bind all comers together. We've seen more than one proposal for what one author called an "unchastity belt"—a sort of elastic band for two. Another

*The Obergs managed to create a second original fish/sex simile in the same book: "Legs get thinner and chests expand, which is good news for both sexes. This happens as a result of fluids rising upward. However, these same fluids also bloat the face, causing the beloved's features to fill out like a blowfish."

concept is the "snuggle tunnel" for anyone who's ever wanted to experience lovemaking in a narrow, poorly ventilated pipe. There's also the 2suit, which would keep a couple connected via Velcro straps. And then, best of all, in the immortal words of engineer and futurist Dr. Thomas Heppenheimer, writing from the glorious 1970s: "One way to enjoy such zero-g delight will be in a space Chevy van."

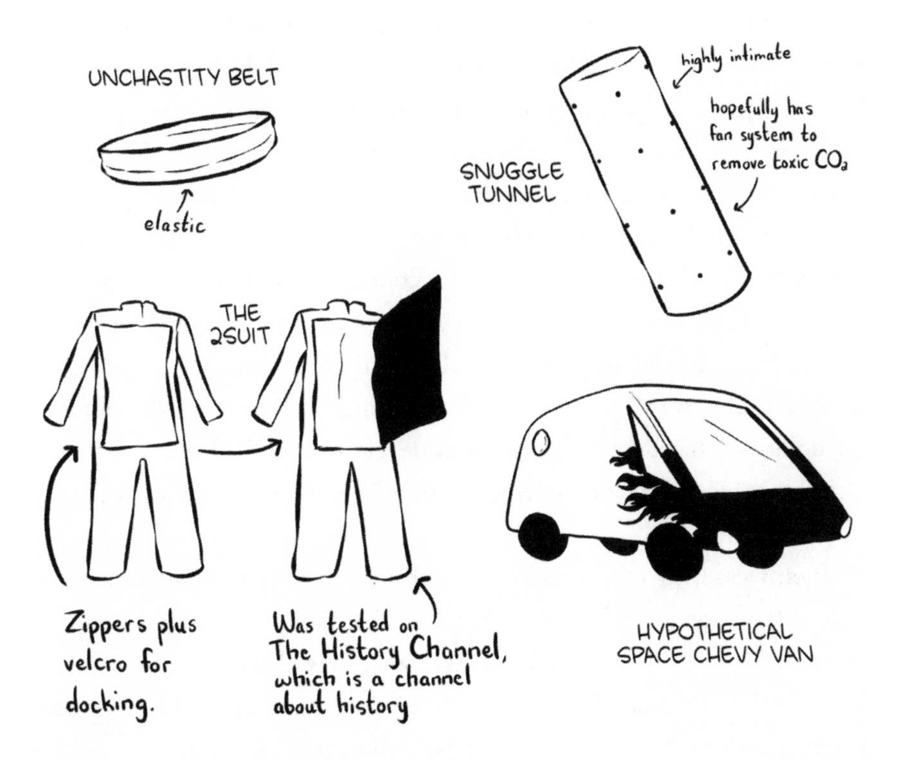

For better or worse, life in partial gravity will be more like life on Earth, only everything will fall more slowly. This will render the snuggle tunnel obsolete, but the Chevy will remain about as desirable as it ever was. The optimal gravity setting will be left as an exercise for future generations.

So much for physics; what about biology? The evidence suggests that the flesh is willing. Space-tourist Dennis Tito reported astro-erections, and Space Shuttle veteran Mike Mullane wrote, no doubt with an eye toward the history books, "I had an erection so intense it was painful. I could have drilled through kryptonite." Kryptonite is not known for its hardness so much as its Superman-withering abilities, but reading between the lines Mullane seems to have been available for mating opportunities.

Desire appears to be present as well: cosmonaut Aleksandr Laveikin said he often missed women back home: "There are sexual dreams, as a substitute." These may have been helped along by artificial means—several accounts refer to a collection of European softcore porn films aboard Mir. Norm Thagard remembers: "There were also some not hard X porno, but sort of soft X stuff, and Veloga would put some of those things on, and Gennady just couldn't bring himself to act like he had any interest in that at all. So Gennady would be sitting there reading a book while Veloga and I were watching this, something like Emmanuelle or whatever."

Other supplements were considered. Cosmonaut Polyakov, who you may remember is the record holder for consecutive days in space, noted at one point there was talk of a "doll which one can buy in a sex shop" being sent up to orbit, though this particular form of self-love apparently never occurred. This was the right choice, according to Polyakov, who opposed sex dolls in space, because a man "may develop a so-called 'doll syndrome,' or in other words, start preferring the doll." Sometimes, people say a lot by saying a little.

In any case, *we* can say with pretty high certainty that onesomes have indeed happened in space. Laveikin said that masturbation was common in orbit. More recently, Scott Kelly was interviewed after a full year in the ISS by Atlas Obscura, and when asked, "Do astronauts masturbate in space," laughed awkwardly, saying, "Can I take the fifth?" According to a paper written by medical experts, these astronauts might have had the right idea: "Little is known about sexual activity in space, and infrequent ejaculation resulting in accumulation of prostate secretions can support bacterial growth."*

But to make a space baby, humanity must move from the onesome to the twosome. Or have we already? Speaking as researchers who tried *really*

---

*Despite a genuine and kind of stupid effort, we failed to find any women astronauts admitting a need to plead the fifth. That could be a genuine behavioral difference or it could simply be a matter of odds—almost all long-term spacefarers to date are men, and most of them haven't admitted to the orientationless breakdance either. It's also possible women are just not talking because they get enough scrutiny of non-career-related behaviors without having this particular subject layered on top.

hard to find an answer . . . we don't know. We don't even agree on the odds. Zach gives it a 5 percent chance. Kelly gives it 75 percent. This has led to a lot of marital strife and the loud and frequent use of the phrase "*Come on.*"

In our experience there are three main camps in this debate:

1. People who think it hasn't happened because there's no direct evidence for it.
2. People who think it MUST have happened given the sheer amount of time humans have spent together in space.
3. People who are pretty sure they read somewhere that it definitely already happened.

People in the last camp are wrong, and we've found they usually refer to a particular shuttle flight: Mark Lee and Jan Davis were both astronauts slated to ride Space Shuttle *Endeavor* in 1992. They had fallen in love during training and gotten married on the sly prior to launch. Newlyweds on a Space Shuttle may be the most plausible case for a successful rendezvous and docking, but listen, a shuttle has about as much living space as a school bus, only with a crew of seven, a sealed atmosphere, and no shower. No doubt this is *someone*'s ideal setting, but if Lee and Davis successfully defiled low Earth orbit, the crew likely would've noticed. We can't categorically deny the existence of a seven-person space-canoodling conspiracy, but we can at least say if we were trying to pull it off, one of the two- or three-person crews on an early space station seems like the easiest place to keep it secret.*

And then there's our favorite space sex rumor: according to G. Harry Stine, an engineer and well-known rocket science popularizer, "clandestine experiments" have been conducted in NASA's neutral buoyancy tank, confirming beyond doubt that "it is indeed possible for humans to copulate

*On the other hand, those crews were almost, though not quite, exclusively male. That doesn't rule out sexual activity, of course, but it's worth noting that most of the Space Age occurred during eras of profound homophobia. Trying to have secret space sex would've been a big enough career risk without the added stigma of doing the wrong kind of secret space sex.

in weightlessness." But, says Stine, an anonymous source informed him it works best with a third swimmer to "push at the right time in the right place." People who executed this particular maneuver in orbit called themselves the "Three Dolphin Club" and even had a membership pin.

"As of 1990," he wrote, there had been "nonscheduled personal activities aboard the space shuttle on *seven flights.*"

One snag in this claim is that dolphins don't actually do that.* In fact, we're not even sure what exactly astronaut number 3 is expected to push and pull on. Visualize it. Or maybe don't. Anyway, the author of the story passed away in 1997. As far as we know, his papers weren't ransacked to find the source of the Three Dolphins story, but given the fact that no other similar claim has surfaced despite apparently nonstop orbital *ménages à trois*, we suspect the only person getting screwed with was Mr. Stine.

Very good. Now that we've finished with the boring topic of space threesomes, we can get back to the fun part: whether all that fussing around in a Chevy van will get you what you really want—*immediate pregnancy.*

A Bun in the Space Oven, *or* Reasons Not to Get Pregnant While Going Around Earth at 7.8 Kilometers per Second

Imagine this: you're informed there's a drug that's really fun, but that results in slow bone loss, major fluid shifts inside the body, renal stones, muscle weakness, dizziness, and eyeball damage. You might still take it if all the cool kids are doing it, sure, but you'd be substantially more careful if you were carrying a baby in your body. Replace the drug with space and you understand why you shouldn't quit birth control in orbit.

Space has potential negative effects at every stage. Sperm and eggs are subject to constant radiation well before anyone dons a 2suit. So is the resulting fetus, and the developing child, and the developing child's gametes.

Microgravity is also frightening. On the one hand, a fetus is in a sort of miniature neutral buoyancy tank, so it's possible the change in gravity regime wouldn't have a huge effect on development in utero. In fact,

*Full credit to best-ever pop science author Mary Roach for being the first to note this problem in her book *Packing for Mars*, which prompted us to read Stine's surprisingly detailed ideas about space sex.

according to the first Google result we could find, you can even do headstands while pregnant if for some reason you feel the need. However, nobody has ever had a zero gravity or low-gravity pregnancy. If any part of the developmental process relies on a consistent Earth-like downward tug, it will be disturbed off-world. Or maybe we'll have to generate artificial gravity by spinning the station. This will be expensive, so an alternative is to require moms to enjoy a centrifuge ride for the duration of pregnancy, possibly making use of this device during labor, which the inventor decided was worth patenting.*

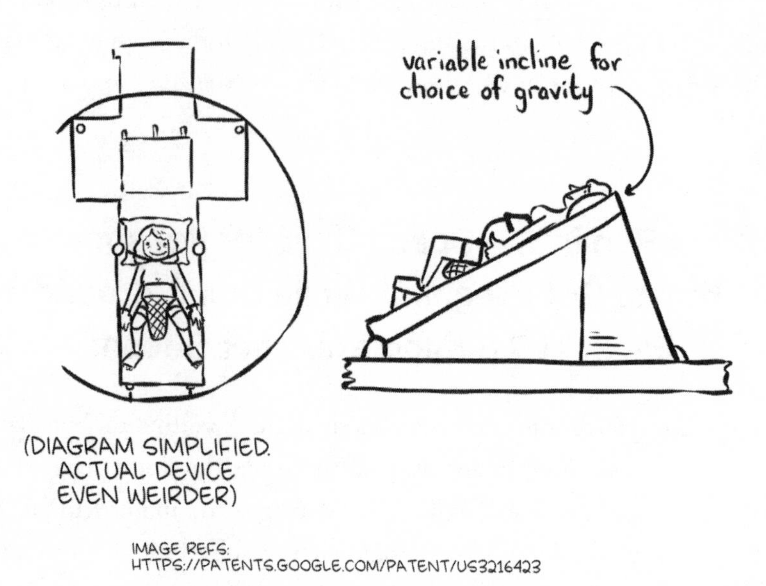

IMAGE REFS:
HTTPS://PATENTS.GOOGLE.COM/PATENT/US3216423

Spinning provides the needed gravity, and moms will never have to wonder whether or not they'll experience morning sickness. And speaking of the privileges of giving life, remember that mom's bones will be weakened in microgravity and perhaps also under only partial Earth gravity. A serious open question is whether birthing is safe with a weakened pelvis. Yep. Bet that mom-centrifuge† is looking pretty darn tempting now.

---

*To be clear, this was originally a very bad idea for use on Earth, but it *could* one day be a very bad idea on Mars. George Blonsky and Charlotte Blonsky, "Apparatus for Facilitating the Birth of a Child by Centrifugal Force," US Patent Number US3216423A, 1963.

†An aerospace engineer friend of ours, Joe Batwinis, showing the usual tendency of engineers toward optimization, noted that if we had artificial wombs, we wouldn't need a momtrifuge, but only a wombtrifuge.

For the baby, the time after birth is especially worrisome. The main current method of preserving bone health in astronauts is exercise, but *you* try getting a three-month-old to conduct resistance training for three hours a day. If you're on the Moon or Mars, possibly you could employ a weighted onesie to solve the problem, but because NASA has yet to spring for a lunar newborn, all we know now is that it'd be really cute. We may at some point have drugs that halt bone loss in space, but whether we're comfortable using them on children is a different matter. What we know about human bones in space today comes entirely from fully developed adults. We have no knowledge about how altered gravity regimes will affect, say, a twelve-year-old girl having a growth spurt.

Another issue at every stage from conception to adolescence is hormones. For instance, there's some evidence of hormonal changes in astronauts, such as lowered testosterone in males. The root cause may simply be stress from riding a giant exploding tube above the atmosphere and then working all day in a high-pressure environment. But we don't know for sure, and once again, whatever the source of the problem, we only know its effects on fully grown adult males, not boys.

Mars in particular may produce hormone problems. We'll get into specifics later, but for now, know that Martian soil is quite high in perchlorates, a class of chemical that messes with thyroid hormones.

Also, what kind of atmosphere are we raising our kids in here? If your settlement is like the ISS, it's one with abnormally high CO_2 levels. That alone would be a novel environment for baby construction, but there may be other issues; when any new equipment goes to the ISS, it has to be carefully checked for what volatile organic compounds it emits. Think about it this way—when a new computer arrives at your house, fresh from the factory, you don't really care if a bunch of weird synthetic gases waft out. You just think to yourself, "Good thing I don't live in a tiny sealed atmosphere!" and go on with your day.

NASA keeps a list of what are called SMACs—spacecraft maximum allowable concentrations. These are how long you're allowed to be exposed to various compounds in a spacecraft. For instance, if you're only exposed for an hour, the atmosphere is allowed to contain 425 parts per million carbon monoxide. Bump that to twenty-four hours and you're only allowed 100 parts per million. If you're staying beyond a thousand days, it drops to 15, which is about what you would find in the kitchen of a house with a gas stove. Why not just always keep things in Earth range? Because it's more expensive. A complaint made by both Scott Kelly and Terry Virts regarding their time on the ISS, which would also be familiar to people who work in submarines, is the high level of carbon dioxide in the atmosphere, which can lead to headaches. Keeping SMACs at a reasonable level is a tricky business, especially when, as one paper laments, "Earth-based [Air Quality Indexes] cannot be extrapolated to microgravity indoor environments."

Now, take all of the above, and add to it that nobody's been in a spacecraft for more than about 1.3 years consecutively, and you can see why, when it comes to safely making babies in space, the answer to most questions is a very *very* nervous shrug.

In fairness to those who are more gung ho, we should remember that in the 1950s, some scientists worried humans couldn't survive in microgravity, period. Maybe reproduction will be similarly benign. That said, one difference is that if we're just trying to verify nondeath, sending up a dog or ape for a little while is a pretty good test. Verifying successful childbirth and development is much more complex. Not only is it harder to test, but if humans *do* have major reproductive issues off-world, it'll be extremely hard to know why. Suppose tomorrow you had a Martian settlement and observed a rate of developmental abnormality that was three times normal. Where do you point the finger? Maybe it's the altered gravity or maybe it's the radiation. Maybe it's the radiation but only in concert with an altered microbiome and the stress of trying to survive in a hostile environment. Maybe it's some ultra-low-concentration SMAC gas nobody's paying enough attention to. And if you're talking about a child conceived and born in space, you also have to think about time—did the problem arise in the parents' gametes, in utero, sometime after birth, or what?

Why Don't We Know More?

Studying reproduction in space is hard to do, expensive, and it's not like the ISS was set up to be a nursery. Also, space agency culture, especially in the United States, has been squeamish about exploring these sorts of questions. This isn't totally unreasonable—nobody really *needs* in-space baby production for a six-person space base, and because these groups receive a lot of public money, we're guessing administrators don't want to have to make public explanations about funds appropriated for zero-gravity-rated sex pipes.

Unfortunately, the work done to date is not the least bit systematic—it's been done on different creatures, in different setups, with different protocols, for different lengths of time. The one thing these experiments have in common is that they aren't run for long enough to be useful for the study of space settlements. If the goal is generations in space, it's not enough to show that conception and birth are possible—we need to know those babies can grow up to have their own babies. We know of no studies on mammals in space where the process was observed from conception through birth, let alone development and conception in the following generation.

Some space simulations can be done on Earth, but it turns out simulating altered gravity is really hard. Note: readers uncomfortable with animal experiments may want to skip the following paragraph.

Some experiments, for example, have been done on small mammals in "hind-limb suspension," meaning their tails have been suspended, causing their rear legs to dangle, arguably kind of simulating weird gravity. It turns out that when you do this to rats their testicles tend to sink into their abdomens, changing the temperature of the scrotum. This can damage sperm. In one experiment, the solution was surgery to keep their testes maintained "intrascrotally." As far as we know, nobody's tried this little maneuver out on humans, but if they did, well, how reliable would you find the resulting data? If the researchers found issues with the sperm, would it be due to the different "gravity" or the presumably quite stressful surgery? Even if you could answer that, would it apply in space?

Experiments in space have similar problems. Launching to orbit means high acceleration with vibrations, followed by floating. If you're a rat, you don't even know why any of this is happening, and stress can produce weird physiological results. For example, early on, researchers thought mice might not experience estrous cycles in space, but it turns out in longer experiments, when you examine the mice before they're brought back down to Earth, they do. Plausibly this is because they've been up there long enough to get used to the new environment and haven't been stressed out by falling back to Earth, but we need to know more.

While we don't have systematic or long-term data, what we can say is that the available experiments indicate that there's not a chance in hell *we* would want to try having and raising kids in space. Experiments show that radiation causes havoc for gametes and that microgravity may damage cell cytoskeletons. In case you don't remember high school biology, these are the structures in a cell that give the cell its shape, kind of like the wooden beams in the walls that hold up your house.

Experiments involving creatures like tadpoles, salamanders, geckos, newts, quail (eggs), rats, and mice have found results like increased rates of abnormality, unusual swimming patterns, unusually large heads, unusually long tails, increases in infant mortality, decreases in oxytocin (an important hormone for birthing and bonding), and in one case an entire rat litter was stillborn due to a single abnormally large pup that wouldn't fit through the mother's birth canal.

But there are also studies where things went well. And there are studies where abnormalities early in development went away a little later. And this is why it's such a problem that no sustained, long-term studies are happening: maybe there is some stage in the reproductive cycle that is irreparably damaged by the space environment, or maybe bodies can compensate for the space environment and be totally fine in the end. We just don't know yet.

And will any of this apply to humans? We don't know that either.

A number of human women have become astronauts and then returned to have children.* They often required assisted reproduction technology.

*Men, too, but male gametes are replenished over time and theirs are not the bodies in which babies are being created.

However, there's a more boring reason for this than space travel: age. The average age at which women astronauts first go to space is thirty-eight, and these women typically wait to have children until after their flights are complete. On the plus side, astronaut moms were not more likely to have spontaneous abortions relative to comparably aged non-astro moms. That's good news overall, but because sexism is the gift that keeps on giving, a grand total of seventy-five women have ever gone to space as of this writing. There just isn't a lot of data to work with.

We wish we could say more, but it's hard to have much confidence in anything pertaining to having babies off-world. We've repeatedly encountered space-settlement enthusiasts who believe that because we don't know whether there's a problem, we should assume it's okay. Given what we know about space medicine and what little we can say from animal reproduction involving space, we think this is insane. Any space-settlement plan that doesn't at least try to account for this stuff should be considered to have a major gap in research.

Mars Ain't the Kind of Place to Raise a Kid

Strangely, in his song "Rocket Man," Elton John didn't mention the potential for pelvic fracture during labor or the need to be centrifuged while pregnant. Such is poetic license. The truth is we don't know if he was right or not, and we don't seem apt to find out soon. If you really wanted to know whether humans can have babies in space, the best experiment would be something like taking over a whole ISS module and dedicating it to a rodent colony that could be observed through a series of generations. The most relevant, highest-quality project on the ISS today is likely MARS—the Multiple Artificial-gravity Research System. Run by the Japanese space agency, JAXA, it is able to simulate gravity. But even that system doesn't simulate all the problems of particular space locations, such as Martian toxins or lunar levels of radiation.

We are not saying that any of this is impossible to solve. But as with space medicine generally, getting the knowledge we'd need to have reproduction in space that is safe and ethical would be a massive, costly, decades-consuming affair, and strangely, among people advocating for vast space settlements in the next thirty years, nobody is doing the sort of enormous spending necessary to get answers.

What we will say is that scaffolding from current knowledge levels to being able to successfully make babies in space is going to be a tricky business, not just in terms of science, but from the perspective of scientific ethics. Adult humans can consent to be in experiments. Babies can't. Electively putting a child in a space environment should only happen after we have a lot of positive data from other organisms. Ideally, this would at some point involve something like a primate center in space, complete with a monkey day-care center and whatever the monkey equivalent of a Chevy van is. This will be very expensive and very time-consuming and of course comes with its own ethical considerations. If there is no urgent reason to settle space, what is the ethical justification for a huge number of experiments on animals, including humans too young to consent?

Tech Fixes

An alternative approach would be to make facilities so Earth-like that we can be fairly comfortable about safety. We'll explore orbital space stations

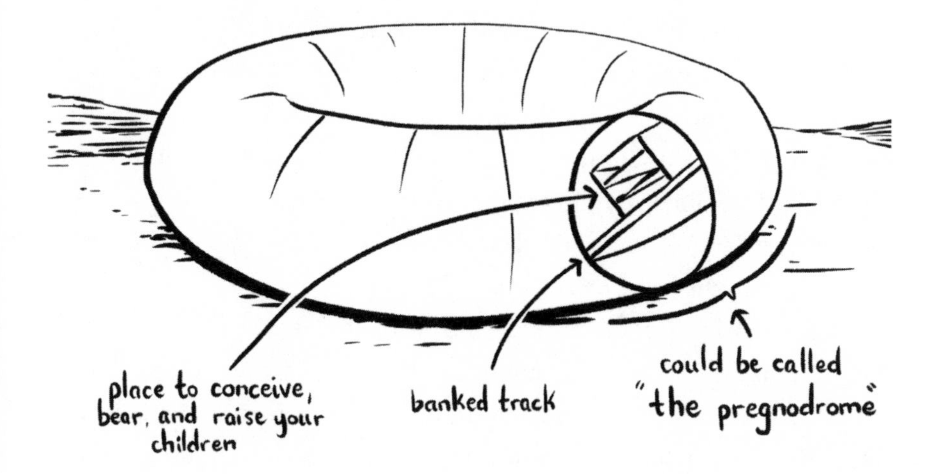

shortly, but one major virtue they possess is the ability to create Earth-like gravity. If we find out that you can't safely bear and raise children on Mars or the Moon because partial gravity creates developmental problems, the solution might have to be a sort of orbital baby-production station. A kind of combination honeymoon suite, day care, and kindergarten in the sky.

For people who want to stay on the ground, there may be alternatives. One author suggested a gigantic banked racetrack on the Martian surface. Pregnant? No problem—just hop on the Earth-gravity Tilt-A-Whirl.

Even if you do have this setup, which we call "the pregnodrome," you'll likely still need shielding against radiation. So it's not just a pregnodrome, it's an *underground* pregnodrome.

This is costly, dangerous, and frankly traveling 40 million miles to live in a heap of dirt is a bit of a disappointment. The alternative is to wait for more technology, but sometimes the desired technology is pretty futuristic even if you're already imagining a pregnodrome. At a National Space Society Space Settlement Workshop Kelly attended in 2021, the Human Biological Response to the Space Environment group emphasized the need for an artificial womb in radiation-proof cladding. But artificial wombs are, and we apologize for this joke, still in their infancy.

Handing pregnancy over to an armored baby-machine may strike you as a bit weird, but it's just the tip of the radiation weirdberg. At the society level, things get stranger still. As we've seen, the risk of radiation is affected by age* and sex. This will have to be accounted for if the goal is successful reproduction. For instance, would it be appropriate to simply keep all developing humans underground, while the elders are allowed to explore the surface? If so, would there be sex differences in when you're allowed to leave the base?

Or suppose humans need full Earth gravity until they're eighteen. What's your responsibility there as a parent? Do Earth and space stations become a kind of spawning ground humans return to for their entire reproductive lives? And note that maybe returning to Earth isn't even an option for you if you've grown up on Mars. If you've lived in partial gravity long enough, then perhaps your bones just can't handle Earth gravity. Or, if space settlements escape the many parasites and pathogens that plague us here on Earth, perhaps an immune system trained in a relatively sterile Martian environment wouldn't be able to keep up with the onslaught of pathogens living on Earth. Again, basic scientific facts may ultimately affect social and political outcomes in ways that are hard to predict based on our limited knowledge.

BioTech Solutions and Their Problems

A commonly proposed solution to a lot of these issues is to perform some sort of genetic engineering. There is at least one book entirely devoted to the question of genetic modifications in space, and it includes proposals that range from prolonging human life spans to altering the ABCC11 gene, responsible for armpit stink in some populations, as a way to improve social harmony. Single-gene edits are possible, but we suspect more complex ideas, like radiation-resistant outer skin or genetic alterations to help bones in microgravity, are a long way off scientifically.

*One of the more uncomfortable reasons age may matter is that if you're older, you have less time to develop radiation-induced cancer. That is, if space radiation increases your yearly odds of getting cancer, one option is to send up people who have fewer years available.

More important, we think the ethical costs are potentially tremendous. Bioethicists already debate whether it's appropriate to develop "designer babies," but it seems pretty straightforward to us that creating new humans adapted specifically to hostile space environments far from the homeworld purely so they can settle it is morally dicey. Just getting to the stage where such a thing would be possible is hard to imagine without a long series of genetic experiments, with questionable medical value, done on babies. If there were some kind of extreme urgency for human migration to Mars, perhaps this sort of thing could be argued for. But there isn't.

When we talk about populations and growth in human settlements, we often talk at a high level, with humans as ecological and economic inputs—as statistics for a settlement strategy. We're nerds—we like stats. We think the aerial view is often appropriate. When we hear things like "statistics aren't people" we adjust our glasses in displeasure. But you have to be careful, because sometimes abstraction can conceal the potential for atrocity.

If your goal is to produce a lot of babies rapidly in an environment apt to create medical problems, the likely result will be a nontrivial population of children not adapted to the harsh and demanding environment. You may have kids who have physical or cognitive disabilities that mean they can't contribute at the level of nondisabled children. You may have kids who, for whatever hard-to-suss-out reason, have difficulty psychologically thriving in a confined unearthly environment. How do space-settlement enthusiasts react to these hard questions? As far as we've found, most don't bother to deal with it. Those who do often possess a surprising level of banality about the likely consequences.

Space anthropologist Dr. Cameron Smith, who we genuinely consider one of the smartest voices in space social science, argues that a space society may favor evolutionary adaptation. As he says, you could have a culture that will endure the "painful transition" required to "allow natural selection to tailor individuals to their new environments." Dr. Alexander Layendecker, one of the best-known researchers on human sexuality in space writes, "if we cease to exist as a species, the deliberation of morality and philosophy issues becomes a moot point. Historically, the human drive for survival has a way of displacing morality and ethics in times of desperation." Perhaps most openly there is Dr. Konrad Szocik, a philosopher specializing in the ethics of space exploration. In a 2018 paper, he and coauthors write, "The idea to protect life at every stage of development may not be suited to a Mars colony." In the same paper, they argue, "We assume that the Martian colony environment would favor these liberal pro-abortion policy [*sic*] because the birth of a disabled child would be highly detrimental to the colony. . . . A Martian community may set new or higher criteria for valuable offspring, and may evolve to favor the preservation of personal and physiological traits more suitable to Martian residents."

Now look, you could argue that these are not moral arguments, but merely descriptions of what might happen. That is, these are not statements about what *ought* to happen, but what *will.* That would be reasonable if, say, we were talking about humans marooned on a small island for generations. They might indeed develop norms and institutions that would have previously been considered morally questionable. The difference is that in the case of a space settlement, we are making a choice with full knowledge of the danger and potential consequences. If someone has a plan to send one hundred people to live stranded in Antarctica, and the plan notes that the Antarcticans might develop a culture that sees morality pertaining to disabled children as moot, your response won't be "neato!" it'll be "DO NOT SEND THEM."

For many people who envision a Martian city, the entire project is wrapped in human aspiration—a chance to make humanity better with a new start. But if the new start requires us to create a new ethical framework that is anathema to most living Earthlings, then what's the point? Note well, this problem isn't just for the settlers and their children. If you accept Daniel Deudney's argument that a large human presence in space gives us a lot of power to destroy ourselves, you should think twice about creating settlements where you anticipate the culture will place a low value on human life.

Going Big

We're sorry to have begun with jokes about the snuggle tunnel and ended with eugenics. Babies make things weird, don't they?

Here's our overall take: everything about reproduction in space is cause for concern. This doesn't mean we can never settle space, but it does argue against certain paths. Many space-settlement proposals call for a bootstrapping approach, in which you start with a minimal base, and then each landing facilitates the next, allowing for exponential growth of the settlement population.* It's possible this would be the quickest way to grow a population, but if there really isn't an urgent reason to get a million people to Mars, there's a good case for a wait-and-go-big approach to settlement. This is an idea we'll return to several times in the book, and the idea is basically this: First, wait until you have the data and technology you need to safely and ethically move forward. Second, if you do try to build a space settlement, you'd be better off in a situation where you can settle a very large number of humans during just a few years. This is a long way off technologically, but the advantages are very clear in a medical context. More population means more specialization—more people specifically doing medical work. It also means that society has enough care workers and redundancy that nobody would ever have to say of a child that evolution needs to run its course on them.

If current technology barely permits survival and only permits natural population growth via throwing conventional moral standards out the window, and if there's no reason to leave right this second, why not be patient? Spend a few decades at least in order to advance the science of human reproduction, as well as every other technology relevant to space settlement. Then, we can send a population large enough and with advanced

*This doesn't have to involve reproduction, but unless you're willing to enforce birth control on millions of people, it probably will.

enough technology that we aren't required to "set new or higher criteria for valuable offspring."

As for Astrid the space settler, we're just going to assume a happy ending on the babies front. The next question, as for so many new parents, will be keeping her wits together.

4.

Spacefarer Psychology: In Which the Only Thing We're Sure of Is That Astronauts Are Liars

In our experience, when the question of space psychology comes up, people tend to fall into two camps: either they think it's a nonissue brought up by pearl-clutching wussbags, or they think that, in the distant and isolated Martian landscape, people will surely go mad in the night. Or anyway, that acute psychiatric problems may be uncomfortably common.

Here, we're going to argue for a third way. Psychology in space is a serious constraint on space settlement, not because of dramatic problems, but because space settlements may be broadly similar to Earth in their rate of mental illness. Given that Earth has lots of facilities and specialists for this sort of thing, space settlements should do likewise, which means once again that a large settlement with a large population is desirable.

It's not surprising that people worry about something like space madness: space settlements will likely be claustrophobic and dangerous. People who've maybe seen a few too many horror movies may wonder whether settlers in such an environment would experience some kind of mental break or even become violent. That's certainly possible and should be planned for, but it's also likely to be quite rare. Humans have been inhabiting space as well as cramped isolated environments like submarines and Antarctic bases for almost a century now, and outside a small number of serious incidents, things have basically gone all right. There are plenty of anecdotes about boredom or malaise near the South Pole, but then people are bored or sad once in a while on the other six continents too. Space will likely be similar.

The one caveat we'll add to this picture should be familiar by now—the data sucks. It's arguably substantially less reliable than even the physiological data. You have the familiar concerns about unsystematic experiments and lack of long-term studies, but now add in the fact, which we will document soon, that astronauts are liars. Especially about psychology. We do not believe self-reported astronaut data. Throw in the fact that we still don't know if there are negative physiological effects on the brain due to long-term exposure to space conditions, and it's reasonable to want a bit more science before we start building a Martian city.

Space-settlement planners can be surprisingly blasé about this stuff. Kelly once attended a space conference talk about a new system for managing astronaut psychology off-world. The idea was clever—place sensors on your astronaut so that you can detect emotional disturbance. If they are disturbed, automatically lock the doors, sealing in the deficient human. Kelly asked whether the speaker thought that astronauts would be cool with this constant surveillance. At that point, the session moderator stepped in to say that, hey, famed astronaut John Glenn wore a rectal thermometer for his entire trip aboard Friendship 7 and nobody heard a peep out of him.

This is literally true, or at least it's true that Glenn's voyage featured that most intimate form of data collection. But it's worth noting that his trip was just about five hours long, not the multiyear time frame of an initial Mars trip, never mind a permanent base or settlement.

This is apparently not an isolated theory of human psychology. We found one paper by a researcher at Johns Hopkins, proposing a computer that would automatically dope astronaut food to maintain social harmony. We think this might be seen as a wee bit invasive, perhaps reflecting the

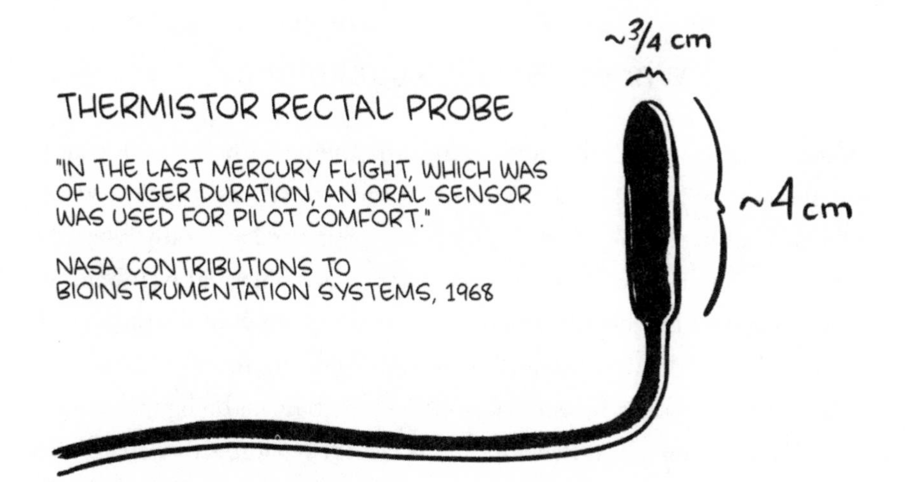

tendency of some in the sciences to see astronauts as something like machines that happen to be made of meat.

But it is possible to go too far in the other direction—to think of humans as overly delicate. One prominent space psychologist talks of an "Earth-out-of-view" effect, where people might suffer serious psychological issues on deep-space voyages after the stark realization that the Earth now appears as just another speck in the sky. This is an idea that as far as we can tell has no basis in observation or theory, however poetic it sounds. And let's remember that while Earth may look like just another dot in the darkness, it's still transmitting Netflix. Our view is that space psychology really matters, and may matter a lot if the long-term data turns out bad, but that the likely case is at least somewhat Earth-like levels of psychiatric issues. However, this doesn't mean space-settlement plans can treat mental health as a nonissue—quite the opposite. We need to produce at least something like the infrastructure here on Earth. What that'd look like on a settlement is hard to say, but we *can* tell you what's done in space right now.

What Is Space Psychology?

Basically, it's keeping astronauts on the job. The term "behavioral health" is sometimes preferred, in part because it avoids any stigma about the words "mental health." It's also apt in that behavior is fundamentally what the concern is. Space psychologists do care about psychological well-being—they sweat specific details about food quality and how birthdays are celebrated—but the ultimate goal is mission completion.

You might wonder why astronauts—elite superhumans doing their dream jobs—have psychological issues at all. The short version is that space on a day-to-day basis isn't so great. It's not the least bit *Star Trek*ky. There are tedious chores: meal prep, cleaning the bathroom, scrubbing mold, taking inventory. Private accommodations are the size of a coffin, funny smells waft from the waste containment facility, the food is mediocre, exercise clothing gets reused for more days than Earthlings would dare. Bits of human waste frequently find their way to airborne status, as do sloughed-off foot callouses from feet that no longer strike the ground. The equipment is quite loud, it's hard to get enough sleep, and historically space stations are pretty crowded. The ISS has about the habitable volume of two average American apartments, but it contains six people, all their equipment, food, exercise stuff, experiments, and every kind of astronaut waste from food containers to feces.

The rectal thermometer is no longer a perpetual presence, but because space stations are science labs, other bodily invasions like blood draws do occur from time to time. Even when they don't involve the extraction of bodily fluids, schedules are packed with experiments, public relations, and maintenance work. All of this is punctuated by occasional emergencies, which have historically included fires, near-drowning in a space suit, and the need to rapidly locate a leak in the space station.

Space as we've known it is a particularly unusual version of what's sometimes called an "isolated and confined environment"—a type of place where humans notoriously experience a certain level of discomfort. The goal of space psychology is to identify problems the environment might

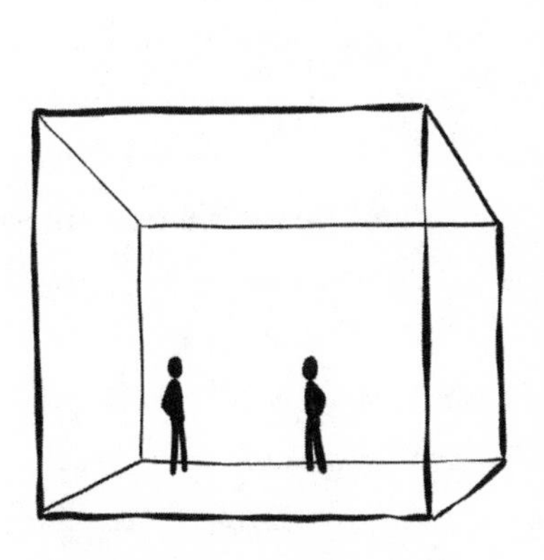

produce and thwart them. The way this is done is via what are called "countermeasures."

Countering the Space Blues

Countermeasures is an overly clinical term for pretty much anything that improves behavioral health. There are essentially two ways to do this—screening out people likely to have problems, then supporting the people who make the cut.

Preflight Countermeasures

Exactly what makes for "the right stuff" has changed over time. As the era of high-flying Space Heroes gave way to the era of long-duration astronauts, the desirable psychological profile shifted. A few quotes will illustrate the difference. The first is from the memoirs of Deke Slayton, one of the original Mercury astronauts, talking about his days as a young pilot:

We spent about a week drinking at night and taking go pills all day long. Looking back, I'm amazed at the amount of drinking and flying we did.

Here's a more recent quote from the memoir of Space Shuttle–era astronaut Dr. Mike Massimino:

Astronauts coming back from space had to splash down in the water, and I hated the water. I didn't know how to swim that well. . . . I was scared of heights, too. Still am. Standing on a balcony four or five stories up and looking over? No, thank you. I didn't like roller coasters, either. They're scary. Hanging upside down? It makes you sick. Who wants to do that?

Space travel, it is clear, has reached middle age.

Although the criteria for selection have changed, the general procedure hasn't—it consists of medical exams, psychological exams, and interviews with the candidate. Different programs have their own exotica—JAXA has candidates fold a thousand origami cranes over the course of a week to show they can maintain high standards while doing monotonous tasks. Russian candidates are dropped out of an airplane and made to solve puzzles tied to their wrists before the parachute opens.

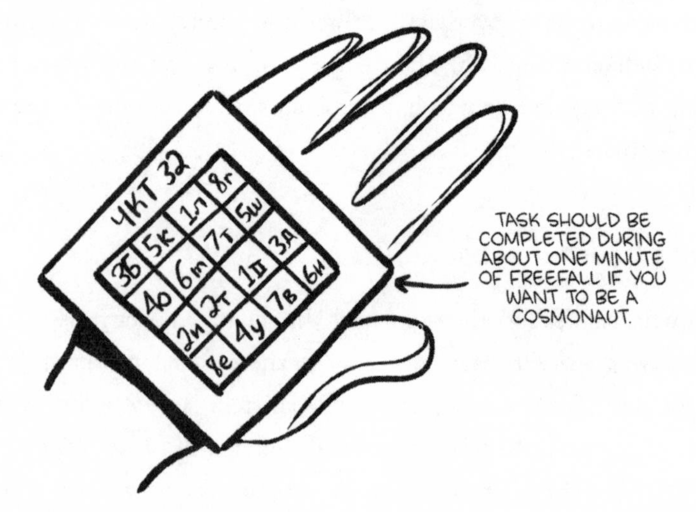

But the basic goal is the same everywhere; the optimal astronaut is a sort of hypercompetent Ned Flanders—reliable, even-tempered, a bit boring, and also the kind of person who could perform brain surgery in a flaming jet if called upon to do so.

Once candidates are accepted, they often wait years—in some cases over a decade—before they ever get to go to space. Much of this is just pragmatic—space flights are few and sometimes delayed quite a bit, but the long preflight period also allows psychiatrists to assess how people work in teams.

These sorts of screening mechanisms are very important, but will not work well for a permanent settlement. You can, of course, screen applicants, but if you want Elon Musk's plan for a million Martians, well, there are simply not a million Chris Hadfields to go around. Standards will have to fall. It's possible that the nature of the project itself will attract the wrong kind of people. The now defunct Mars One program, which proposed to build a Mars settlement for reality-TV purposes, successfully signed up thousands of people for a one-way Mars trip. They never went, of course, but it's worth noting that many of these people had families, including children, that they were willing to leave behind. Not to be too judgmental, but if the goal is a new society, well, these people sound a wee bit antisocial to us.

Another source of subpar candidates will be your children. One of the most effective ways of screening astronauts is age, because a number of psychiatric disorders are most likely to manifest in the late teens and early twenties. A forty-five-year-old may need SAG spectacles, but you typically know what you're getting psychology-wise. In a space settlement, humans will be made locally, but it is notoriously difficult to get a newborn to fill out a psych questionnaire or conduct an interview with a professional. In the long run, screening will not be an effective means of dealing with human psychology. Settlement psychology will be much more about management than selection.

On-Flight Countermeasures

One of the most frequently cited countermeasures is the chance to communicate with loved ones back on Earth. When cosmonaut Valentin Lebedev was feeling down during his 211 days in orbit, mission control noticed and brought in his son to play the piano for him during a flyover. Lebedev recognized the player at once and tears came to his eyes.

Things today are less dramatic. For most of space history, contact with home was limited. Lebedev's call had to be snuck in during a brief open communication interval and prioritized over mission information. Today, the ISS has internet. Perpetual access to email, video, voice calls, and psychological counseling services.

Whether this sort of care will be available in a settlement is a matter of size and location. The Moon, for instance, is close enough to Earth that you can have something like a live conversation with a psychiatrist. Mars is not. Although early Mars trips, or even Mars bases, may not want to spend the mass to bring in a professional psychiatric staff, a full-size settlement should. At the very least, they should cross-train some of their people to have the psychiatric skills to spot problems before they get worse. As for friends and family back home, well, that'll be hard. At greatest distance, it takes forty minutes for a signal to get from Earth to Mars and back. Communication will be more like letter writing than full conversation.

For NASA astronauts, more important than family, friends, or thoughtful guides through psychological distress is the fact that sometimes William Shatner calls. As has Will Smith, Stephen Colbert, the pope, and best of all, Fabio. Astronauts are allowed to request this sort of thing, and NASA does its best to make it happen. Tragically, it will not be possible to speak live with Fabio from Mars settlements.

Day-to-day amusement is also pretty good these days. As late as 1997, Dr. Michael Foale scored points with his Russian compatriots aboard Mir by setting up a VCR and live-translating *Total Recall* and *Crocodile Dundee*. Today's ISS has Netflix, and according to Terry Virts, "the Russian psychologists uplinked *50 Shades of Grey*, in Russian." Not exactly high art, but then again if the goal is population growth, there are worse choices.

Astronauts also regularly receive little gifts with resupply. In one recent case, a scientific refrigerator was sent up preloaded with ice cream—no doubt a welcome break from typical space fare, which often consists of lukewarm food packets. Resupplies also bring fresh fruits and vegetables, which provide benefits beyond taste and nutrition. For Viktor Patsayev's in-space birthday on Salyut-1, he celebrated with a fresh lemon and onion. Anatoliy Berezovoy and Valentin Lebedev craved fresh onions so much they once ate the miniature ones meant for an experimental cultivator on Salyut-7. For Foale, the scent that brought him home was fresh apples. Resupply also brings little surprises for holidays and birthdays. Especially beloved are care packages from friends and family. These are often very simple things—Dr. Shannon Lucid recalled Mir cosmonauts appreciating perfumed letters from the women in their lives. Lucid herself was pleased to receive Jell-O and a big bag of M&Ms.

Ironically, for a distant space settlement, fresh apples may be easier to acquire than M&Ms. As we'll explore later, you basically can't have space settlements without creating an ecosystem to grow food and clean air. Or anyway, you can, but the price difference between growing an apple and flinging it to another planet is likely to remain substantial.

One fact of human psychology we found even in old polar exploration diaries was a deep concern with celebrations and with food. Crews on those monotonous trips often celebrated whatever holidays they could lay their hands on as an excuse to bring some novelty into the world. Today's ISS crews receive themed care packages for things like birthdays and holidays. Mars can only be reached once every two years, but it might be smart to mimic the ISS and send up impossible-to-get Earth goodies for distribution on special occasions.

In-space farming also likely provides a psychiatric benefit. Partially this is just the love of fresh food, but it's also the sights and smells of nature in an otherwise artificial environment. One way to think about onflight countermeasures is that they are ways to recreate a little bit of Earth far from home. Cosmonaut Anatoliy Berezovoy loved cassette recordings of nature during his 211-day flight in 1982, recalling, "We had recordings of sounds: thunder, rain, the singing of birds. We switched them on most frequently of all, and we never grew tired of them. They were like meetings with Earth." Terry Virts described how the Russian psychology team sent up a set of recordings called "Sounds of Earth"—ambient noises such as ocean waves or the inside of a café. These recordings were beloved. For an entire month, Virts went to sleep to the sound of rainfall, and at one point the station crew agreed to play it simultaneously on the ship's laptops for a whole weekend.

There is an entire field of literature dedicated to these sorts of space ergonomics, including detailed analysis of exactly what foods astronauts like,* whether they're sleeping enough, and what they're apt to complain about in private diaries. Astronaut behavioral health is not perfect—there are complaints, there are arguments, and sometimes this has risen to the level of astronauts disobeying safety orders, as when the crew of Apollo 7 refused to wear helmets during descent due to head colds. But all this is just regular human behavior stuff. Psychologically things have been pretty good, even boring, in space.

This may not be what you've heard. In our experience, people often report half-remembered tales of astronaut madness. If something about space life really does cause major psychiatric episodes, we'd like to know. So, does it?

Incidents of Space Madness . . .

. . . are hard to come by.

Look, this is a pop-science book. We're doing our best to keep the particulars of in-space psychiatry interesting. We would love to tell you that one time Chris Hadfield went mad and attacked colleagues with

*And what they don't like. Dr. Jack Stuster had astronauts keep journals, one of which noted: "I have had many programmatic fights with the Russians about the inequality of Russian/US foods (supposed to be 50/50). They have always had more Russian on board, using the rationale that people like Russian food better. That is bull____."

razor-sharp fragments of astronaut ice cream or that one time Sally Ride tried to commandeer the Space Shuttle and drive like hell for Jupiter. No such luck. Spacefarers *have* had plenty of issues on the ground, but space has been surprisingly mellow. We searched and searched, and as far as we can tell, the space-madness stories that get repeated tend to be either wild exaggerations or just straight-up wrong.

To give one example, there's a supposed quote from cosmonaut Valery Ryumin, shared endlessly on social media that goes, "All the conditions necessary for murder are met if you shut two men in a cabin measuring 18 feet by 20 and leave them together for two months." This sounds pretty bleak. The problem is that those aren't really Ryumin's words. He was quoting from literature. Here's the full statement he made: "O. Henry wrote in one of his stories that if you want to encourage the craft of murder, all you have to do is lock up two men for two months in an eighteen-by-twenty-foot room. Naturally, this now sounds humorous. Confidentially, a long stay even with a pleasant person is a test in itself." Substantially less terrifying. And in fact, Ryumin was a model cosmonaut, receiving a variety of awards and promotions over a long career. And as far as we know he never murdered anyone.

Memoirs and oral histories do reveal issues here and there. John Blaha, who spent four months as a guest on Mir, experienced moderate depression. Dr. Jerry Linenger had a sudden, terrifying sensation of rapidly falling while performing a spacewalk, but he pushed through. Perhaps the scariest story is that of Dr. Taylor Wang, a shuttle payload specialist, who brought a complex scientific instrument to space only to have it break. Mission control said he couldn't spend the time to fix it, and after some back-and-forth, he told them, "Hey, if you guys don't give me a chance to repair my instrument, *I'm not going back*." Fortunately for Wang, nobody called his bluff, because as he noted, "if you tried to hang yourself with no gravity, you'd just dangle there and look like an idiot."* In the end, nobody

*The full quote: "I was relieved, because I hadn't really figured out how not to come back if they'd called my bluff. The Asian tradition of honorable suicide, *seppuku*, would have failed, since everything on the shuttle is designed for safety. The knife onboard can't even cut the bread. You could put your head in the oven, but it's really just a food warmer. You wouldn't even burn yourself. And if you tried to hang yourself with no gravity, you'd just dangle there and look like an idiot."

went bonkers: the crew worked together, taking over Wang's other duties so he could fix the machine.

It's worth taking a moment to note that these stories are objectively boring in any other setting than space. Something about orbit makes the dullest facts seem like high drama. Aboard the Skylab 4 mission, there was an argument about scheduling, which is sometimes referred to as the Skylab 4 "Mutiny." Without going into details, it went something like this: the crew felt overscheduled, had a discussion with mission control, changed the scheduling protocol, and ended up completing more than the scheduled workload. If this is mutiny, astronauts suck at mutiny.

Try to imagine a friend describing the equivalent situation in an office on Earth. It'd be a long story about grumbling with the boss and then compromising on schedule changes. If you were still awake at the end, you'd probably side with the boss. Yet if it happens in orbit, it becomes the stuff of legend. One case, frequently cited in scholarly literature on space psychology, involves a cosmonaut pathologically obsessing over a potential toothache. We managed to track down the reference about this from 1985, and the actual story was of a man who was kind of worried about getting a toothache, had a dream about it, woke up and thought he really did have a toothache, but then felt fine by the morning. It's hard to see this having serious psychological implications. Meanwhile, another man, cosmonaut Yuri Romanenko, actually had an exposed tooth root in space, and instead of doing pop-science authors the courtesy of going mad, he simply toughed it out for the duration of the voyage.

As we kept checking these space stories, hoping against hope for just one measly instance of space madness with which we could regale our readers, we started to wonder if any dramatic space psychology situations had ever occurred. The answer is a full-throated kinda maybe.

Moderate, Qualified Cases of Potential Space Madness, Sort Of

In our research covering life in space stations from 1971 to today, we found precisely two stories in which a psychological problem may have resulted in a shortened mission:

In 1976, cosmonauts Boris Volynov and Vitaly Zholobov were on orbit in Salyut-5. They didn't have great rapport. This happens from time to time, but in the case of Salyut-5, conditions conspired to make things worse, beginning on day 42. Here is a description of what happened from a 2003 paper by space historian Dr. Asif Siddiqi:

> *As the crew were working, the station's alarm suddenly went off; simultaneously, all interior lights turned off and several onboard systems simply died. At the time, the station was passing over the night side of the Earth. In the darkness, with the loud shrill sound of the siren, the crew were totally confused. Within seconds, however, they first turned off the alarm, only to hear dead silence, i.e., it seemed that all of the station's systems had shut down. Volynov immediately transmitted an emergency message to ground control: "There's been an accident on board."*

Moments later, as ground control tried to figure out what was up, the cosmonauts drifted out of communication range. The station had lost attitude control and life support—silently, darkly, it turned above the shadowed Earth. Only over the course of two hours were the men able to get the station back to normal, at which point the mission was cut short and the crew came home early.

We've found various reasons reported for why the mission was cut short. In one telling, there was a bad smell aboard ship, which was concerning enough for the crew to leave. In another, the bad smell was purely psychological, since a later crew went to Salyut-5 and smelled nothing amiss. In more dramatic versions, Zholobov was overwhelmed by interpersonal issues, or by the vastness of the cosmos itself, and suffered a mental break. We are 100 percent positive we would've snapped under these conditions, but given how generally tame space psychology stories turn out to be, we're betting the true story is the least exciting one: the men were grumbling at each other and the ship was having major issues, so the mission got called off early.

Our one other example comes from cosmonaut Aleksandr Laveikin—the same guy who said all spacemen masturbate back in the chapter on space babies. His trip to Mir was cut short in 1987. Accounts about what happened differ, even when they all come from Laveikin. The agreed-upon fact is that heart monitors found Laveikin to be experiencing cardiac irregularity. After a battery of medical tests found nothing wrong, the condition reappeared during a spacewalk. At that point, doctors decided to bring him home. According to Laveikin, he was surprised and horrified by this decision. He later recalled, "I felt well during the flight, I felt great throughout. But you can't argue with the doctors."

However, decades later, Mary Roach interviewed Laveikin over a few glasses of whiskey for her book *Packing for Mars*, and Laveikin admitted that the mission was tougher than he'd expected. The work was hard, the station was noisy, and he'd had motion sickness. He ended up depressed and suicidal, making an observation that is apparently common among spacefarers: "I wanted to hang myself. Of course, it's impossible because of weightlessness."

Once again, we can't know exactly what happened. Both Laveikin and his commander, Romanenko (of exposed-tooth-root fame), apparently had serious friction with mission control, so things were at least not behaviorally ideal. But, if we assume the psychiatric worst case—that Laveikin had serious mental health issues, which led to heart issues, which caused him to be evacuated—the good news is that it'd be the only mission to space that ever got cut short entirely due to a psychological reaction to life in space without some major external trigger.

We document these cases mostly to clear up any idea that psychology has so far shown itself to be the major barrier in space activity. At least so far, screening and on-flight countermeasures have been incredibly successful. Nothing like those two incidents has ever occurred on the ISS.

The other available data paints a similarly rosy picture.

One analysis of every single reported in-flight health issue in all shuttle flights from April 1981 to January 1998 found a total of thirty-four events of "behavioral signs and symptoms," giving an incidence of about one for every three person-years of spaceflight. That's amazingly low when you consider that the thirty-four events include things like "anxiety" and "annoyance." We had thirty-four anxiety and annoyance events just editing this chapter.

According to NASA, there have been no reported psychiatric disorders on either the shuttle or the ISS. That is, ISS occupants may have felt depressed, annoyed, or anxious, but never—not once—has the issue risen to the level of diagnosis.

Based on all this, you might be tempted to conclude that at least the very first space settlers should have no major issues. We don't have super long-term data, true, but almost all the data we do have is pretty darn comforting.

It's a nice thought. It is. But here's the thing: *astronauts are liars.*

Carrying the Pants-Fire: The Rich History of Space Fibbing

When Eileen Collins, later the first female space pilot, went for her NASA psych interview, she became concerned. Having filled out an innocent-seeming psychological questionnaire, she was looking forward to her hour with the psychologist.

That was before he told her she'd gotten a number of questions "wrong." "Tell me about your hallucinations," he said. When Collins expressed confusion, he told her that for the question "When I walk down the street, I see things other people don't see," she'd written "True." Surprised, she tried to clarify: "When I am with my husband, for example, I'll notice flowers, landscaping, and people's clothing, while he has no clue they're there. He looks at other things." The psychiatrist replied, "Well, that question is designed to find out if you hallucinate."

Pay attention to how she responded. Not "there's been some confusion here." Not "this is a bizarre definition of hallucination." What she said was this: "Sir, with all due respect, someone who wants to be an astronaut is not going to confess to unusual mental issues."

Indeed, Collins was participating in what she no doubt knew was a grand tradition that spanned generations and continents. Valentina Ponomareva, backup pilot to the first woman who ever went to space way back in 1963, said in an interview many years later that the women in the program had cheated on their medical exams. As she said, "the [male] cosmonauts of the first group treated us very well; they cared about us, they helped us, they taught us how to deceive physicians and how to pass tests easier." Michael Collins, Moon-chauffeur to Armstrong and Aldrin, kept secret that he got claustrophobic in space suits, which he called "nasty little coffins," even though he had to conduct spacewalks during missions. Nor have the years diminished the tradition. Terry Virts, in his 2020 memoir, recalls pretending to be happy whenever talking to psychiatric counselors, regardless of his personal life or private emotional state.

Just reading through memoirs of astronauts and astronaut candidates who were willing to confess the truth, we found lies (and lies of omission) told to medical staff about: color blindness, height, chest pains, back pain, a suspected heart attack, suspected bone cancer, a severe inner-ear problem that resulted in disorientation and dizziness, one's entire childhood, whether or not one had vomited in one's own gloves during test pilot training, whether one had vomited upon docking with Skylab, and in at least one case a man physically removed pages from his own medical records to hide information from flight surgeons.*

Why do astronauts and pilots lie to the medical professionals tasked with keeping them healthy? A simple bit of game theory: if you're allowed to fly when you walk into the doctor's office, you can only come out still allowed to fly, or grounded. There are no benefits for truth-telling about anything negative. Doctors and psychiatrists know this game, and because

*We hasten to note that the list is made somewhat longer by repeat offenders.

of it a natural antagonism exists between flyers and medical people.* Or, as veteran astronaut/physician Joe Kerwin reportedly told NASA psychologist Al Holland: "Son, you gotta understand, the crews won't be happy until the last psychologist has been strangled on the entrails of the last flight surgeon."

It is simply impossible to read the personal accounts of astronauts and trust self-reported data. Mike Mullane, who removed medical pages from his file and also lied about his childhood, wrote, "We would have lied about a wooden leg or a glass eye. *You find it* would have been our attitude." Glass eyes are hard to lie about. Psychological state is not, especially for elite professionals with careers on the line. If you are planning a space settlement and basing your ideas of astronaut psychology on the reports of actual astronauts, you are making a terrible mistake.

Looking Elsewhere

So here's our problem: we want to stick at least a few thousand people somewhere in space, and ideally quite a few more. We don't have long-term data, and the data we do have comes from the professional class of liars known as astronauts. Is there an alternative? Kind of. Researchers look at what are called analogs and sims to try to get space insights while remaining on the ground.

*In fairness to the astro-duplicitous, our research turned up only one case where lies negatively impacted a mission. In 1985, Commander Vladimir Vasyutin had to return early from Salyut-7 because of an apparent "acute inflammatory disease." There are good reasons to believe he knew about this problem prior to flight but kept it a secret. More interesting for those of us concerned with cosmo-mendacity was the reaction of fellow cosmonaut Yekaterina Ivanova, whose flight was canceled by Vasyutin's early return. She was furious. Not because Vasyutin had lied about his fitness for flight, but because he broke the unspoken code of aviators and spacefarers: "One fundamental rule—if you're asked how you feel, you cheerfully answer fine, even if you can barely stand on your feet . . . But Vasyutin didn't cope and because of all of this I was once again grounded." That is, lying to medical staff is fine. What's forbidden is to *stop* lying. Bart Hendrickx, "Illness in Orbit," *Spaceflight* 53 (2011): 108.

Analogs

Many researchers believe the best place to look is situations on Earth that are similar to space. Indeed, there are entire books on the topic of "space analogs," and they tend to focus in particular on Antarctica and submarines—both places where humans live for extended periods confined in built structures, with harsh conditions outside.

The basic findings from these settings are broadly reassuring. The internet has plenty of questionable stories about stabbings over chess matches in Antarctic bases, but at the statistical level things are pretty okay. People at Antarctic research stations do get depressed and have the kinds of psychological struggles you'd expect from being stuck inside with a small group of people in a cold environment where the sun doesn't shine. But these problems rarely rise to the level of a clinical diagnosis, and sometimes studies even find people at research stations have better average psychological outcomes than the general population back home. This is partly due to prescreening with psych exams, and partly due to countermeasures similar to ones used in space. It's also likely due to self-selection—people aren't chosen at random to go to the South Pole, and those who decide to go may be looking forward to eating canned food for several months of perpetual night.

In the case of submarines we find similar results. The available data is limited, but studies report low rates of psychological disturbance, and in some cases even lower rates of psychological disturbance compared to the population they come from. Again, this is likely down to self-selection, screening, and countermeasures.

We would happily tell you more about these settings, and frankly we did a lot of reading about them that was objectively interesting. Topline findings: submarines are surprisingly gross and Antarcticans have a surprising amount of sex and sometimes run around outside naked. In terms of applying this data to space settlements, where running outside naked will be less attractive and more asphyxiating, the basic picture is a good one. Like in space, self-selected prescreened pros typically do fine.

However, this data isn't perfectly applicable because both analogs are too small in population and too short term. There are no permanent Antarcticans and certainly no permanent submariners. There's a world of difference between needing to deal with harsh conditions for a specific limited time and needing to stay forever and raise children.

Sims

The one other data source that might help us guess about space-settlement behavioral health is what are called simulations or "sims." Basically, you stick people in a controlled setting that is space-like in some way and you observe them. For instance, people may live in a desert, trying to survive on only what's inside the base, only going outside wearing simulated space suits, all while monitoring their psychological state.

These settings regularly feature in media reports, and there are documentaries and books about them. We have friends who participate in them. However, we are skeptical of how applicable sims are to space activity, beyond solving small ergonomic matters. We're not alone in this—some scholars believe the simple fact that sims carry no risk of death makes them useless for analyzing actual space psychology. How accurately can you simulate life on Mars if all the subjects know they can leave at a moment's notice and go to McDonald's? Also, with only a few exceptions, most sims run for weeks or maybe months, but not the years necessary to really understand the psychological toll of living with a small number of people in an isolated and confined environment for a long time.

The other thing is that, to be honest, some of this data is collected in scientifically questionable ways. For instance, in researching Mars-500, the most involved space sim ever conducted, lasting 520 days with a crew of 6, we found a paper coauthored by someone who was also a participant. This is a major scientific no-no, because the "subject" of the experiment knows in advance what the most interesting findings would be. Even if we give them the benefit of the doubt that they're trying to make honest observations, it's hard to believe the results weren't biased in some way.

At the Intersection of Ignorances

So our alternate data sources aren't great, we don't have very long-term data, and astronauts are liars. Where does this leave us? The one good thing we can say at the end of a mountain of research is that although the data are mediocre, they at least all point in the same direction. Prescreened professionals in harsh environments, including space, don't seem to have any special mental health problems. Unless there's something lurking beyond the 1.3 year barrier, we should assume space-settlement psychology will not be wildly different from anywhere on Earth. That said, acute psychiatric episodes in space, while not necessarily more common, might be substantially more dangerous.

Dealing with Major Psychiatric Issues in Space

Although missions have occasionally been cut short, there's never been a need for an emergency evacuation in space due to a psychological issue. However, it could be done on relatively short notice if needed. This is what happens on Antarctica and in submarines. If someone has a medical or behavioral emergency, the typical solution is an airlift to somewhere they can get proper professional care.

But what if the person is a risk to themselves or others now, and evacuation isn't an option? Well, we can tell you what they do on the ISS, thanks to the "International Space Station Integrated Medical Group Medical Checklist," which includes a section on dealing with "acute psychosis."

Instructions: unstow the drug subpack, the gray tape, the bungees, and towels. Talk to the "patient" so they know you're "using a restraint to ensure that he is safe." Tie wrists and ankles with tape, reserving the bungee for the torso. If the head needs to be restrained, stick a towel below the neck and use some more tape.

The patient is now offered tranquilizers and sedatives orally, but if they refuse then the drugs will be given through intramuscular injection. The medical officer is instructed to remain with the patient, carefully monitoring and recording vital signs. A similar protocol is advised in case of suicidal behavior.

Nothing like this has ever been done in space. If it happened on the ISS, conceivably you could keep the person sedated for the duration of an emergency trip home. And perhaps the Moon is close enough that an emergency trip to Earth within a few days wouldn't be out of the question. For a settlement on distant Mars, you'll need ways to deal with psychological emergencies locally. This suggests that pretty early on in the settlement process you'll want to be sure you have psychiatric care, medicine, and for the particular case of people who've become dangerous to themselves, you may need some way to safely confine them until the problem has been solved or they can be moved back to Earth. This will not be a trivial problem—thanks to the unforgiving nature of orbital mechanics, in the worst-case scenario, confinement would be over a year, followed by a launch to orbit and a six-month transit to Earth. This is something that could be handled in a large-enough settlement that included specialists and appropriate facilities. If it happens in a one-hundred-person outpost, you've got a real problem.

Psychology and Space Settlements

The big picture with space psychology is that so far things have been good, but we shouldn't expect perfection. We think it's likely psychology isn't as big a barrier to settlement as, for instance, reproduction. If you just go by movies and novels, you might get the impression that humans in isolation wig out pretty fast. In reality, the history of polar exploration furnishes plenty of examples where crew have been trapped in desolate places with dwindling food supplies and have been reasonably psychologically steady, at times even genteel.

That said, the lack of long-term data on cognitive effects is worrisome. In terms of our wait-and-go-big approach, that would be a big reason to wait. We'd like to see much longer-term data that either rules out cognitive effects or tells us how to mitigate them. Given how isolated and dangerous a Mars settlement would be, any kind of slowly progressing mental health deterioration could turn into a nightmare.

The combination of day-to-day needs and the risk of occasional acute problems means that any settlement needs a support plan for the entire range of psychiatric possibilities. That's easy to say, but it's a huge constraint. In a world where people are permanently living off-world and raising children, there has got to be a way to take care of mental well-being locally. That's not just about people and buildings; it's also about medicine. It's going to be a long time before the entire psychological pharmacopeia can be produced anywhere but Earth, and modern drugs aren't designed to be shelf stable for years while also being exposed to space radiation. All of this is much easier if we wait until the science and technology are in place to send big facilities and large populations, while also having some regular means of allowing people to go back to Earth.

For our space settler Astrid, and her kids, we'll assume everyone is psychologically well adapted to having the Earth out of view. The whole family is physically and psychologically fantastic, and so the question becomes where to settle.

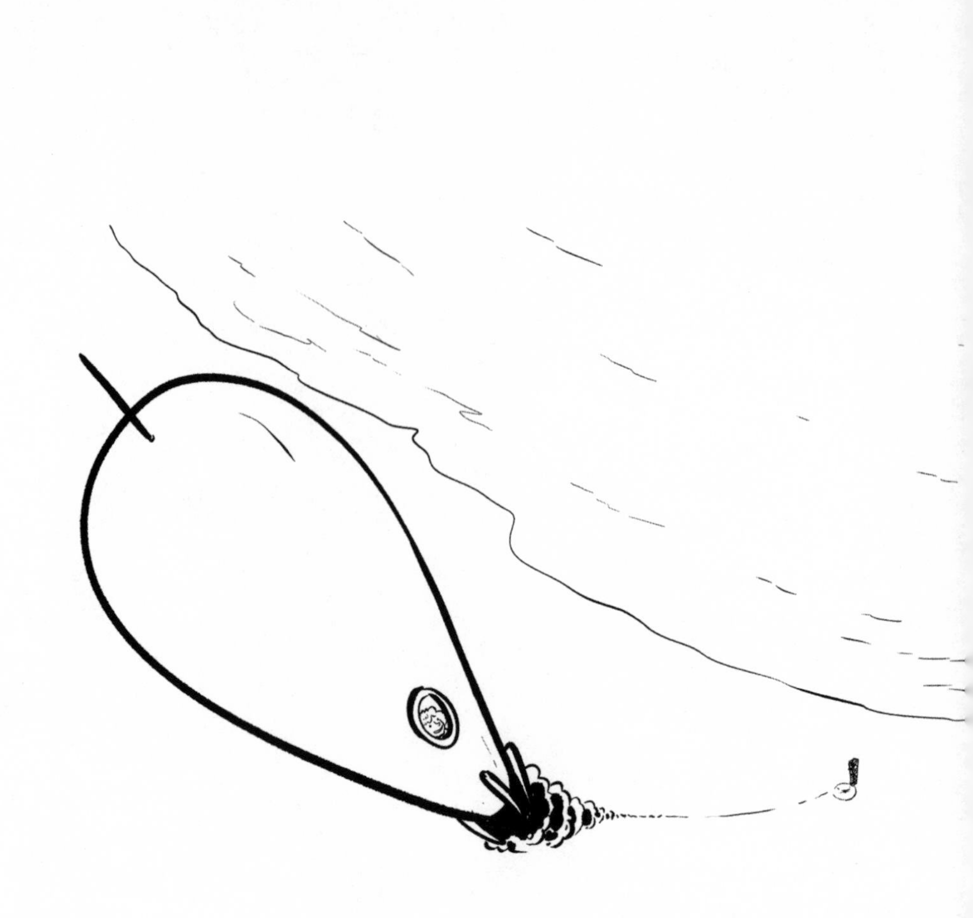

Nota Bene

ROCKETRY GOES TO THE MOVIES, OR, SPACE CAPITALISM IN DAYS OF YORE, PART 1

Hermann Oberth was born in 1894 and died in 1989. He was the only founding father of rocketry who lived long enough to see Armstrong's "one small step for a man."

Oberth was something of a prodigy. As a boy in Romania, before he'd learned any advanced mathematics, he noticed an error in Jules Verne's *From the Earth to the Moon*. Verne has his spacefarers fired from a 275-meter-long cannon. Young Oberth calculated that this would produce 23,000 times the normal force of gravity—enough to have liquefied the main characters in chapter 26. This was both precocious intelligence and, for a future rocket scientist, quite a high level of concern for human welfare.

As a young man, Oberth wanted to go into the physical sciences, but his physician father insisted he pursue medicine. Oberth obeyed, but was undeterred in his spacefaring interest. As a young medic in World War I, he took drugs to degrade his sense of balance, then sunk himself under bathtub water, breathing through a tube. Why? He wanted to simulate the orientationless world of space to see how a future rocketeer might function. Don't try this at home, but it turns out you can medicate yourself, bathtub scuba dive, and still keep your bearings. In fact, this was probably one of the less dangerous things to do in World War I.

One imagines Oberth would've preferred somebody else be the guinea pig in his experiments, but in those early days there was very little money for rocketry. When his father relented, Oberth pursued his space studies in earnest. Without funding, he focused on theory work, using his wife's money to publish several editions of his *The Rocket into Planetary Space*, sparking something of a mania for rockets in Germany. But although he briefly had a chance at venture capital for a rocket project, he was thwarted when a professor insisted Oberth's ideas were not feasible. Without an academic post or funding source, Oberth took up a job as a country village teacher, corresponding with other early space geeks while writing pop-science in his spare time.

Then, one day in 1928, he got a telegraph from the world-famous movie director Fritz Lang. Lang was working on a space travel film called *Frau im Mond* (*The Woman on the Moon*) and wanted it to be technically accurate. Oberth packed his things and headed for UFA Studios in Berlin.

Oberth ended up in charge of a sales gimmick, wherein he would be tasked with building a thirteen-meter-tall rocket that would reach the stratosphere as part of the film's premier. Appreciate for a moment how absolutely insane this is. Oberth was a theorist who taught physics, not an engineer. It'd be kind of like asking Einstein to build an atom bomb from scratch in a few months and promising him a loft, some helpers, and a little funding.

The first step for Oberth was to start messing around with liquid rocket fuels. He was the kind of person who'd take disorienting drugs and go underwater for the sake of science, but if he had known from his vantage point in the late 1920s how many people—including friends of his—would die fiddling with rocket fuel, he would surely not have done what he did next.

If you've ever wondered what happens when you take liquid oxygen and pour gasoline on top, we can tell you that in at least this one case, you get exploded across a room, burst an eardrum, and have your left eye damaged. Then, if you're enthusiastic enough about rockets, you get right back to work.

With a deadline in six weeks, and no time to build a wind tunnel, the only way to test his rocket design's aerodynamics was to drop a model of it down a smoke stack. According to an account by Oberth's friend and contemporary Dr. Boris V. Rauschenbach, UFA come up with a brilliant work-around: they took a photo of the falling rocket model, turned it upside down, and sent copies out as promo.

Oberth never finished the rocket. In fact, he had a nervous breakdown and headed for home before the premier. UFA, in proper studio style, made a fortune off the marketing without delivering the goods, and Oberth went back to his country village, fated to play the role of a theorist while others built the great machines of the early Space Age. This would be his last brush with space capitalism.

PART II

Spome, Spome on the Range: Where Will Humans Live Off-World?

IN 1835, THE NEW YORK NEWSPAPER *THE SUN* CLAIMED LIFE ON THE Moon had been glimpsed by Sir John Herschel, LL.D., F.R.S., &c.,* who while on expedition in the Southern Hemisphere made use of a "telescope of vast dimensions and an entirely new principle." Seeing the Moon up close, Herschel was astonished to find much *better* life than we have on Earth—a Valley of the Unicorns, horned bears, bipedal beavers, a race of humanoid bats soaring above teeming riverbanks.

This story was widely circulated, but it turned out to contain some inaccuracies, and is now remembered as the 1835 Moon Hoax. It was neither the first nor the last time otherwise intelligent people were fooled into thinking space would be friendly for the living. A long-running theory that

*Our best guess here is that this means "Doctor of Laws, Fellow of the Royal Society, etc."

persisted well into the twentieth century claimed the Martian surface was lined with canals, built by aliens husbanding their diminishing water supply. In the days before radio, there were proposals to signal Mars using light, fire, or carving shapes into a forest. As late as 1964, one Mars mission proposal called for astronauts to land and "investigate life forms for possible nutritional value." But when the Mariner 4 probe arrived in Martian orbit a year later, the truth became apparent—there's no sense carving "Pardon me, Martians, but what's your nutritional value?" into the local woods because nobody is on the receiving end. The same goes for the Moon, for Venus, and as far as we can tell for everywhere in the solar system except this pale blue dot.

Why? Because space is terrible. All of it. *Terrible.* Even photos and videos taken on the actual surfaces of the Moon and Mars can be deceiving. They don't look half bad—sort of like dusty deserts with rolling hills. Not exactly bearnicorns and man-bats in Happy Valley, but not entirely uninviting either. To see these images properly, you need to know what they can't show you. The Moon isn't just a sort of gray Sahara without air. Its surface is made of jagged, electrically charged microscopic glass and stone, which clings to pressure suits and landing vehicles. Nor is Mars just an off-world Death Valley—its soil is laden with toxic chemicals, and its thin carbonic atmosphere whips up worldwide dust storms that blot out the Sun for weeks at a time.

And those are the good places to land.

Venus, with its oven-heat, ocean-bottom pressure, and sulfuric acid clouds has been compared unfavorably with hell. Mercury, naked to the nearby Sun, has equatorial temperature swings of over 600°C. The farther-out planets take years to reach with current technology and from their vantage the Sun is dim and distant. Some moons of Jupiter or Saturn may harbor life in warm subsurface oceans, but even if those depths are toasty enough to permit life, they are shielded from human eyes by kilometers-thick crusts of ice.

The question for a would-be space settler isn't "Where's the good place?" It's "Where's the survivable place?" Options are limited. Space is big, yes, but space-settlement sites are few. Until we reach some extravagantly more advanced level of technology, there are only two worlds we can hope to settle—the Moon and Mars. Mars has about the same land surface as Earth, but that's only due to the lack of any oceans. The Moon, similarly unencumbered, has about as much land area as 1.25 Africas.

The only other serious spome possibility would be the construction of a massive space station—a very hard project that nevertheless has prominent advocates, and which we'll explore shortly.

Consider this a travel brochure on regions above the atmosphere. We'll see the places you might go, what amenities are available, what your accommodations might be like, and if there's an elevated risk of death, we won't dwell on it.

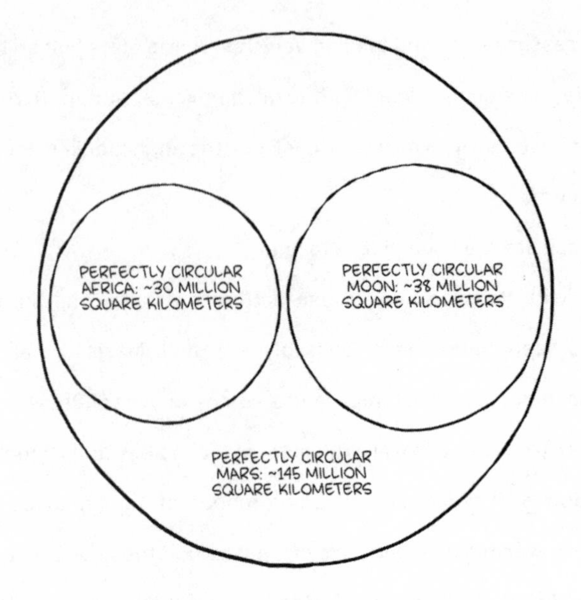
PERFECTLY CIRCULAR AFRICA: ~30 MILLION SQUARE KILOMETERS
PERFECTLY CIRCULAR MOON: ~38 MILLION SQUARE KILOMETERS
PERFECTLY CIRCULAR MARS: ~145 MILLION SQUARE KILOMETERS

5.

The Moon: Great Location, Bit of a Fixer-Upper

Buzz Aldrin once referred to the Moon's surface as "magnificent desolation." "Magnificent" can be debated, but desolation is scientifically accurate. If you build a Moon settlement, the defining feature of life will be things lacked.

The Case Against the Moon

The Moon has almost no carbon on its surface. What little carbon can be found has been deposited by solar wind and space-object impacts in minute quantities over billions of years. This is a problem, because humans just can't get enough of the stuff. We are about 20 percent carbon by mass. Plants are worse. Trees, for example, have a dry weight consisting of about 50 percent carbon. Plus, the Moon's surface is low in other important stuff, like nitrogen and phosphorous. Life as we know it literally cannot construct itself with what the Moon offers. At the moment, there are precisely six small caches of high-concentration carbon on the Moon—the

Apollo Moon landing sites, where the Heroes of the Space Age left behind a grand total of ninety-six bags of feces, urine, and vomit. Sadly, you are legally forbidden to use these precious historical artifacts.*

Outside of small regions we'll get to shortly, the Moon is also dry. There is *some* water bound up in the lunar surface, but then, if we wanna be technical, which we do, there is water bound up in concrete. In fact, going by the latest estimates, concrete is comparatively moist. We did a back-of-the-envelope calculation and estimated you'd need to cook all the water out of six tons of lunar soil to get the three kilograms of water you need daily to survive, not including cleaning, showering, and the occasional water balloon fight. However, we were told by lunar surface expert Robert Wagner of Arizona State University, "That looks like you're using some of the wetter materials."

The Moon also lacks a radiation-thwarting planetwide magnetosphere. Nor does it have a thick blanket of atmosphere, which is handy when you want to breathe. It's also handy because it provides protection against radiation and meteors.

The lack of protection also affects the ground itself. Earth's surface has wind and water and the general sloshy squishiness of a living world. The Moon's surface is the result of billions of unhealed wounds—violent impacts from space objects large and small. The heat of these strikes fuses the surface while shattering what came before, and this fuse-and-shatter routine happens over and over for eons. Add to this the regular fracturing that comes with extreme heat followed by extreme cold, and the result is that the Moon is coated in "regolith," from the Greek roots meaning "blanket of rock." Nasty little jagged bits of stone and glass.

So, that dirt you're baking your three kilograms of water out of? It's not exactly easy to work with. Harrison Schmitt of Apollo 17 reported allergy-like symptoms arising from inhaled moondust. Some habitat researchers fear that too much inhaled regolith over a long enough period will result in

*Those poo bags are of historical value and scientific interest. Under current space law, that poo is US government property and is protected by NASA's 2011 "Recommendations to Space-Faring Entities: How to Protect and Preserve the Historical and Scientific Value of U.S. Government Lunar Artifacts." So if you've ever dreamt of growing a lunar vegetable patch in Neil Armstrong's long-lost bowel movements, we're sorry, it's time to move on.

something like silicosis (also known as stone-grinder's disease), in which repeated microscopic lung scarification makes breathing extremely difficult.

It's not good for the equipment either. As John Young said during Apollo 16: "Houston, this dust is just like an abrasive. Any time you rub something, you can no longer read it. And that's what's happened to our RCUs and our . . . (pause) every piece of gear we've got. In other words, it's a mistake to rub something to clean it off."

Regolith is a constant nuisance that shouldn't be underestimated. The lunar surface is electrically charged, meaning it clings like fresh laundry. This is directly bad for machinery, but it also causes temperature dysregulation. Everyone knows black is the most awesome color for a space suit, but you never see one because white is the color that reflects sunlight. This is important because sunbeams reach the lunar surface without the mitigating influence of air. But the static cling of regolith means suits that don't get cleaned slowly take on the dark plastery gray color of the Moon, making them absorb more heat. Once the coating is thick enough, it can also act as an insulator, which can cause a totally new problem—equipment that needs to radiate heat from human bodies cannot. Humans haven't been on the Moon long enough for this sort of thing to be a big problem, though it has made life annoying.

Robots have had it worse. It's thought that the Soviet lunar rover Lunokhod 2 ("Moonwalker 2") eventually died after it acquired a heat-retaining patina of regolith, ultimately cooking until it couldn't function.

On the plus side, you'll come to really appreciate lunar sunrise, mostly because it arrives just once every two weeks. Lunar nights are Earth fortnights—two weeks light, two weeks dark. Combine this with the lack of oceans and atmosphere to moderate climate and you get regular equatorial temperature swings from -130°C to 120°C, with temperatures as low as -250°C recorded in a crater on the South Pole. That's bad for equipment, bad for humans, and two weeks of darkness is decidedly undesirable if you're trying to generate solar power, which you probably will be.

Is there a positive tradeoff for all this? Historically people have been willing to endure all sorts of hardships in exchange for a return on investment or eternal fame. Eternal fame is a substance with diminishing returns, so here we'll focus on commodities.

Moon dirt fetches a pretty high price if you can get it back home, but in a future where spacefaring becomes common, gifting your sweetie a vial of glass and stone dust will lose a bit of its romance. If we're talking about local minerals that you might refine and ship back to Earth for a profit, the Moon harbors nothing worth getting. The economics of spacefaring are changing, but in order to be exportable, for the foreseeable future any goods mined in space for sale on Earth need to be high value, low mass, and fairly easy to acquire. Nothing on the Moon matches this description.

You may have heard otherwise. A surprisingly large number of books that discuss Moon settlements mention something called "helium-3," a valuable helium isotope. We have been forbidden by our editors to go on a ten-page rant about helium isotope economics in a pop-science book, but if you want to see a nerd hyperventilate, buy us a beer and ask about it. In short, helium-3 is more common on the Moon than Earth, yes, but still extremely diffuse. We're talking parts per billion. One estimate suggests it takes 150 tons of regolith to produce a single gram of helium-3. In other words, you likely have to process square kilometers of the lunar surface to get a decent amount. What's it good for? In a single sentence: it's good for a small set of medical applications and for a futuristic type of fusion reactor, which would work great except that we can't build it yet, and almost nobody is trying to build it, because it's far harder to make work than more typical kinds of fusion reactor we *also* can't build, and which use a far cheaper, more plentiful fuel, and anyway helium-3 is a by-product we already make with a well-known nuclear power source called a heavy-water reactor. There. We'll stop now because we're getting spittle on the screen.*

So, that's the Moon: hotter than a desert, colder than Antarctica, airless, irradiated by space, lacking in carbon, and with no minerals valuable enough to sell back home. It's not obviously gold rush territory. Oh, and what are the long-term health effects of living in one-sixth Earth gravity while inhaling glass? Your guess is as good as ours.

The Case for the Moon

But you should see the location!

Other than the Sun, which is hard to land on, the Moon is the only

*If you would like to join us in being a nerd hyperventilating over Moon isotopes, we recommend the recent paper: Gerrit Bruhaug and William Phillips, "Nuclear Fuel Resources of the Moon: A Broad Analysis of Future Lunar Nuclear Fuel Utilization," *NSS Space Settlement Journal* 5 (June 2021).

place in the solar system that stays put with respect to Earth. Our cosmic roommate remains just about 385,000 kilometers away, perpetually.

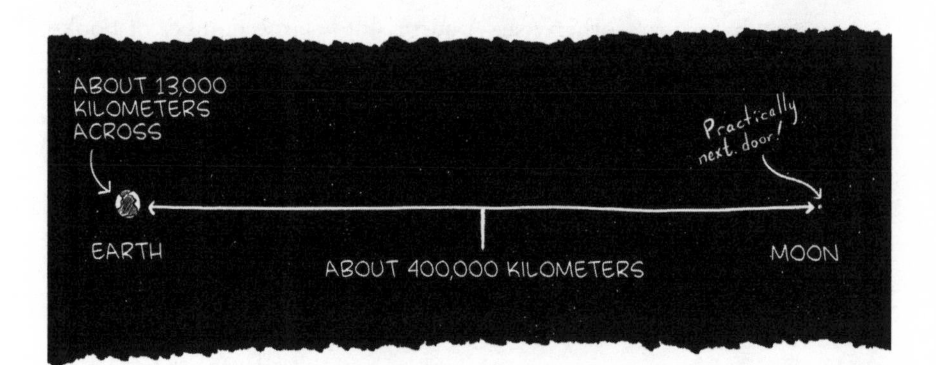

The Moon is a cosmic day-trip. Easy access, easy escape. That means astronauts can receive fresh resupplies on a regular basis. It means signals only take about one second each way, allowing nearly live communication in case of emergencies. It means construction equipment could be remotely operated from Earth—no fancy autonomous robots required.

The Moon is also an excellent place for rocket launches. From an energy perspective, the hard part of spacefaring isn't traveling a long distance, but getting off-planet in the first place. On a trip to Mars, most of the propellant you will ever use gets burned up reaching a stable orbit above Earth. Once orbital, a relatively modest use of propellant will sling you elsewhere.

You can imagine space as a sort of giant air-hockey board with deep wells pitted in its surface. Once you're out of the well, or spinning on its rim, it's pretty easy to get where you're going as long as you have the time to get there. Earth is a deep gravity well, the Moon much less so.

Throw in the fact that the Moon hasn't got a pesky atmosphere to slow down launches, and you've got a great platform for throwing things into space, at least compared to Earth. You could plausibly even set up a mass driver—basically a roller coaster to space, no rocket required—something nearly impossible on Earth.

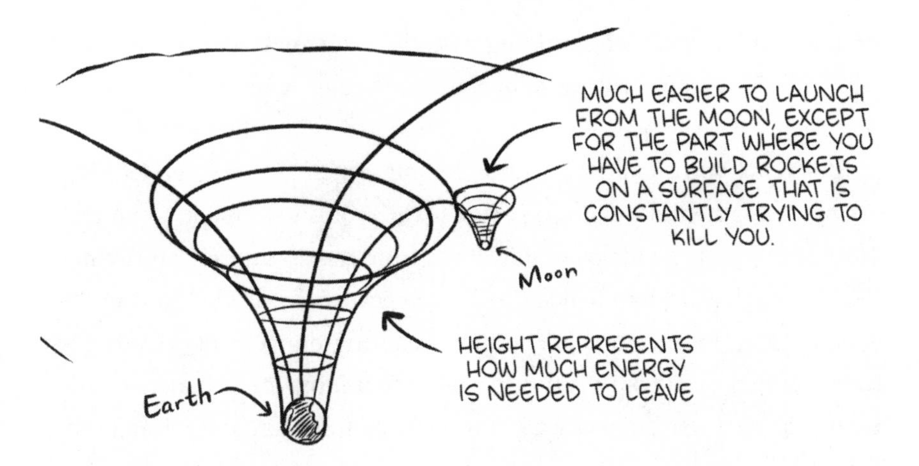

Based on figure in Marilyn Dudley-Flores and Thomas Gangale, "Manufactured on the Moon, Made on Mars—Sustainment for the Earth Beyond the Earth," AIAA Space Conference and Exhibition, Pasadena, CA, September 14–17, 2009, AIAA-2009-6428.

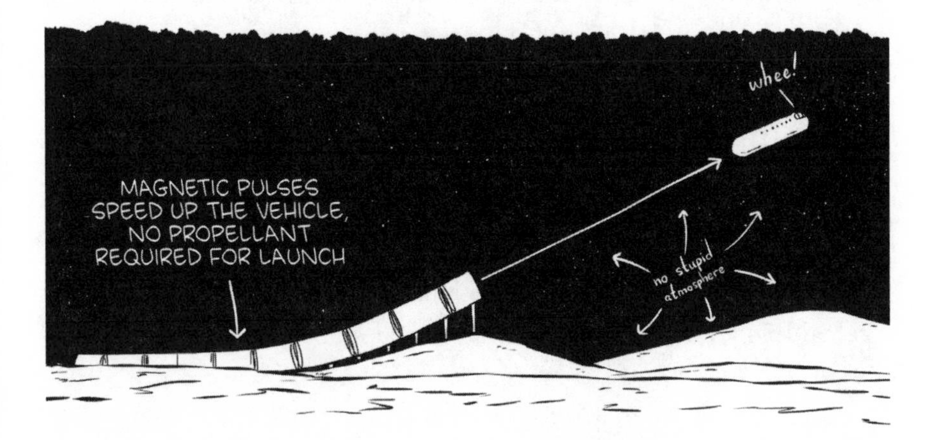

To be clear, setting up a launch facility on the Moon would be very hard. A lot of your mass will have to be expensively sourced from Earth, especially at first. But, at least in principle, some construction materials can be generated lunar-locally. The Moon's surface is high in silicon, aluminum, magnesium, iron, and titanium. Silicon has all sorts of applications, from windows to photovoltaic solar panels. Aluminum, iron, and titanium are all excellent building materials. Magnesium is easy to work with due to its low melting temperature, though it has the downside of reacting explosively with oxygen, a popular gas among humans.

But note, you should be a little nervous when you hear geeks like us saying you can build something because the requisite elements are present. Noting that you can make photovoltaics from metals and silicon is kind of like noting that you can build an airplane because the dirt below your lawn contains aluminum, iron, and carbon. Focusing on elements can disguise some serious complexity. Working with titanium, for instance, requires tremendously high temperatures in specialized furnaces. The silicon needed for solar panels begins life as jagged dust or compressed stone. Trying to work with lunar iron is kind of like trying to build steel beams from a pile of rust. It is literally possible, but very hard, and it's worth noting that people never talk about Earth resources this way. You never hear anyone say "We should mine for copper here, because there is literally any amount of copper here." The moral of the story: beware of nerds ticking off lists of mineral content, unless they also happen to be carrying a futuristic limitless energy supply, perhaps involving helium-3.

Although there are some useful materials on the Moon, settlement prospects are fairly bleak pretty much anywhere on the lunar surface. But not everywhere. The Moon has a few tantalizing locations worth considering for settlement. These are what we'll call the Moon's "Premium Real Estate." Of special interest are three lunar features: Lava Tubes, Peaks of Eternal Light, and Craters of Eternal Darkness.*

The Moon's Upper Crust

Lava tubes are a particular type of cave, also found on Earth and Mars. They form in various ways, but a common one goes something like this: as lava flows, its outer layer cools and hardens, similar to how the top of a river turns to ice when it's cold enough. This "crust" acts as a sort of

*Technically, today's term of art is "Permanently Shadowed Regions," and not all of them are craters. But we're sticking with the pretty language.

thermos, keeping the lava inside nice and hot and flowing. Eventually, when all that molten stone has gone on its merry way, the crust stays behind as the roof of an enormous cave with high arched ceilings, almost like an underground cathedral.

NOTE: THIS IS ONE OF SEVERAL PROCESSES THAT CAN CREATE LAVA TUBES. FOR ALL THE GRITTY GEOLOGICAL DETAILS, SEE SAURO ET AL., "LAVA TUBES ON EARTH, MOON AND MARS: A REVIEW ON THEIR SIZE AND MORPHOLOGY REVEALED BY COMPARATIVE PLANETOLOGY," EARTH-SCIENCE REVIEWS 209 (2020), 103288.

The Moon no longer produces lava flows, but it once did. And, thanks to the Moon's low gravity, some of these tubes reach titanically alien proportions—perhaps more than ten times larger than anything comparable on Earth, enclosing over a *billion* cubic meters of space. Substantially roomier than the ISS's 388 cubic meters. Weirdest of all, recent evidence suggests that at least in some cases lunar caves may have temperatures that are reliably Earth-like, in the range of 17°C. We'll talk about habitat design in part III, but for now understand that these caves may provide a prebuilt hole in the ground with protection against radiation, micrometeorite impacts, and wild temperature swings, while also allowing early settlers to leapfrog past many of the difficult first steps of inhabiting the Moon.

There are a few potential catches here. Lava tubes will be difficult to access safely due to their sheer size. Moonwalking is hard enough without having to moon-spelunk. More important, we just don't know that much about them, and their stability remains a hard-to-assess question. Scientists are working on models to try to determine features that make for stable tubes, but Wagner informs us that "the strongest evidence for lunar lava tubes' size (and, indeed, presence) is observations of collapsed segments." That's not the most reassuring information imaginable, but recent work demonstrates that there are at least some pits around today that are found in photos from the Apollo era. And also, these things are really, really old and *some* of them haven't caved in yet. We wouldn't phrase it that way if we were selling lava tubes on Zillow, but if a cave is substantially older than life on Earth, it can probably be expected to stick around a little longer.

And then there are the Peaks of Eternal Light and Craters of Eternal Darkness. This sounds like the description of a sixteen-year-old's emotional states, but in fact it's a business opportunity, and perhaps one day a source of geopolitical tension. As we said earlier, the Moon gets two weeks of light and two weeks of darkness. However, just like on Earth, at the poles the day-night cycle gets weird. Starting in the early nineteenth century, scientists speculated that if you had a celestial body at the proper angle to the Sun, parts of it will get perpetual sunlight. Of particular interest were peaks atop the lunar poles, which appeared to be permanently grazed by sunlight.

EARTH IS TILTED 23.5 DEGREES WITH RESPECT TO THE SUN, SO WE GET SEASONS

THE MOON IS ONLY TILTED 1.5 DEGREES WITH RESPECT TO THE SUN, SO SOME PEAKS ARE ALMOST PERPETUALLY LIT UP

Recent data confirm that this is nearly the case. They're called the Peaks of Eternal Light, but Peaks-of-Pretty-Much-Eternal Light would be more accurate. In particular, parts of the rims of the North Pole's Peary crater and the South Pole's Shackleton crater are illuminated more than 80 percent of the time. This is tempting stuff for spomesteaders: First, if you're using solar power, you can get energy on a consistent basis instead of dealing with two-week-long stretches of night. Second, because you're perpetually *grazed* by light, rather than alternately blasted and deprived, you can get a less volatile and more mellow temperature. Well, mellow for the Moon. For example, a ridge between the Shackleton and de Gerlache craters has an average summer temperature that hovers around -70°C, which is merely about 10°C colder than the average temperature in Antarctica. In the interior. In the winter. Not exactly Cancun, but the view is nice.

Then there are the Craters of Eternal Darkness, which are also at the poles, and which are substantially more inviting than their name suggests. Because light hits the poles at such a sharp angle, you get craters where part of their interior has never seen the light of day. This is easy to visualize if you imagine a giant coffee mug sitting at the North Pole. The rim of the cup and part of the interior will be lit, but as long as the cup is deep enough, there are regions where light will never penetrate. Perpetual darkness, which means perpetual cold.

Why do you want permanently cold domains of shadow? Because thanks to their extremely low temperature, some of them appear to have held on to water ice. Neither rover nor human have visited these sites, so they are imperfectly characterized, but the water in them likely comes from sources like comets that crashed into the Moon, and possibly long-ago lunar volcanism. This water would've moved around the lunar surface, in some cases for millennia, before finding itself trapped inside these exceptionally cold regions. There it remained, confined in darkness for eons, crystalline and mysterious, until a Blue Origin spacecraft landed nearby and Jeff Bezos needed a fill-up for his hot tub.

Mind you, he won't have an easy time. At these temperatures, ice is more like stone than the stuff in your freezer. It also contains other compounds, like methane, hydrogen sulfide, and ammonia. These are potentially valuable chemicals, but are also toxic and will have to be separated out before anyone takes a drink. Some of these chemicals contain precious carbon, but sadly, even after harvesting the Apollo poo bags, your combined carbon total won't be nearly enough to start the farm.

Still, if you can get a base on one of these craters, with their combination of perpetual darkness and (almost) perpetual light, you've got solar power and water. That means almost everything in space. See, your friend H_2O is the jack-of-all-space-trades. You can drink it. You can crack it into oxygen and hydrogen, then use the oxygen for air. And in the right forms, hydrogen and oxygen can be reacted back together for rocket propellant or fuel cells. As long as you have a lot of energy, water means survival, mobility, escape, and if there's a market for Moon-based rocket propellant, it may one day mean business.

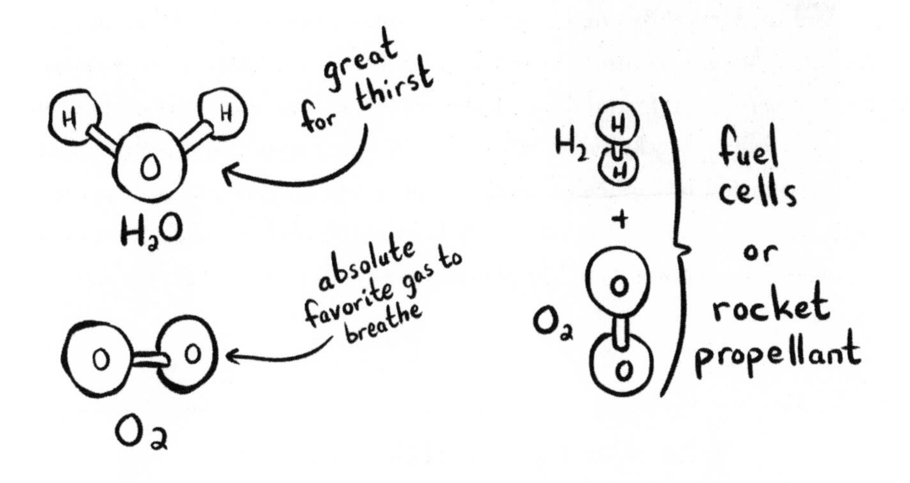

The main downside of the Moon's Premium Real Estate is that, if it proves genuinely valuable, we might fight over it. The legal regime governing these places is vague, and you'll learn about it later, but the major issue we want to raise here is that these resources are finite and small.

Taken together, using optimistic numbers, the Craters of Eternal Darkness only make up 0.1 percent of the lunar surface. They also don't have that much water—perhaps as much as 100 million tons. That sounds like a lot, but it's about the weight of one tenth of a cubic kilometer of water. So, the total amount of water hidden in Eternal Darkness may be roughly equivalent to 10 percent of the volume of Sardis Lake. You know, Sardis Lake? A manmade lake in Mississippi? We hadn't heard of it either, but it looks nice.

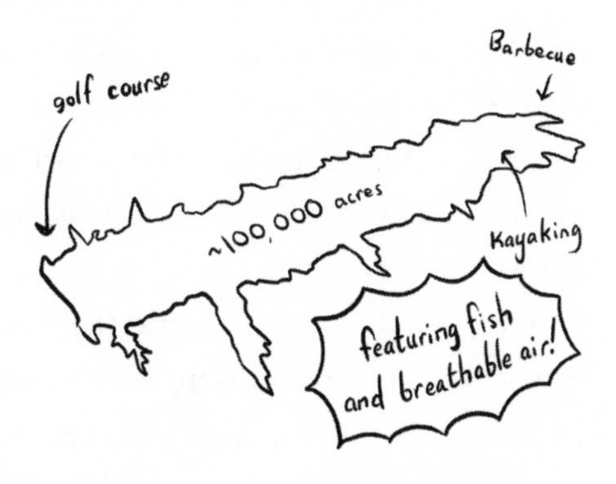

They have bass. There's a whole restaurant. And it has one other advantage over water-laden Moon craters—it gets replenished over time periods shorter than eons. If the water is mostly used as part of a human habitat, and you get very good at recycling, it can go a long way. If the water is used for rocket fuel, once the burn is over, it's gone.

And the Peaks? By one estimate, they are one one-hundred-billionth of the lunar surface. Doing a little math, that's less than two tennis courts. You can spread out farther if you raise your solar panels up on platforms or if you're willing to take even less perpetual light, but the point is that while the Moon itself has an area a bit larger than two Russias put together, the especially good parts cover substantially less ground than Liechtenstein. If there ever is a scramble for Moon land, these will be the places we scramble for.

Who Wants to Settle the Moon?

Almost nobody wants to settle the Moon as an end in itself. Because most of its value is positional, and because it lacks the resources to sustain life, the Moon is mainly enticing as a stepping-stone elsewhere. This gives us at least two reasons to set up facilities on the Moon.

First, we could use it as a sort of giant spaceport. Earth is nice, especially on weekends, but it's hard to launch here. If we can just get the right setup, the Moon is the ideal place to gas up old spacecraft or launch new ones. As space visionary Krafft Ehricke reportedly said, "If God wanted man to become a spacefaring species, *He would have given man a Moon.*"

The second reason to settle the Moon is practice. Before permanent outposts in Antarctica were built, human beings had been trekking in Antarctica for over fifty years. That success was in part possible due to knowledge gained from people who'd lived in the Far North for thousands of years. We don't have anything like this level of working knowledge for the Moon. We barely even have veteran Moon travelers. The twelve Apollo moonwalkers spent a grand total of less than a month on the Moon, and as of this writing, only four of them are still alive.

For all its deficiencies, the Moon is close, which spares us from a lot of the logistical complexity of a trip anywhere else. The Moon is a place to master all the problems we've described above—low-gravity medicine, countermeasures for long-term mental health issues in space, dust mitigation, robotic construction, *baby* construction, and a thousand other issues, some of which haven't yet been anticipated. A place to paddle a canoe before we set out on the sea.

This appears to also be the view of large space agencies and launch corporations. As NASA's international "2020 Artemis Plan" for returning humans to the Moon says, "The sooner we get to the Moon, the sooner we get American astronauts to Mars." China is also targeting a human lunar landing for the 2030s and is teaming up with Russia to do it. Jeff Bezos's Blue Origin is interested in Shackleton crater's water, sunlight, and minerals. NASA is teaming up with SpaceX in the hope that their "Starship" can take astronauts to the Moon, and SpaceX also has a private contract to send tourists on a lunar-orbit vacation.

One day, if the Moon is settled enough to have a large manufacturing operation, it may become a lot easier to explore or even settle other parts of space. Because it is far easier to launch from the Moon than from Earth, creating propellant or rocket parts on the Moon could in principle be far cheaper than boosting them from Earth.

But all these things are second-order considerations—the Moon is probably only valuable for settlement if *somewhere else* is even more valuable. If humanity decides, either with public or private spending, that sending humans deep into space is worthwhile, the utility of the Moon may justify a full-on settlement. But while this possibility is compelling to many space enthusiasts, most of them see the Moon not as the goal, but as a stopover en route to somewhere better. And usually, the better location they're talking about is Mars.

Astrid and company haven't yet made up their minds, so we're just going to give them a travel brochure to hold on to as we consider the most important alternative.

"But think of all the ways it *isn't* trying to kill you!"
MOON!

6.

Mars: Landscapes of Poison and Toxic Skies, but What an Opportunity!

> Send old men to Mars because they're going to die anyway.
>
> —John Young, astronaut

We're going to argue that Mars is a good bet for space settlement, but first let us be clear: by earthly standards Mars sucks. Sucks more than Scott Kelly's pants. Mars is nowhere near being a Plan B home for humanity anytime soon. Consider a worst-case climate scenario. The oceans have swollen ten meters higher, drowning New York City and Boston. Low-lying countries like Belgium and the Netherlands have been swallowed up whole. Heat waves make parts of the Southern Hemisphere uninhabitable as the planet is ravaged by floods, droughts, wildfires, and massive tropical cyclones. More than half of the world's species die, coral reefs become bleached skeletons, freshwater sources from snowpack melt away or are fouled by rising seas, tropical diseases make their way into formerly temperate climates. Crops fail, people starve, and violence breaks out as over a billion climate refugees beat against the closed gates of the comparatively livable North.

That planet? Eden compared to Mars or the Moon. That Earth still has a breathable atmosphere, a magnetosphere to protect against radiation, and quite possibly still has McDonald's breakfast. It's not a world we would *like* to inhabit, but it *is* the one world in the solar system where you can run around naked for ten minutes and still be alive at the end.

What makes Mars beguiling to space settlers is not its current state, but its potential. On Mars, you have, at the level of chemistry, most of the stuff you need to stay forever. And the most basic stuff like carbon, oxygen, and water is easy to acquire, at least by the stingy standards of space. Mars is thus a place we can not only survive on, but expand into. With enough time and effort, Mars at least holds out the *possibility* of a second independent home for humanity.

But it sucks.

The Case Against Mars

Like the Moon, Mars is covered with dead regolith. There is some weathering from the blowing wind, but not enough to prevent jagged particles. And Mars has something extra—the Martian surface is poisonous. Perchlorates, a class of chemicals found in trace levels on Earth, make up 0.5 to 1.0 percent of Martian surface soil. Exactly how bad this is depends on your perspective. More optimistic space-settlement geeks will tell you that the perchlorates can be easily reacted into oxygen. But we should note that perchlorates are a pretty nasty chemical. At high doses, they cause thyroid problems by competing with the iodine ions your body needs to produce certain hormones. This is probably not good, especially for developing fetuses and children. So we lean toward the less optimistic take. When people talk about space, they are often weirdly generous to conditions they would never accept on Earth. Imagine you're planning to have kids and are looking to buy a homestead. How do you feel about a real estate agent saying, "It's a great location, but I should mention the surface contains high levels of chemicals that are dangerous for children. And those chemicals can get taken up by edible plants, so if I were you I'd convert those perchlorates to oxygen before you put in the veggie patch."

Martian dust is also more active than moondust. In 1971, as the first Mars orbiter, Mariner 9, approached its target planet, something happened that surprised the scientists. The red surface of Mars seemed to resolve itself into featureless smoothness—apparently a flat disk where a sphere was supposed to be. It turned out the entire planet, except for bits at the poles and the tall volcanic peaks, was enveloped in a single massive dust storm.

Impressively, from the perspective of human discomfort, these dust storms occur even though the atmosphere is quite thin—just about 1 percent of Earth pressure, almost entirely made of carbon dioxide. The net result here is that if you step outside, you still die about as quickly as you would on the Moon, but also from time to time the sky is blotted out by killer toxic dirt.

This will keep the humans indoors. Unfortunately, their outdoor equipment, such as solar panels, will become less useful when coated with toxic regolith. Even without dust storms, photovoltaics won't work as well on

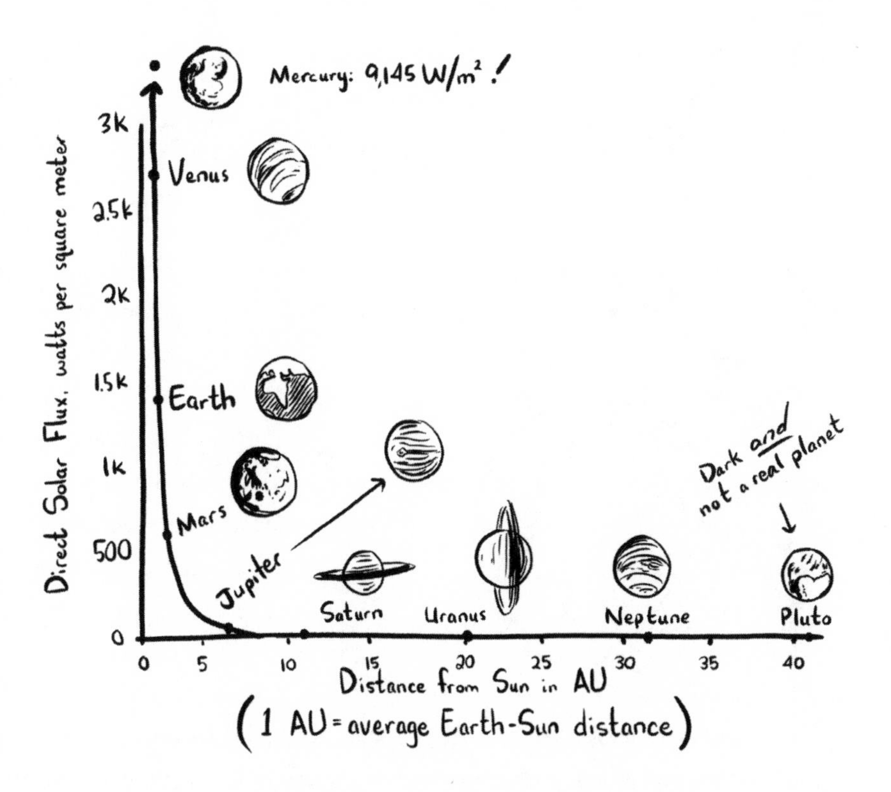

Based on figure in David Buden, Nuclear Thermal Propulsion System *(Lakewood, CO: Polaris Books, 2011).*

Mars as they do in the equivalent latitude on Earth. Light emanating from the Sun obeys an inverse square law. Every time you go twice as far from the Sun, you have one quarter the brightness. Out at Mars, you get less than half the sunlight per area that Earth and the Moon get.*

And that takes us to the biggest problem for Mars settlement: distance. Without some exotic propulsion system, your trip inbound is going to take about half a year each way. We're talking about six-month voyages in a tight ship, minus fresh apples or live calls with Fabio. You will not be packing light for this one. Six months of food, six months of water, six months of undies and toothpaste, assuming no redundancy.

If something goes wrong, getting home will be extremely difficult because the trip to Mars already has to be quite fuel efficient. Proposals for trips to Mars typically look like this:

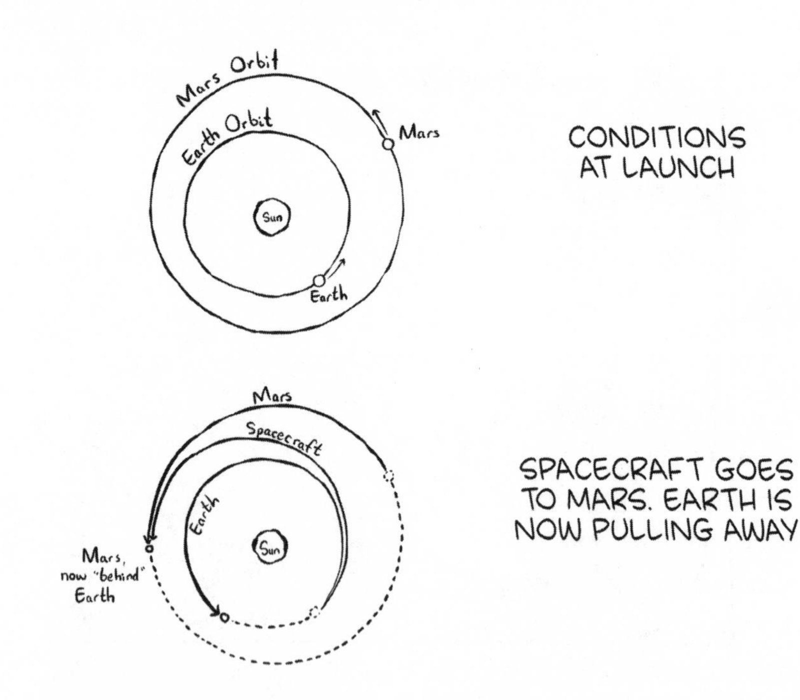

Simply put: you leave Earth at high speed, slowed down over time by the pull of the Sun. Having used up your outward momentum just as you

*Though, not counting the toxic dust storms, you'll do a bit better thanks to the thin atmosphere.

arrive, you burn a little propellant to switch to Mars orbit. Now, make sure to settle in and get comfy, because Earth has raced ahead of you in its comparatively short journey around the Sun.* Even for an initial Mars trip, that's a good two to three years total spent away from Earth.†

Once you're partway into your trip, you cannot go home until Earth and Mars are about to sync up once more. This is risky. When the service module for the Apollo 13 mission experienced an explosion en route to the Moon, part of why the men survived is that physics provided a very short "free return trajectory" back home with a minimal use of propellant.

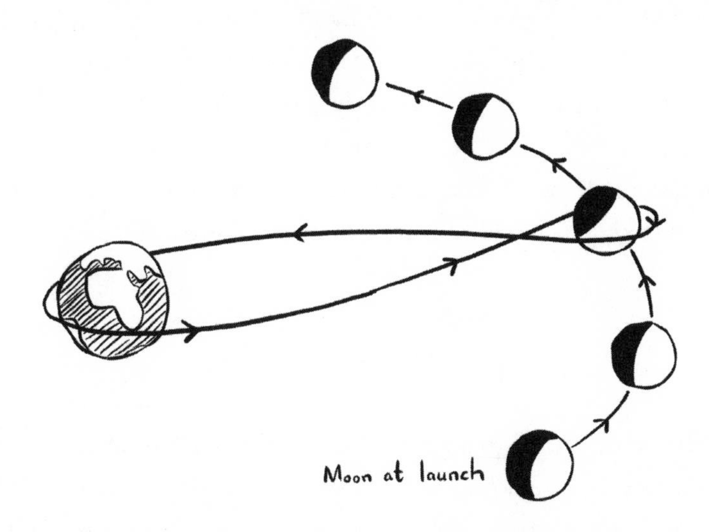

The Martian equivalent would take longer than a year. If something goes wrong on the surface of the Moon, there's some world in which you can be rescued by a ship from Earth or an escape pod home. If something goes wrong on Mars, you're likely on your own.

You can't even get a real-time phone call to help with surgery or repairs. At the greatest Mars-Earth distance, a signal takes twenty-two minutes each way. At shortest, three minutes.

And like on the Moon, there don't appear to be minerals with economic

*Theoretically you could burn a massive quantity of propellant to make a beeline for Mars, but that route is doubly inefficient. You use way more fuel to get fast, then use way more fuel at the end so you don't beeline right past Mars. Even if we had a Moon base supplying ample propellant for cheap, you still might not do this, because by efficiently using that larger amount of fuel, you can get more stuff on the spaceship.

†There are a number of different proposals for Mars landings, but all of them involve a multiyear voyage.

value for export to Earth. The proposals we've read sometimes talk about deuterium, a hydrogen isotope found at higher concentration on Mars. This is even less plausible than helium-3 on the Moon because it's farther away, worth less, and is readily available on Earth. Others talk about finding rare elements on Mars, which in fairness hasn't had several millennia of being mined by humans. But we don't know if such elements are easily accessible, and even if we did, getting a rare-element mining operation running on Mars won't be happening anytime soon. For these reasons, even among enthusiasts, return on Mars investment plans tend to focus on services rather than goods—things like tourism, scientific research, and media sales. Getting to Mars may require beautiful science and elegant engineering, but paying the rent? Only reality TV can do that.

So that's Mars. Most of the problems of the Moon, plus toxic dust storms and a half-year flight each way. Why then do so many settlement advocates favor it as the ideal second home for humanity?

The Case for Mars

Okay, the location isn't great, but the gardening opportunities are tremendous. Once you've removed toxins from all the soil, that is. But listen, Mars has all your favorite elements: oxygen, hydrogen, carbon, nitrogen! Huge amounts of water are locked up in Martian ice caps. Even if you head pretty far south there appears to be plentiful water under the surface, and there is even a little in the atmosphere too. This is substantially better than baking water out of stone or fighting Jeff Bezos for a frozen lake of ammoniated H_2O.

And sure the atmosphere is 95 percent carbon dioxide, but that's only toxic to humans. Plants *love* CO_2. They use it to build themselves and then emit free oxygen! Also, there's the Sabatier process. You may not have heard of it, but Mars enthusiasts know it by heart. If you recite the reactants out loud at a space-settlement conference, someone will reply with the products.*

*We wrote this originally as a joke, but are informed it has actually happened at a National Space Society conference.

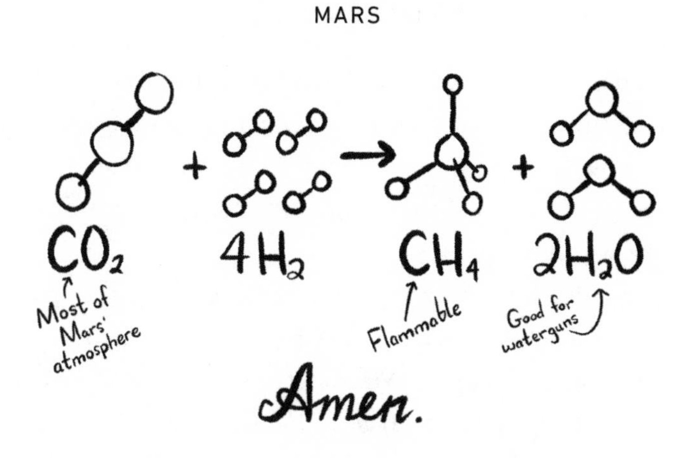

Carbon dioxide and hydrogen in, methane and water out. Water is convenient when you want to not die in three days, but methane is handy over the long term. If methane sounds familiar, that's because it's a common constituent of flatulence. Contrary to widespread belief, methane is odorless. Also, it combusts when reacted with oxygen and can be used to power rovers, habitats, and in liquid form it's a fine rocket propellant.* If you're an early Mars explorer, one of your main activities may be running a chemical reaction to stockpile methane for the return voyage. Or if you can pull it off, machines may arrive first and stockpile methane for you.

None of this will be easy, but it'll all be far easier than the lunar equivalent. While most of the Moon's oxygen is bound up in stone and much of

*So is hydrogen, but hydrogen requires ultra-extreme-cold temperatures to stay liquid. At atmospheric pressure, a mere -162°C will keep your methane liquid. Hydrogen must be held at -253°C, which is close to absolute zero.

its carbon is bound up in poo bags, Martian oxygen and carbon float in the sky. Although the Martian surface is still poor in some elements, they are needed only in trace amounts. Elements like potassium, boron, and manganese—stuff you could ship from Earth much more easily than shipping a farm's worth of carbon.

The climate is also surprisingly decent, at least by the standards of space. The typical behavior of a non-Earth planet encountering a human is to cook it, freeze it, or crush it. Mars certainly does have its freezy parts: the planetwide temperature average is -65°C, and the poles get as low as -140°C in winter. For comparison, Earth's record low temperature is -89°C, experienced at Vostok Station in Antarctica in 1983. However, toward the equator at summertime, Mars gets around 21°C—room temperature. Add in the Earth-like 24.7-hour day, and it's practically home, notwithstanding the endless lifeless horizons and the poison storms that shroud the world in darkness.

Despite its distance from Earth, Mars does have some positional value as a launch spot. We're skeptical of the asteroid-mining business, but if one day humanity is capable of processing asteroids at a profit, Mars is near the main belt asteroids, and its 40 percent Earth gravity and thin atmo-

sphere will make rocket launch much easier than on Earth. Or if you can pull it off, Mars's tiny moons would be even better.

In some visions, this creates a three-point exchange—high-tech goods go from Earth to Mars, which ships food and other low-tech resources to people working out in the asteroids, where valuable raw inputs like metals are mined then flung to Earth, where hopefully there are some very very detailed rules about flinging 100-ton pieces of dense metal at the cradle of humanity.

Mars doesn't have premium real estate in the same way the Moon does—but mostly because there are *lots* of good spots. Different proposals focus on different areas: Should we go to the cold poles where water ice is plentiful or the warmer more equatorial regions where we may need to dig for a drink? Or should we go to the lava tubes, which though not as big as the biggest lunar lava tubes, still tend to be bigger than the ones found on Earth?

Or should we go looking for the one other thing Mars may have that would be so valuable that it could justify vast spending of the sort required to develop a near-term settlement—alien life? As late as 1968, Arthur C. Clarke could write in his book *The Promise of Space* that the "evidence for the growth of vegetation is impressive." It later turned out that the seasonally expanding dark areas on Mars were not plant life, but dust storms. Hope was rekindled with the Viking probes in the 1970s, which in one experiment applied a sort of nutrient soup to a container of Martian soil, whereupon it immediately produced chemical signs of life. The meaning of this has been debated since, with the dominant view being that the reaction, especially given its high speed, was chemical, not biological. Subsequent Martian missions have failed to detect life, but have found plenty of evidence for a warm, wet Martian past, suggesting that at least the *conditions* for life as we know it were once present.

If life managed to hold out somewhere on Mars, lava tubes could've provided the last redoubt. If so, we might at last have a chance to encounter alien life and check it for nutritional value. The tricky thing is that while Martian microbes might actually have tremendous value in a way

Martian resources might not, the risk of accidentally killing the only alien life we've ever met could give us a good reason not to settle Mars. Or anyway, to settle only with extreme caution.

Who Wants to Go to Mars?

Just about everyone who wants to do space settlement. Mars is the most commonly suggested place to settle in space, basically for two reasons: First, it has what life as we know it needs to survive. Second, everywhere else is far far worse. While the solar system is quite big, the places that are even remotely friendly to human existence are small. If you think of the Moon and Mars being your overwhelmingly best bets for Human Civilization Part 2, Mars represents 80 percent of the available land for settlement.

In the more distant future, Mars could potentially be terraformed, meaning that its climate could be altered to be more human friendly. Not everyone thinks terraforming will be possible, and the proposals in favor of it tend to be, shall we say, a bit dramatic, calling for fleets of nuclear weapons to be detonated at the poles or for redirected celestial objects to create a similar effect using only speed and mass. Why do this? To convert all that ice into water vapor—a potent greenhouse gas. By this means, it's conceivable that Mars could be tuned to be warmer, wetter, and more inviting to oxygen-producing plant life. In the fullness of time, this might make for a world where humans could go outside without a pressure suit. We don't intend to explore this too deeply, because the technology won't be available for a very long time, and perhaps just as important the international law consequences of redirecting nuclear weapons and giant space objects to permanently alter the climate of the only settlement-worthy planet are, let's say, interesting. But if you buy our earlier argument that wait-and-go-big might be the right path to populating space, terraforming Mars would in some sense be the ultimate version of that.

So, the Moon and Mars are your best options. But as we discussed in our look at space medicine, it's at least possible that partial gravity will create major long-term physiological issues. If so, the next best option is likely to be gigantic rotating stations built in space.

7.

Giant Rotating Space Wheels: Not Literally the Worst Option

Some people think the giant balls of matter colloquially known as "planets" and "moons" are simply not the way to go for settlement. Like, have you looked at planets? They're so wasteful. There's this whole middle part of Earth you never even visit. Does that inner 6 billion trillion tons of mass really *spark joy*? Sure, you get a magnetosphere and atmosphere to stop radiation, but all you really need for that is some shielding. And sure you get gravity, which is nice when you're walking around, but that can be generated artificially just by spinning a million-ton space wheel. Why bother trying to fix up nature when you could start from scratch?

In short, because it's harder than the other already-really-hard options for space settlement. This is unfortunate, because huge rotating space stations are probably the most compelling visuals in the history of space-settlement concepts—often depicted as sweeping artificial countrysides where views of forests and streams are cut through with open windows to the speckled blackness of space. Not everyone loves these pictures, though. In our experience talking to space architects, the phrase "I *hate* those" was used, because they gave the public a thoroughly misleading idea of what's plausible.

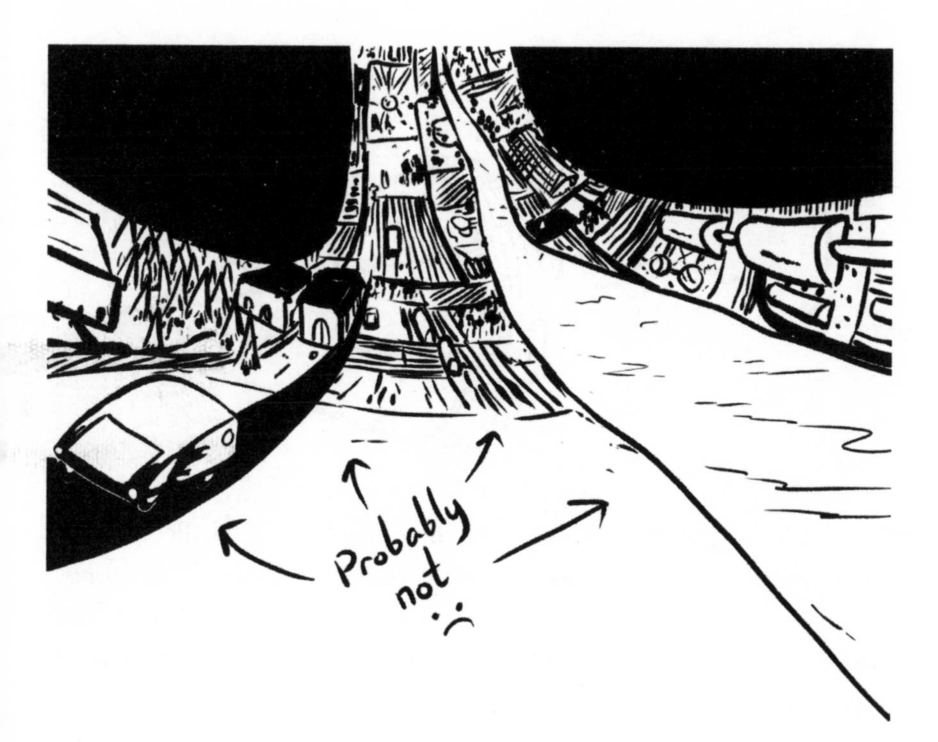

The Case Against Open Space Settlements

Any near-term design for such a station will likely be built from parts created on Earth. But Earth is a pretty deep gravity well to rocket out of, and typical proposals for open space settlements require millions of tons of material. Let's suppose you can launch fifty tons of stuff per rocket—about as much as the largest rockets ever launched from Earth. A million-ton space station would require twenty thousand such rocket launches. For this reason, proposals generally require the material to be harvested from the Moon or from asteroids, then fired to an in-space construction site.

The basic setup then is to build a technically complex launch facility off-world, then create a catcher's mitt to receive hunks of mass that are not especially promising as industrial inputs, and then convert all that into the most complex built structure humanity has ever contemplated.

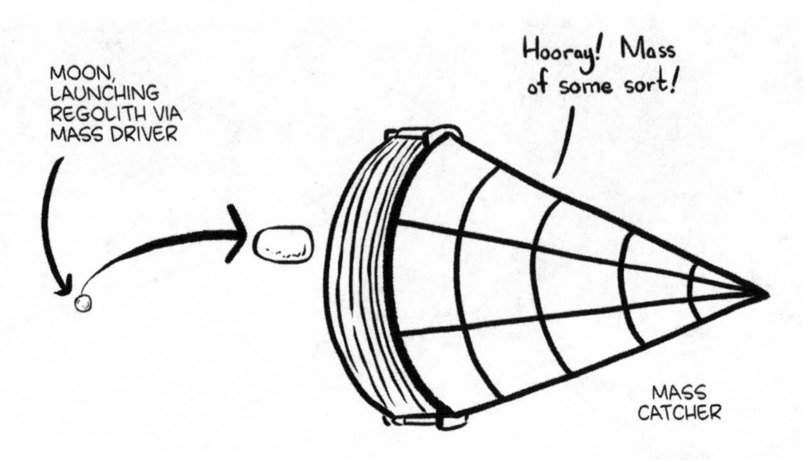

That sounds hard. Can we make these things a lot smaller perhaps? Probably not. In order to walk around the rim of a spinning wheel without getting nauseated, the wheel needs to be really really big. If it's not clear why, consider an extreme case: you're in space, and the radius of the wheel is precisely the same as your height.

Imagine the wheel is spinning so quickly that your feet push into it with the force of Earth gravity. However, your head is at the center of the wheel, experiencing far slower rotation—something close to zero gravity. Effectively, the top of you is floating and the bottom of you is planted. Thus, the middle of you decides to puke.

So you need a bigger wheel, but how big? At two rotations per minute, you need a diameter of 450 meters to hit one Earth gravity. If you up the rotation to four per minute, you can pull it off with about 112 meters in diameter.*

But how is a human who is spinning around at four rotations per minute going to feel? Honestly, we don't really know. The studies that have

*You could also do a more low-rent version by having an incomplete rim, possibly even just a pair of living spaces connected by a tether, but now you're really giving up on the big space-station settlement dream.

been done often use small sample sizes, don't last very long, and frequently are done on people who are known to be less susceptible to motion sickness. Plus, they are mostly done on Earth, and some data from Skylab

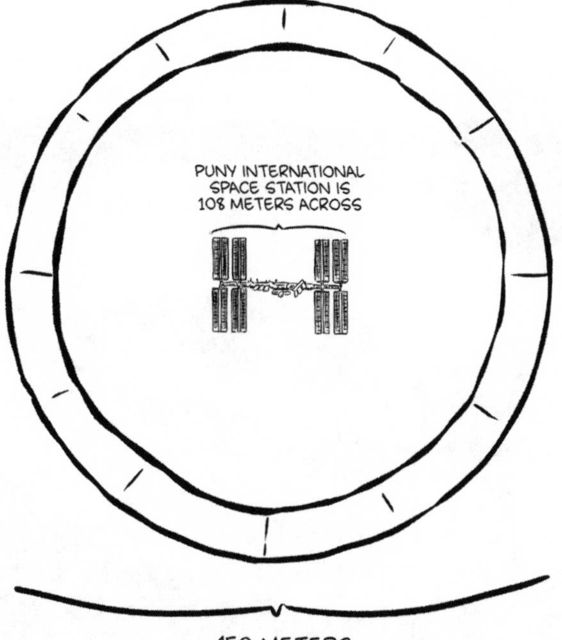

suggests that rotating on the ground may be more sickness inducing than rotating in space.

Even if we assume the best case, where you only need 112 meters diameter, that's still far more ambitious than anything we've ever built. The ISS, which cost over $150 billion to construct, is about 112 meters long at its greatest length, but doesn't include a huge habitable wheel around its rim. Even with the recent cost reductions for launch, this is going to be pricey.

And smallness creates its own problems. A small wheel has to be pretty clever to avoid what you might call the "washing machine effect." Take a washing machine, put a heavy towel to one side and start it spinning. Inevitably, you hear a worrisome KA-CHUNK KA-CHUNK KA-CHUNK. This is what happens when a spinning object is unbalanced, and it'll be extra embarrassing if the wheel is a tiny bubble of life in a hostile void. The solution is to counterbalance the source of the KA-CHUNK. For instance, you can have a hydraulic system that shifts water around to maintain the right mass distribution.

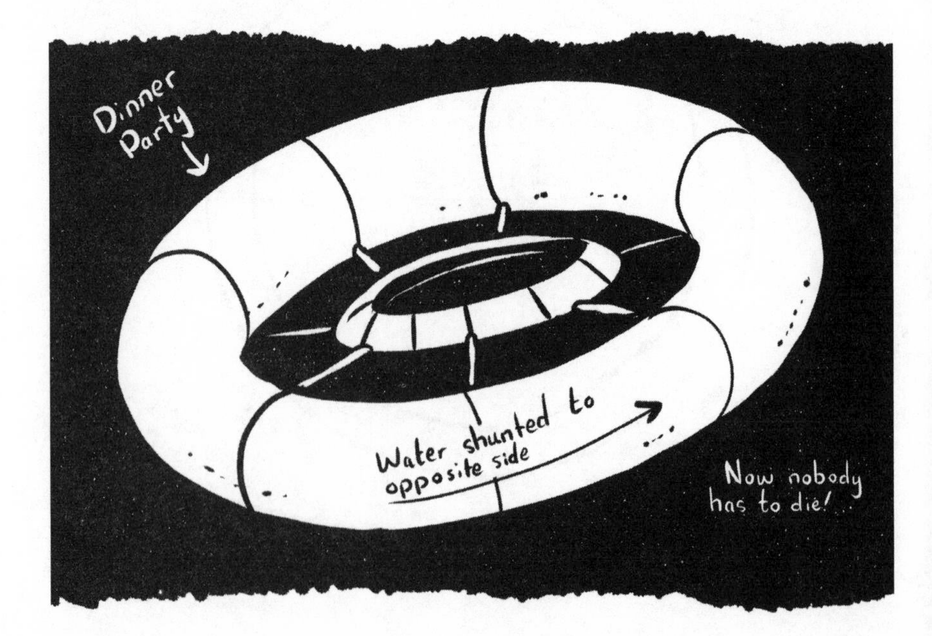

This strikes us as terrifyingly vulnerable to Murphy's Law. As one paper notes: "A major shift of weight (all the crew to dinner at one sitting) would require a programmed and controlled counter-movement of ballast." Let's hope the pipes don't malfunction for Friday night potluck—you'd hate to have Mrs. Sanderson's baked ziti ripped out through a hole in the world. The best solution here may be to give up any idea of a small station—the bigger and more massive the wheel, the fewer problems are created by the puny motion of humans within. But, well, now we're right back to an enormous and enormously complex project.

There are other issues we could get into, but the upshot is this: even a lot of the very basic aspects of rotating space wheels are difficult and dangerous and wildly expensive. While it is literally true that there is enough of the right stuff in space to build these things, it's kind of like saying we can build solar panels on the Moon. What seems easy to the physicist may not work so well for the engineer, or the investor for that matter.

Even if we could pull all this off, there remains a great big "why bother" standing in the way. In order to build a 400-meter wide wheel in space, you have to assume we've already got the technological level necessary to build launch facilities on the Moon or to field great big asteroid-trawling spacecraft, not to mention outer space factories capable of converting high-speed cargos of regolith into orbital suburbs. But if we have that level of know-how, why not just use that capability to remain on the Moon or Mars, where all this stuff we're launching is just sitting around?

Is there any justification for building these things?

We should start by saying there are a lot of bad cases often made. We've seen repeated claims that an advantage of space stations will be the ability to completely control things like temperature, light, and weather. This sounds compelling until you realize it's also a thing that happens in, like, buildings. Another notion is that we need more land because after all there's no new land being created for the teeming masses of Earth. This sort of argument goes back at least to the 1920s, but gained more prominence with the rise of the environmental movement in the 1960s and '70s.

One can dispute what the optimal population of Earth is in terms of consumption, but it's pretty clear that *land* for people to live on is not our biggest problem. As of 2018, Japan had 8.49 million abandoned homes, and in some rural regions tax incentives are offered to people willing to move into them. A quick Google search locates eight towns in Canada alone offering free land if you'll pretty please just move in. Antarctica, which is substantially more inviting than anywhere in space, is 40 percent larger than Europe and has a population under ten thousand during the busy season. Additionally, we *are* making more land. At least, if you're willing to count artificially constructed surfaces on which humans live, which you should if you're going to count space stations. We couldn't find numbers on the total amount of living space in buildings added to Earth per year, but as an example of scale, the world's tallest residential building is New York City's Central Park Tower, finished in 2020. It has a floor area just under 120,000 square meters—about twenty soccer fields worth of new living, leisure, and shopping space, in a single building, and incidentally with perfectly controlled weather and climate. No rotating space wheel required. Meanwhile, although population is growing, world fertility rates are declining, and the United Nations's *World Population Prospects 2022* expects population to reach its peak in the 2080s.

A more nuanced version of this argument would say that space stations don't just make new land, they make new biosphere, sparing our overtaxed planet from humans and human-made pollutants. This is possible in principle and may happen one day, but it's not coming in time to save us from major environmental issues like climate change. Even if it becomes possible to build all this stuff in an environmentally neutral way, you'd have to get around 80 million people into space per year just to keep the population stable. That's about 220,000 people *per day*. For reference, Central Park Tower contains about two hundred condominiums. Assuming our space-station settlers will have less roomy accommodations and accounting for how some of that space is for things like shopping and hotel space, let's bump it to a thousand condominiums and assume five people per unit. With these numbers you'll want to boost about sixteen thousand Central Park Towers to space per year just to keep Earth's population from growing. Oh, and that's not counting how those people will need on-site farms, shopping, and space Chevy van services.

Space stations are incredibly cool. If we could take a vacation in any space-settlement concept, we'd always rather a giant space toroid than the dusty wastes of Mars or the Moon. But when you start looking at the details, in-space settlements are basically doing space settlement on hard mode.

The Case for Open Space Settlements

There are a few good arguments for space stations under particular conditions.

First, babies. Since you are now an expert on space sex and its consequences, you are aware of the concerns about life in partial gravity. In a world where the Moon's and Mars's partial gravity are bad for human flourishing, we might have to await the coming of space-wheel technology before we can settle space. That still wouldn't necessarily argue for space stations as the default mode of off-world life, since they could just be something like orbital nurseries. Not exactly the vision of *Star Trek*, but anyway a lot cuter.

Second, there might be utility for these things if you assume a developed space economy is already established. Space stations have a very shallow gravity well, meaning spacecraft could come and go without costly fuel expenditures. Also access to multiple gravity regimes inside the station might provide all sorts of side benefits. In a rotating space cylinder, the artificial gravity level depends on your position in the wheel. Without getting into the math, the basic deal is that gravity is highest at the rim, diminishing as you approach the axis.

The most awesome result here is that people could fly down the middle of a cylindrical space station, but there would also be practical appeal for manufacturing. Spacecraft construction might be easier in a zero-gravity facility. Or, you could set up a very-low-gravity facility, so you can easily manipulate very heavy objects* without having to deal with the floating

*Mind you, that doesn't mean it's totally safe. Objects in zero gravity still have inertia. If a giant steel ball presses you against a wall, you still go splat in zero gravity.

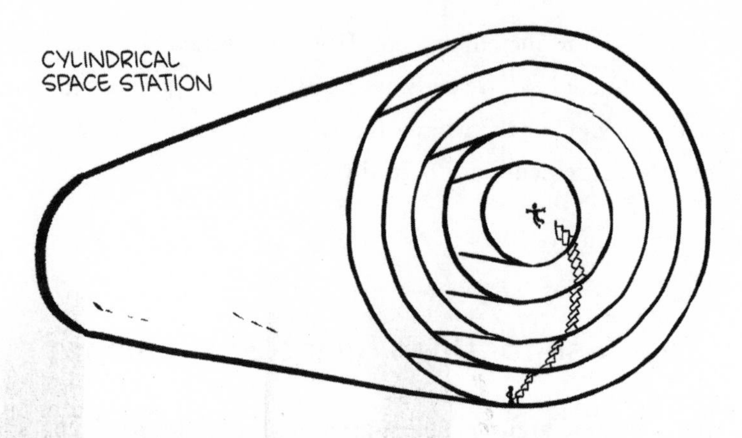

AS YOU CLIMB TOWARD THE AXIS, EACH DECK HAS PROPORTIONATELY LESS "GRAVITY." HALFWAY TO THE AXIS REDUCES YOUR WEIGHT BY HALF, WITH PURE MICROGRAVITY ALONG THE AXIS.

debris of zero gravity. Zero gravity may also provide benefits for certain manufacturing processes, such as crystal formation, though we should add that these benefits have been promised since at least the 1970s and haven't yet materialized in any large-scale way.

Third, giant space wheels would make trips to Mars or the Moon substantially more luxurious. Also safer, since the artificial gravity would protect you from the microgravity issues you typically experience in space vehicles. However, in our experience, "very wealthy people would have a lovely time" is not a typical argument for these projects. In any case, even if we need a lot of transports, that doesn't argue for a full-on space settlement.

Honestly, that's about the best we can say for space stations. They're popular among space geeks, and they have for decades been the main pursuit of the National Space Society. But unless partial gravity is a showstopper, we don't see rotating space stations as a settlement priority for a long time.

Many of the arguments made for space stations have their historical origins during a particular period of the 1970s when it appeared that environmental degradation was going to cause imminent worldwide famines, and when the price of space launch appeared to be falling rapidly. Also, renewable tech that we take for granted, like cheap photovoltaics, massive wind turbines, and advanced battery storage, were still decades away. If you combine all those constraints—resource calamity in the extreme short term, space access very cheap, and no other options—there might maybe have been a case for huge solar-power-collecting space stations with factories and artificial human habitats. But the widespread famines never came,* renewables got cheap, and space travel has remained relatively costly, even accounting for recent changes. Today, if you're going to move a bunch of humans off Earth, space stations are unlikely to be a priority.

But hey, there are worse options.

*That's not to say there weren't any famines. However, the ones that did happen weren't on the predicted scale of hundreds of millions of people, and in addition were frequently caused by war, not consumption outstripping resources. Contrast this with Dr. Paul Ehrlich's statement in the prologue of 1968's *The Population Bomb*: "The battle to feed all of humanity is over. In the 1970s the world will undergo famines—hundreds of millions of people are going to starve to death in spite of any crash programs embarked upon now."

8.
Worse Options

The Moon, Mars, and space stations are the most common proposals for space settlement, but not the only ones. However, everywhere else is so very much worse. Here, we provide the alternatives, ranked and ordered by increasing awfulness.

Asteroids

The asteroids in the belt are even farther out than Mars, meaning solar power is limited. Also, despite what *Star Wars* told you, many asteroids aren't solid potato-shaped rocks, but "rubble piles." Zero-gravity rock and dust are not great as a landing surface. Also they aren't very close together. If you're parked on one asteroid, you likely can't even see another with the naked eye.

Ideas for using asteroids to promote settlement tend to focus on the potential to make money by harvesting asteroid resources. We are skeptical. Reading proposals, you hear about hundreds of trillions of dollars' worth of minerals in the asteroid belt at current prices. Setting aside the fact that reducing the scarcity of materials lowers their prices, it's not clear we can even get asteroid materials at a profit. It does you no good to know the asteroids are worth $700 trillion if it costs $700 trillion and ten cents to get them to market. After all, if you're willing to just ignore the cost of acquisition, you're really better off digging on Earth. Earth contains about 10^{23} tons of iron. If we assume a value of $100 a ton, that's roughly a bajillion zillion hojillion dollars' worth of iron, and that's not even counting the gold, the silver, the diamonds, and all the friends we'll make along the way.

Even if *someone* makes enormous profits in space, there is no plan for an equitable distribution of space-acquired goods. At a recent meeting of the Space Generation Congress, where the young future leaders of space gather to hobnob, an African delegate raised the point that the economies of some African countries are dependent on mining, and might be devastated by an influx of space minerals. How much the reader cares about the distribution of purported space riches is a matter of political ideology, but it's important to realize that when it's said that "we" can access the riches of space, the "we" tends to go unspecified.

The more serious proposals for an economics of asteroids call for nabbing the ones that have very special qualities. An ideal asteroid should be on a trajectory that takes it near Earth. Note, this means it's not one of the majority of asteroids, which are found in the main belt. The asteroid should be moving at low relative speed as it comes close to us. And it should be one of the rare asteroids with a high concentration of valuable metals, like rhodium and platinum. A recent estimate suggests that the total number of known slow-moving, near-Earth asteroids containing at least a billion dollars' worth of platinum-group metals is around a few dozen. We talked to Dr. Martin Elvis,* whose name is Elvis, and who is also an asteroid-mining expert, and he said he expected the number to go higher, but that the ultimate goal would have to be mining the asteroids in the main belt beyond Mars.

*We are just leaving a footnote to declare our love for Dr. Elvis, who not only tolerates but insists on our Elvis jokes. Also, nearly as important, he gave us excellent notes on this section even though he disagrees with parts of our perspective. If you want to learn more about asteroid resources, check out Elvis's 2021 book *Asteroids.*

Even if we can do this, don't visualize a giant hunk of platinum—we're still talking about ore with only a few grams per ton of precious metals. This is your best bet for asteroid money, and right now it's not very good. More important for our purposes, it offers no reason to create a settlement.

If we ever do have a large human presence among the asteroids, it will likely depend on a large human presence somewhere better. Platinum-bearing accessible asteroids are in short supply, but more homely materials are available. If we just want nearby slow-moving asteroids with water, there are about nine thousand known so far. And because they're relatively small, kicking these asteroids to a Moon or Mars base, or even a space station, to provide things like drinking water will take far less energy than shipping it from Earth. There could one day be significant asteroid-mining operations, but it's hard to see why you'd ever want a large population living there. Indeed, asteroid-settlement proposals are rare, with the typical idea being to use asteroid resources to supply people living in the less-crappy regions of the solar system. But we've already described those, so let us move on to the *more*-crappy regions of the solar system.

Venus

Venus's average daily surface temperature is over 450°C—hot enough to melt lead. You won't mind because you'll have already been crushed by an atmospheric pressure more than ninety times Earth's. That's assuming you survived the sulfuric acid clouds on the way in. On the plus side, the thick atmosphere will provide your remains ample protection against radiation.

We haven't found too many proposals for a Venusian habitat, but there are a few ideas for a floating base in that thick atmosphere. It turns out that there's a slim shell of the Venusian sky that has human-friendly temperatures and pressure, low radiation, 90 percent Earth gravity, and access to atmospheric carbon dioxide. Location, location, location!

If you're somehow not impressed by the idea of a life spent dangling above *and* below hell, you should talk to the nice people who proposed a project called Cloud Ten. Their plan is to use all that atmospheric carbon dioxide to grow bamboo and kombucha, out of which to build small cell-like habitats.

Look, if you've read the footnotes, you know we're the kind of people who read technical documents about legal aspects of Buzz Aldrin's feces.

We are not qualified to advise you on any lifestyle choice whatsoever. But if you dream of a home constructed of bamboo and kombucha, surely someone in Northern California is prepared to accommodate you at a lower price.

Mercury

Mercury is like the Moon, but nudged much closer to the Sun, with the result that the average day-night cycle produces temperature swings from -180°C to over 425°C. In fairness, the temperatures are a bit milder at the poles, and like with the Moon, we think there may be ice there, permanently frozen in deep craters.

But Mercury is not a popular settlement choice, outside of a few wacky proposals—the best of which involves settling the terminus. What's the terminus? The little region where day meets night. The Goldilocks zone where you're not frozen to death or cooked alive. The one downside is you have to stay safely in the penumbra as night makes its way around the planet. The good news is that the Mercurial equator is a mere 15,325 kilometers around—well under half the circumference of Earth. Better still, the Mercurial day is a leisurely 4,222.6 hours long. So all you have to do is move every bit of human civilization on the planet 86 kilometers every

24 hours forever, and you get to stay alive.* Or, anyway, at least not killed in particular by the cold or heat. You'll still have the radiation, lack of air, and constant regret over your life choices.

The Outer Solar System

One thing the worlds beyond the asteroid belt have in common is darkness. Once you're much past the asteroid belt, the Sun is almost useless as a settlement power source.

The planetary landing sites aren't so great either. There are a few tantalizing moons spinning around Jupiter and Saturn—worlds like Enceladus and Europa that likely have warm subsurface water. If your goal is to find alien life in the solar system, these are great candidates. If your goal is to *be* that alien life, you're better off closer to home. Using current methods, doing the trip safely is going to take you years, possibly decades depending on where you're going. Until we have some very futuristic propulsion technology, these places will remain the magisterium of robotic space probes.

Other Suns

The nearest star, Proxima Centauri, is about 4.2 light years away. If we assume you're going as fast as the Parker Solar Probe, which *Guinness World Records* says is the fastest-ever spacecraft, that's about 8,000 years.

We can't do this. Look, your only option for interstellar travel using anything remotely like near-term technology is to build a ship inside of which a human civilization can survive and reproduce for four hundred generations without killing each other. Does that *sound* like something humans can pull off?

*We are, in fact, being somewhat unfair to Mercury. You could shorten your around-the-world trip just by living closer to the poles. However, if you've chosen to live somewhere that is pizza-oven hot if you stray too far from home, you've already demonstrated limiting reasoning ability.

If you'll allow a little sci-fi technology, maybe we can have ultra-long-term hibernation. Actually, come to think of it, this is probably more plausible than the four hundred generations of harmony. That said, you'll still need to build a spaceship with no major technical malfunctions during the next eight millennia. Go ahead and splurge on the nice paint.

If you're willing to allow a *lot* of sci-fi technology, the fastest rate you could get there would be a bit over four years, traveling at light speed and somehow not being obliterated as you encounter small interstellar objects at velocities normally reserved for tiny particles. We are not futurecasters, but we're willing to bet that well before lightspeed spaceships we'll all be uploading our brains to Amazon or whatever, and we can just beam ourselves to the next star if we want to.

Faraway stars with exoplanets are neat. Maybe one day we'll find signs of alien life. Maybe we can even ask it for its nutritional value. But the odds of humanity as we know it showing up to greet those aliens? Just about zero.

Space: Quite Bad

If you take just one concept away from part II, it should be that space is different from Earth. Space is different, and attempts to compare spacefaring to earthly exploration tend to make it sound easier than it will be. Sailing a galleon around the world is damned impressive, but it's still an extension of swimming or sitting on a floating log—stuff humans and other animals do naturally.

No other species we've ever encountered has intentionally made its way to space, likely because space combines just about every bad environment on Earth, plus a few curveballs like ultra-extreme temperatures, poison-soaked soil, and endless horizons of charged jagged glass. Space settlement is not impossible, but it will be damn hard.

Astrid, as you see, has decided to go to the conveniently local Moon. Our goal in part III is to figure out how to have her survive.

Yes please!

Nota Bene

SPACE IS THE PLACE FOR PRODUCT PLACEMENT, OR, SPACE CAPITALISM IN DAYS OF YORE, PART 2

Given their vast cost and battlefield potential, in the years after Hermann Oberth exploded himself, rocket funding would not come from movies or venture capital. It would come from governments and militaries. But when the spigot of Cold War money began to run dry, things changed.

As the USSR limped into the late 1980s, the Soviet space agency found itself in want of cash and willing to be a little less communist in the pursuit thereof. By the late '90s, Russia had pushed so far into commercialization, some of their activities made even Americans cringe.

Emblematic of this change was a 1997 commercial using footage filmed aboard Mir, the largest space station ever constructed at that time. It was shot by Aleksandr Lazutkin, featuring Commander Vasily Tsibliyev squirting little spheres of milk into the air and gobbling them up with as much enthusiasm as he could muster.

In the commercial, ground control loses contact with Mir, causing serious concerns before they manage to restore the connection. In the flush of their success, they ask the commander if he needs anything.

The commander replies, "A glass of real milk."

"Real milk in space? It's never been done!" says the shocked head of ground control.

Tsibliyev looks out the window at the great blue ocean below and asks, "Hmm . . . why not Israeli milk?"

This was meant to be a rhetorical question, but it turns out the answer was cancer. Tnuva, the Israeli manufacturer of the milk, was in trouble at that time for putting a suspected carcinogen into their milk. Also, for lying about putting a suspected carcinogen into their milk.

However, in the commercial, Tnuva's white-lab-coated scientists spring into action, rapidly producing that one beverage every Russian man wants to have on hand for a celebration—shelf-stable milk. The drink is sent up to Commander Tsibliyev, who Pac-Mans the delicious white spheres. An attractive, prodigiously blond woman of unknown vocation swoons from mission control. *Fini.*

Incidentally, Tsibliyev was made to reshoot a scene from the commercial, because its director on the ground felt he hadn't been smiling enough.

From the late '80s through today there have been countless shameless promotions interfacing uncomfortably with the final frontier. We can't document them all, but here are a few highlights:

1990—Tokyo Broadcast System pays to fly journalist Toyohiro Akiyama to Mir aboard a rocket covered with advertisements for Japanese companies, including Unicharm, maker of disposable hygiene products.

1996—Pepsi sends a 1.2-meter-long inflatable fake Pepsi can to Mir, which cosmonauts have to take on more than one EVA (extravehicular activity), aka a spacewalk, aka the most dangerous activity available in low Earth orbit, and which doesn't typically involve gigantic novelty soda cans. Incidentally, although both Coke and Pepsi have made their way to space, they aren't terribly popular with astronauts, because burping in zero gravity is a dicey proposition.

2000—Pizza Hut plasters their logo on a Russian rocket. One year later, they make the first pizza delivery to orbit. They had earlier looked into projecting their logo onto the Moon itself, but changed their minds after being told the laser projection would need to be the size of Texas and would cost hundreds of millions of dollars.

2001—NBC announces a new setting for its popular reality TV show *Survivor* called "Destination Mir." It was never filmed, not least because Mir was deorbited in 2001. Which leads us to this:

2001—Taco Bell sets up a target at sea, offering a free taco to everyone in the United States if Mir hits the bull's-eye when it deorbits. Mir, perhaps as a final act of defiance, misses.

Other space-promoted products include Cup O'Noodles, Rold Gold Pretzels, and Radio Shack. Honorable mention goes to KFC's "Zinger" sandwich, which went where no chicken had gone before during a ride on a high-altitude balloon.

Star Trek may have given you the impression that the future of space will be free from crass consumerism, but the best available evidence suggests commerce is just getting warmed up.

Pizza Hut was thwarted in its attempt to laser an ad into the Moon, but PepsiCo (Pizza Hut's owner) was not entirely deterred in their pursuit of turning the heavens into a billboard. Their Russian branch recently worked with a local start-up called StartRocket, with plans to launch gigantic orbiting billboards made of Mylar.

After word of this plan leaked, causing a predictably negative response, Pepsi changed its mind about ruining the sky. But if they had gone ahead with it, there wouldn't be a strong legal case to stop them. US law allows private companies to put logos on their rockets and payloads, but bars them from putting ads in space that can be seen by the naked eye. However, legal precedent going back to the first satellite launch in 1957 says satellites can move freely over all countries. So in principle, if you want to blot out the stars over America with a Mylar sheet that says "American Beer Sucks," you just have to launch from somewhere other than the United States.

The United Nation's Committee on the Peaceful Uses of Outer Space has shown interest in organizing to stop obnoxious space ads, but no concrete steps have been taken. Given more pressing issues, like space weaponization, this is hardly surprising. As they note in a 2002 committee report: "the view was also expressed that there was a question as to the priority nature of such a recommendation." Fair enough.

Our perspective in this book is that in the wildly alien environment of space, human nature will remain decidedly earthy. So while we don't know whether your grandchildren's grandchildren will inhabit underground Martian caves or floating cells of kombucha skin in the Venusian skies, we can be certain that wherever they are in the cosmos, Ronald McDonald will find them.

PART III

Pocket Edens: How to Create a Human Terrarium That Isn't All That Terrible

SO FAR WE'VE DEALT WITH A LOT OF STUFF SPACE DOES THAT IS WEIRD. Gravity, radiation, crazy temperature swings, and so on. The basic solution to all these problems is to avoid them if you can. What that means specifically can vary, but will always be some sort of bubble world that has to perform all the functions of Earth's biosphere, only in miniature.

Ecosystem design is another one of those weird space-settlement problems that gets relegated to the margins while being a major practical barrier to off-world survival. Much like human reproduction in space, ecosystem design is a wickedly complex scientific problem that is only lightly funded despite being fundamental to any space-settlement project. Perhaps because it involves uncool stuff like growing mangos and recycling poop. Or maybe because it'd be really expensive without any clear geopolitical clout for whomever builds it.

Just speaking for ourselves, though, if we were sent to a Mars base, we would like not to die. So as our last stop before we get into the laws governing space, we consider some more homely matters—food, waste, and how to garden. Having accomplished that, we just need to shield all of it from instant death in space.

9.

Outputs and Inputs: Poop, Food, and "Closing the Loop"

Space settlers will have to receive shipments from Earth for a long time, but at least for basic things like food and water, they'll need to strive for as much independence as possible. Potato salad starts to get expensive when it has to be boosted out of Earth's gravity well, flown across the void, then gently deposited outside a Martian airlock. Also, growing local produce means a higher chance of not dying if a shipment doesn't make it or arrives contaminated.

However, becoming self-sustaining in space means managing waste with an unearthly level of sophistication—something human science is still learning about, and which we'll explore at the end of this chapter. But before we get to that, we present a brief tour of how eating and excreting are actually done in space, so you know the two ends of the loop you're trying to close.

Outputs: The Inevitable Space Toilet Discussion

There have been a lot of space toilets, but outside of launch and landing, there have been two major ways to go number one and two.

In the heroic early days of space travel, the solution was an elongated plastic baggie with adhesive and a little finger-shaped divot.

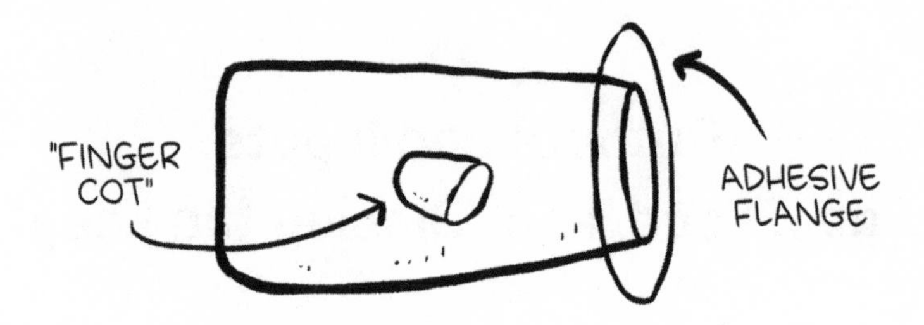

The adhesive is there for obvious reasons, but why the finger-shaped divot? To nudge waste in the right direction.

Nobody liked this. In fact, during Gemini 7, Commander Frank Borman, who had a reputation for being, shall we say, anal retentive, decided he would attempt not to poop for the entire two-week orbit. This quest was somewhat facilitated by NASA's use of what are called "low-residue foods." But, fourteen days is a lot of time to accumulate residue. By day 9, Borman had to face the inevitable. He turned to his lone crewman, Jim Lovell, and said "Jim, I think this is it." Lovell, known for his calm demeanor and sense of humor, joked back, "Frank, you have only five more days left to go here." Frank did not make it five days. Incidentally, this is how big a Gemini was:

2.55 CUBIC METERS OF HABITABLE VOLUME

(FOR COMPARISON, A TESLA MODEL 3 HAS AN INTERIOR VOLUME OF ABOUT 3.2 CUBIC METERS)

By the early space-station days, this method had been replaced, essentially, by a very specialized two-headed vacuum cleaner. One head for liquid waste, which is fairly easy to deal with. One head for solid waste, which must be netted, bagged, and later packed down. And don't visualize some sleek metal-and-plastic disposal unit. As ISS Commander Peggy Whitson noted in an interview, "After it starts getting full, you have to put a rubber glove on and pack it down."

On the plus side, partial gravity will restore the traditional relationship of humanity to waste, in which once it leaves the body it does not fly. That said, if there is any takeaway from space toilets of old, it's that they have always been complex and finicky. NASA's gigantic Human Integration Design Handbook includes detailed analysis for things like the maximum bowel movement mass and frequency, and detailed specifications about how astronauts must be able to poop and pee at the same time, and yet over the years, astronauts have felt compelled to create a variety of neologisms to describe the phenomenon of airborne feces. "Floaters," "escapees," and worst of all, "brown trout." In the original Space Shuttle waste system, when the poop was suctioned down, a device one astronaut called "the slinger" thwacked it against a cold surface, effectively freeze-drying it. But containment was flawed, allowing tiny freeze-dried poop particles to enter the cabin. Aboard today's ISS, the tradition lives on in different form: Tim Peake reported an astronaut who accidentally lost "a decent-sized portion of metabolic waste," only for it to turn up two weeks later in "a small gap near the return air filter."

As recently as SpaceX's 2021 "Inspiration4" mission, which sent four intrepid private astronauts on an orbital voyage in especially cool space suits, one major problem was the busted toilet. As of this writing, details are scarce, but apparently something went wrong with the suction system, leading to the delightful *USA Today* headline: "Elon Musk Says Inspiration4 Crew Had 'Challenges' with Toilet, Vows Bathroom Upgrades."

While space-settlement sanitation will be easier, it will also present a new challenge. Historically, poop is something astronauts just dispose of, hence the tempting yet forbidden lunar poo bags. This won't work on a space settlement. Remember, you're trying to create Eden, and the truth is that human waste is substantially closer to garden perfection than the dead soil of other worlds. But so far, no solid waste has been recycled in space. The closest equivalent is the recycled urine and moisture on the ISS, which is turned into potable water. Or as the Americans call it, "yesterday's coffee."

Given the utility of waste off-world, a likely setup will involve some sort of composting system for human outputs. This is well-understood technology on Earth, but rarely employed in places with low gravity and a sealed atmosphere.

Inputs: Space Food—Bad, But Not as Bad as It Used to Be

Tang sucks.

—Buzz Aldrin, second man on the Moon

Food in space is subject to a lot of constraints. It has to be nutritious. It has to come in a convenient easy-to-prep container that doesn't leach chemicals into the food. It has to be shelf stable for as long as possible. It has to produce no crumbs or other little food bits that'd find their way into the ship's atmosphere or equipment. And ideally it should do all that while being tasty and varied in both flavor and texture.

Complicating matters, you may not have a kitchen. Not a real one, anyway. Mir and Skylab both had refrigerators, which allowed for delicacies like filet mignon, ice cream, and Jell-O. But refrigerators take up a lot of space and energy. The ISS only got one in 2020, though according to rumor some of the scientific cooling fridges have been used for drinks. However, most food is either room temperature or hot. But never piping hot and certainly not freshly fried or roasted—activities that might off-gas undesirable chemicals into the sealed atmosphere. Heating is only done via hot-water injection or in a convection oven that can only get to about 75°C. Space settlements will almost certainly have more Earth-like kitchens, but they won't quite be identical. If you have a stove at home, very likely it comes with a ventilation hood. In space settlements, the atmosphere will have to be recycled. The desire of humans to have cooking procedures like those from Earth will have to be balanced against the costs of scrubbing the atmosphere.

The environment of space also reportedly makes food taste less flavorful. This may be a result of the fluid shift creating sinus pressure similar to a cold, or it may be that in zero gravity smells don't waft up into your nose, or it may be something about the artificial atmosphere.* Whatever the reason, astronauts often lust for piquant condiments, such as salt, pepper, Tabasco, and mayonnaise. And, of course, taco sauce. Salty, zesty taco sauce is so beloved by astronauts that for about a week in 1991 it became the first form of currency specific to outer space. On shuttle flight STS-40, taco sauce went on everything. Pilot Sid Gutierrez recalled, "Although I didn't do it myself, I observed crewmates putting taco sauce on Rice Krispies in the morning." Around day 8, STS-40's Commander Bryan O'Connor realized the crew's rate of taco sauce consumption would soon outstrip the taco sauce supply. According to Gutierrez, the commander "secured all the remaining taco sauce and divided it equally among the crew members. Thereafter taco sauce became the medium of exchange. For example, if it was your turn to clean the latrine, you could pay someone a taco sauce or two to do it for you."

However bad Rice Krispies *picante* sounds, it's an improvement over the old days. The very first space foods were pastes inside tubes, a style of meal originally designed for on-the-go jet pilots. During the 1960s, American space food became weirdly abstract. A variety of food cubes were produced, often encased in an oily, gelatinous substance, including "cheese cubes," "toasted bread cubes," and the intimidatingly nonspecific "red cubes." Partially this was just engineering run amok,† but there was

*One experiment aboard Salyut-6, which went by the name of "taste," tried to better understand this phenomenon by applying an electric current to the cosmonaut's taste nerves.

†We found one story about Sidney A. Schwartz, a scientist at Grumman Aircraft, who combined "flour, corn starch, powdered milk, banana flakes, and hominy grits," then "baked them in a hydraulic press at 400 degrees Fahrenheit under 3,000-pound pressure." That's about 200°C and about 1,360 kilograms. The resulting product was "a grainy brown slab as tough as tempered Masonite that could be cut on a bandsaw without splintering or drilled for bolts and screws." And, said Schwartz, after being "pulverized with a tiny grinder," and "soaked for a few hours in water," it becomes edible and "tastes like breakfast

some logic to it. If you want to efficiently pack a lot of food in a form that will not produce crumbs, bite-size cubes with an outer coating make sense. The downside was that nobody liked them. According to an account by food scientist Dr. Paul Lachance, astronaut Wally Schirra suggested they fly with only the bite-size food. The food scientists realized this wouldn't be a great idea and suggested the astronaut and some buddies try living off them for a few days. "And they all gave it up, because it was coating the throat . . . and they got tired of biting these hard chips."

Much of early space food was freeze-dried or dehydrated. Removing water was more valuable back then because the first human spacecraft got power from fuel cells, which generated pure water as a by-product. Simply use some power, open a baggie of mashed potato flakes, and indulge. As the space station era began, things got a little nicer. The menus on the Russian side sound positively fancy, with Salyut-1 having sausage, chocolate, coffee, cheese, cookies, meat, and fish.

The Space Shuttle, which was the locus of US space activity from the early 1980s until the ISS went up, also had fuel cells to provide water and only orbited for about two weeks at a time. So stocking food was a bit easier, with many of the early rations just being military MREs—"meals ready to eat."* Still, there were major developments at that time, including the very first space tortillas. The noble tortilla first went to orbit in 1985 when mission specialist Mary Cleave and payload specialist Rodolfo Vela, the first Mexican astronaut, brought fresh ones. Tortillas are ideal space bread because they pack well, create no crumbs, and are really just edible plates. Shelf-stable tortillas have remained the astronauts' bread of choice ever since. On the Russian side, a more technical approach has been taken. Cosmonauts eat tiny squares of bread specially engineered to be low crumb. Because of their size, they are jokingly referred to as Barbie bread.

Modern space cuisine is mostly dehydrated food, freeze-dried food, shelf-stable "intermediate moisture" foods like dried fruit and M&Ms,

---

cereal topped with bananas. I rather like it." Richard Foss, *Food in the Air and Space: The Surprising History of Food and Drink in the Skies* (New York: Rowman & Littlefield Publishers, 2016), 161.

*These are not always fine dining, we are informed, and have a variety of joke nicknames, such as "Meals Rarely Edible" and, due to their frequently low fiber content, "Meals Requiring Enemas."

and food that has been thermostabilized or irradiated to keep it from rotting. Reading through the literature on space food and the descriptions by astronauts in memoirs, the impression one gets is that the food is decidedly okay—basically the sort of thing you'd eat if you were living in a fallout bunker stocked by someone with a degree in food science.

One sometimes comes across gustatory freaks who actually enjoy space food, like Charlie Duke, who couldn't get enough of the "ambrosia"* packed for him. Dr. Mike Massimino, despite growing up in an Italian American family, loved thermostabilized spaghetti and meatball packets. However, love for space food is rare, and many astronauts lose weight during their missions.

And what of Earth's most beloved drinks—wine, beer, and liquor? There has been at least one tipple on another world. It happened in 1969, when Buzz Aldrin took Communion in the *Eagle* lander. That's one small sip for a man, one giant leap for man wined.

NASA officially prohibits alcohol on the ISS, but booze has gone to space many times. Soviet/Russian space stations were substantially more permissive, in a way that occasionally caused culture shock. Dr. Jerry Linenger was surprised to find bottles of cognac and whiskey hidden in the gloves of a space suit on Mir.† Non-Russian/non-American flyers have perhaps been more classy—the French *spacionaute* Patrick Baudry brought Chateau Lynch-Bages 1975 during his Salyut trip.

Orbiting under the influence is, if not quite a tradition, a behavior with a long pedigree. We've been told by sources that'll remain anonymous that the atmosphere emanating from the Russian segment of the ISS has been known to carry a hint of ethanol from time to time. Will space settlements be likewise? At least one author counsels against a Martian distillery anytime soon. Science fiction writer Andy Weir said in the preface to *Alcohol in Space* that fans often ask whether his heroic Martian survivor and potato farmer Mark Watney could've made liquor. The answer is no—not if he

*Zach, who grew up in the American South, was familiar with this stuff. Basically, imagine a variety of canned fruits in syrup, plus bits of sweet gelatin and tiny marshmallows, all combined with more sugar and essentially any kind of fat—mayo, whipped cream, sour cream, whatever. No doubt even tastier in an irradiated shelf-stable packet.

†Armenian cognac was reportedly the drink of choice among cosmonauts.

wanted to live. "It takes almost eight kilograms of potatoes to make a bottle of vodka . . . almost a week of meals for our poor stranded astronaut."

It's likely that a future space settlement will still make use of a lot of these preservative technologies, both because they'll supplement whatever we can grow off-world, and because they provide a taste of home. After two years of locally grown organic vegetables, a sleeve of slightly space-irradiated Oreos may start to look pretty tempting. People who study space psychology report good food as one of the most important factors in day-to-day well-being—an idea also found in books from the era of polar exploration. Also, by routing that food into humans, we effectively create a source of additional soil for our space farm. However, the more food we can grow on-site, the more efficient, resilient, and potentially permanent our space settlement becomes.

A Farmer on Mars: The Fresh Future of Space Food

Cosmonaut Valentin Lebedev never liked gardening on Earth. But in orbit, he fell in love with tending green life in an artificial world. As he said, "A tiny leaf opened up and it seemed to fling open a bright window out into the world."

Plants clean the air while creating more organic matter on lifeless worlds. They also provide nutrients like vitamin C, which are hard to keep shelf stable. Lack of vitamin C produces scurvy, and while that would give the operation a charmingly piratical flair, settlers would probably prefer not to have bleeding gums, wiggly teeth, and wounds that won't heal.

We know plants can be grown in space stations so long as you supply water, atmosphere, nutrients, and a grow bed. The ISS "Vegetable Production System" and the later "Automated Plant Habitat" have successfully grown plants like mizuna mustard, kale, lettuce, dwarf wheat, and Chinese cabbage. As with humans in space, keeping plants alive isn't a simple procedure, but the good news is that they don't just immediately up and die on you.

However, saying "I can grow food as long as there's water, atmosphere, and nutrients" is essentially just saying the growth medium is not actively harming the plant. Lunar regolith may not even clear this low standard. The first experiment growing plants in actual lunar regolith was only published in 2022, using *Arabidopsis thaliana* seeds. The soil was supplied with nutrients, water, and a plant-friendly atmosphere. Plants did grow, but poorly. Compared to those in comparably nutrient-crappy Earth soil, plants grown in old Apollo mission samples had a reddish color, usually taken to indicate stress. Further molecular examination also found they had a physiological stress response. They were also smaller than the Earth-soil plants.

Okay, but we knew the Moon was a fixer-upper. How about Mars? You may have read a popular article or two saying that experimenters successfully grew plants in "Martian simulant soil." What's typically left out is that Martian simulant soil doesn't precisely simulate Martian soil. At least not at the chemical level, since it doesn't start with any perchlorates in it. Simulant soil is a particular product that captures some of the texture of Mars soil, but is really just Earth soil that is the closest match to what we think Mars soil would be. Nobody has ever grown a plant in Mars soil, but most likely, Mars dirt will require an involved process of cleansing and fertilizing and seeding with microbes, and even if it works out, we'll need to validate that the plants themselves are safe for human consumption. Otherwise, the solution may be hydroponic or aeroponic gardening, where nutrient-laced water or air replace soil.

By the way, lighting will be a problem. For reasons we'll discuss later, you probably can't use glass as your exterior. If you're on Mars, the light levels are already substantially lower than on Earth, and occasionally the sky is blotted out with dust. Likely you're using an artificial light source, or perhaps piping light from the surface via fiber-optic cables, or both.

Also, just as with humans, we don't know for sure how microgravity and space radiation will impact plants. A recent mission by the European Space Agency called Eu:CROPIS* tried to answer the question by growing plants in a small satellite, but was thwarted by a software

*As an obscure trivia fact, note that the "Eu" is not for Europe, but "Euglena," for the microorganisms used in the study.

malfunction—an excuse we have ever after used for the death of our own houseplants. More successful was China's Chang'e 4, which sprouted cotton on the far side of the Moon. It did pretty well until the sub-Antarctic lunar night overtook its container, freezing it to death. But hey, so far so good.

So much for gardening. Can we ranch? Yes and no. As a general rule, the larger the animal, the less efficient it is at converting fodder to meat. At least initially, cows are out. Pigs are out. Chickens are okay, but squirrel is better. In fact, why not hamsters? Because there's a better option: insects.

Insects—what's not to like? They reproduce quickly, don't take up a lot of space, eat food scraps, are high in protein, and you never name one Wilbur and fall in love with it. Proposed bug-protein sources include crickets, silkworms, mealworms, hawkmoths, drugstore beetles, termites, and flies. Whether you enjoy these is a matter of taste. Or possibly a matter of treachery. According to a 2020 editorial in the journal *Foods*, "Several consumer studies have concluded that hiding insects in traditional foods can increase people's willingness to eat insect-based foods."

Slightly more sci-fi options include bioreactors that make meat from cells. A number of start-ups are already working on this sort of technology, and at least in theory you could create the meat of cows, pigs, goats, and other large animals without having to bring along two of each kind.

One expert we spoke to wanted to take things even below the cellular level—he suspected plants might be grown for spices, but that otherwise we should create our meals from fundamental food building blocks, like fats and amino acids. Others thought we should just keep it simple and go vegan in space. Not everyone's favorite option, but it might complement your Venusian kombucha hut. Just don't eat the walls.

Closing the Loop: Building a Space Ecosystem and Then Not Dying Inside It

For a completely closed-loop ecology, we'd like as much as possible to be recycled: your breath, the moisture your body gives off, urine, feces, flaking skin cells, and whatever other effluvium is currently radiating from your mortal coil.

In an orbital space station, you'd really want to get as close to 100 percent as possible. On the Moon or Mars there can be substantially more tolerance, though you'll still have to be mindful about those elements that are hard to replace locally, such as carbon on the Moon or phosphorous on Mars, and nitrogen almost anywhere.

Other than the creation of "yesterday's coffee," we don't have a lot of experience with in-space recycling. The ISS is big for a space station, but likely far too small ever to have a self-contained ecosystem. We say likely because you're never going to get anywhere close to the plant-to-human ratio of Earth with six people in such a small container. That said, it'd be quite valuable to know in principle how small you *could* take one of these systems. For the moment, we don't know. Closed-loop ecologies are complicated and not a lot of money has gone into them.

The earliest experiments on humans in closed loops were the BIOS-1, 2, and 3 systems built in the USSR starting in 1965. They originally tried using pure algae to keep the men alive. In case you've forgotten high school biology, algae are basically plant sludge. They eat carbon dioxide, exhale oxygen, and are a nutritious source of calories if you don't mind eating the same horrible meal every day forever.

Later BIOS experiments introduced plants like wheat, beetroots, and vegetables, and ultimately were able to control CO_2 at a reasonable level. A major problem they had, which will be quite relevant for space settlement, was that it took a lot of work—20 percent of the crew's time was spent maintaining the system, despite the fact that BIOS was never even close to a completely closed loop. Solid waste was not recycled, and meat was imported into the experiment. Apparently, a vegetarian diet was never even considered, on the basis that "Siberians must have their meat!"

A near completely closed system would only be accomplished in the 1990s, and never duplicated afterward. This is the strange, but we believe undervalued, tale of Biosphere 2.

Biosphere 2: Contrary to Legend, Not an Unmitigated Calamity

Earth is "Biosphere 1," so naturally Biosphere 2 is a 3.14-acre nearly airtight greenhouse in Arizona containing the most complex sealed ecosystem ever created by humans. It was born through a strange partnership between a billionaire and an enigmatic counterculture technologist group called the Synergists. They weren't quite a cult, but they were at least, let's say, cult-*adjacent*. This created issues. Their charismatic leader, John Allen, had space in mind when he designed Biosphere 2, but his training and crew-selection notions weren't exactly NASA's. To find the right people, he sent the candidates on a weirdly 1990s eco-adventure through the Outback and on the high seas. No doubt his explanation would be more involved, but it sure seems like a major criterion for selection was "John Allen thinks you're cool."

Socially, things didn't go great. Early in their stay in Biosphere 2, the crew of eight began fighting, ultimately splitting into two hateful factions, each consisting of two women and two men. With more than a year to go, they were no longer on speaking terms. They stopped dining together or even making eye contact. Jane Poynter, one of the eight "biospherians," recalled a day in which two different members of the opposing faction spat on her, separately, in a sort of coordinated saliva strike. Biosphere 2 had been set up with lessons for a future settlement in space in mind, which apparently stayed on the minds of the participants. As Poynter said of this rift: "Perhaps on Mars, with the safety of home at least forty-eight million miles away, we would have been able to pull together. But then again, perhaps not. I wondered if people are really meant to be enclosed in small spaces, even as large, beautiful, and varied as Biosphere 2. The human species, after all, did not evolve indoors."

There were also particular reasons why relations broke down, mostly having to do with a dispute with the equivalent of mission control. However, the deeper issue was likely the screening process. We have trouble imagining a crew made of Chris Hadfields doing this sort of thing.

There were also problems with food production. Shortly after the experiment ended, crew member Sally Silverstone published a book of recipes they'd used, called *Eating In: From the Field to the Kitchen in Biosphere 2.* Opposite page 1 is a photo of the eight "biospherians," smiling awkwardly on the first anniversary of their stay, each looking substantially more gaunt than your typical chef. In fact, they were starving.

The half-feral chickens they'd hoped would be especially robust had failed to lay eggs regularly, and the unusual breed of pigs they'd selected had been reluctant to eat scraps, sometimes preferring to eat the chickens instead.* Insects and microbes attacked their crops, withering several high-protein high-calorie species. Ultimately, they were forced to eat unripe bananas and an unpalatable type of beans normally used to feed farm animals, and that were only usable after cooking for an hour and hiding them in non-terrible food. None of these people had been exactly corpulent prior to entry, but the men lost about 18 percent of their body weight and the women about 10 percent. This was despite at one point eating seed intended for future planting.

The ecosystem suffered too. The designers of Biosphere 2's ecology used a technique that may be used in space one day, called "species packing." The idea is you start with a lot of species, many of which are ecologically redundant, on the assumption that some will be wiped out before balance is achieved. This indeed happened, but they ended up packing more than they intended. Proposals for space settlements sometimes note the possibility of precisely selecting what life forms come along with the project. In the one big experiment where this was tried, at least two stowaways came along: cockroaches and bark scorpions. In case you don't have a sense of what bark scorpions are like, the Arizona-Sonora Desert Museum supplies what they call "Extra Fun-facts," including "This is the only species of scorpion in Arizona that is truly considered as life threatening."

*According to *Eating In: From the Field to the Kitchen in Biosphere 2*, the chickens were a cross between "jungle fowl"—basically the ancestor of chickens, presumably accustomed to tougher fare than chicken feed—and the Silkie breed of domesticated chickens. They weren't great, laying only about one egg a month each, though possibly they would've done better on a higher-calorie diet. The pigs were originally going to be Vietnamese pot-bellied pigs, which were supposed to be small and high in fat. But at the time they were also a popular pet, which caused anger from activists, so the "Ossabaw feral swine" was used. They ate chickens and wanted starchy food over scraps. Thus, their final contribution to Biosphere 2 came during Thanksgiving and Christmas 1992.

The biospherians also nearly lost control of the atmosphere. In a functioning closed loop, CO_2 and O_2 do a sort of dance as animals convert oxygen to carbon dioxide and plants do the reverse. But at one point the O_2 started falling even as CO_2 kept rising. This was not supposed to happen, and the crew became breathless and lethargic as they tried to do their work, frantically adding new plant life wherever they could steal a patch of sunlight. It didn't help. After an acrimonious debate, the leadership team decided to pipe in fresh oxygen. What happened? The choice of high-organic-matter soil meant that microorganisms pulled a lot of oxygen out of the atmosphere, releasing CO_2. Then, some of that CO_2 was absorbed by structural concrete, effectively locking a huge amount of oxygen out of the system.

So why do we say it wasn't a calamity? For one thing, it more or less worked.

Eight people entered, eight left. The problems they had could have been rectified for future experiments simply by selecting better animals and modes of pest control and by having concrete coated with epoxy so it wasn't actively suffocating the humans. Possibly also by adding a question about spitting on coworkers as part of the screening process. Some errors were downright trivial. Several high-calorie fruit trees like mango and avocado were simply not fruiting by the time the experiment began, but would have been for subsequent crews. At one point, the crew were mildly poisoning themselves because they had no idea how to cook the taro they were growing.

Biosphere 2 was originally intended to do consecutive two-year stints on an ongoing basis, but it was canceled partway through the second round, due to financial issues and infighting.* If it had continued, likely we would have a lot more data on how to maximize production and safety. Assuming these problems could be solved, could the whole system have been ported to Mars?

Nope.

*As an irrelevant but too-weird-to-skip coda, Chris Bannon and his brother Steve Bannon, later director of the 2016 Donald Trump presidential campaign, were brought in to take over the financially strained experiment as it careened toward its demise.

Most important, the structure would have to be changed. On Earth, the air pressure inside Biosphere 2 was relatively close to the pressure outside. Still, the system required two huge "lungs" to make sure the expansion and contraction of the atmosphere due to temperature change was accommodated.

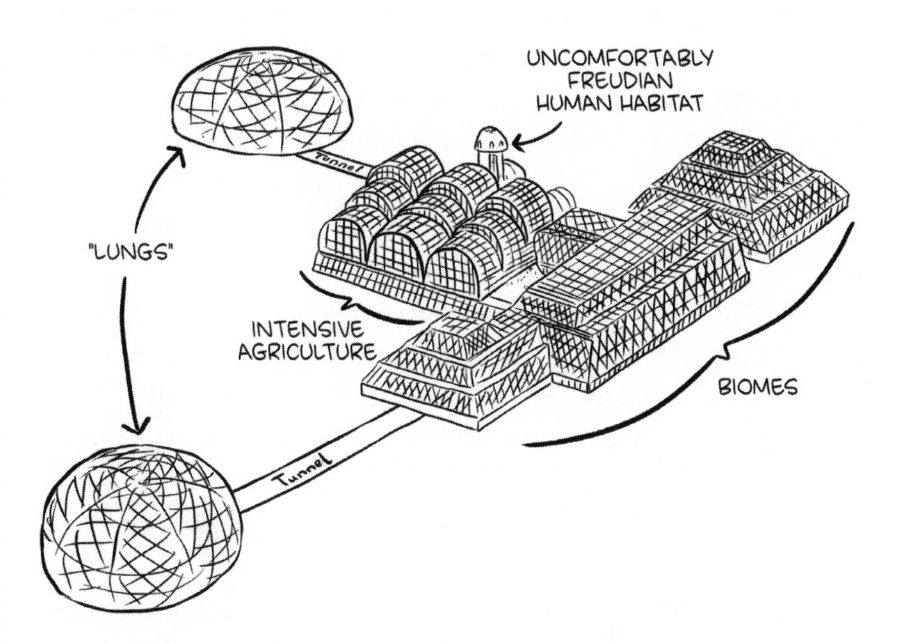

This will be trickier in the near vacuum of space. On the ISS, windows have four panes of glass each, the thinnest of which are a little over one and a quarter centimeters thick, and the thickest of which is over three centimeters. It may be possible to duplicate something like this over several acres in space, but it will not be cheap. A likely alternative will be to bury the whole thing underground, having regolith supply the outside pressure. But now you absolutely have to pipe in or generate light by some means. That's not great, but then again you're likely going to heap regolith on your home anyway to deal with radiation and temperature swings.

Biosphere 2 would likely also require efficiency improvements to work on Mars. Remember, if you want bread inside the biosphere, you start with seeds planted in the ground. You have to grow wheat, take it to a threshing machine, grind the seeds, and then start the normal process of baking. If you'd like cheese on your toast, you'll have to be a goatherd and a cheesemaker. Effectively, you're subsistence farming on a tiny plot of land that sometimes requires technical system maintenance. The result in Biosphere 2 was an average workload of eight to ten hours a day, five and a half days a week. This is despite not doing all the chores of a space settlement, including running a power plant, doing construction, manufacturing, and at some point, taking care of children. A long-term space settlement will benefit from quite a bit of division of labor, but that'll be tricky if everyone has to be a full-time farmer just to not-quite-starve.

Kelly once talked to a group of researchers who work on closed-loop ecologies and asked them what they thought of Biosphere 2. A common view was that it was cool, but too ambitious. For the cost, which was around $300 million to $400 million in today's money, better science could've been accomplished using a series of small facilities, scaling up as you go. Because we didn't have such a methodical approach, Biosphere 2 remains a sort of tantalizing proof of concept. It doesn't *prove* that we can build ecologies on other worlds, but it strongly winks at the possibility. How well it would do trying to scale the population through five orders of magnitude is a different question.

There have been other, smaller-scale experiments since. For example, Japan's Closed Ecology Experiment Facilities (CEEF), in which the system successfully supplied most of the food for two human "eco-nauts" as well as their goats, but had difficulty supplying the needed levels of oxygen. China operates the Lunar Permanent Astrobase Life-support Artificial Closed Ecosystem, or just the PALACE. Clearly, their acronym game is already equal to NASA's. The Lunar PALACE has had crews of three do partial seal-ins and had good success recycling water and clearing carbon dioxide out of the air with plants, particularly when the crew included two female members, who were smaller than the males on the crews. The European Space Agency has a very methodical ecosystem project called the Micro-Ecological Life Support System Alternative, or MELiSSA, but they aren't including humans just yet. These are the right type of systems to enhance our knowledge of closed-loop ecology, but none are nearly on the scale of Biosphere 2, which itself wasn't nearly on the scale needed for space settlement.

Space Composting! Not as Cool as Space Travel, but Very Important

We would love to be able to share a more definitive science of, say, how to put a hundred humans in a glass bubble and have them live indefinitely. We don't have it. In some sense this is the ultimate proof that space agencies are not oriented around space settlement.

Biosphere 2 cost around a tenth of a percent of the ISS. People can debate what the point of space stations is, but if they're part of a project leading to human settlement off-world, this is not a good allocation of resources. For the same cost as the ISS, five hundred biospheres could've been built. Better yet, a decades-long sequence of experiments could've been run in order to create what we really need—an extremely detailed computer model for how to design sealed worlds. What life forms to use, how to arrange them, how to make them as efficient as possible, and how to maximize human population and well-being inside. Oh, and there'd be the side benefits of deeply enhancing our knowledge of ecosystem sustainability and our ability to create food efficiently in environments without hospitable climates—both potentially relevant in the coming decades.

A major reason governments pay to stick tiny habitats full of humans in orbit is national prestige—looking smart and strong in front of the world community. Blasting humans in a rocket to the Moon is substantially more impressive than creating detailed reports about how to turn poop and food scraps into wheat. But if the goal of space agencies is eventual space settlement, this sort of thing should get a lot more funding, even if it generates a collective yawn among the public. If we cannot supply on-site agriculture, space settlement only works with constant vast shipments of food. This may be possible on the Moon at tremendous cost, but would be very risky on faraway Mars.

More concerted efforts in closed-loop ecology would also provide sociological insights. Just as we don't know enough about closed loops, we don't know enough about how humans handle life in these bubbles. That'd be useful because we the apes bring our own chaos to ecology. Earlier, we cited Andy Weir, who said booze would be a bad idea in space because it would destroy valuable calories. He's not wrong biologically, but psychologically he may be missing an insight. Not to second-guess Mr. Weir, but in fact the calorie-deprived biospherians *did* make booze. In order to brew

ADAPTED FROM SALLY SILVERSTONE'S RECIPE IN EATING IN: FROM THE FIELD TO THE KITCHEN IN BIOSPHERE 2

up wine they used several of their reliable calorie sources: bananas, rice, and beets.*

Here, we could accuse the biospherians of behaving irrationally, but we believe the picture is a bit more rich. It's not irrational to seek comforts. Living in a sealed bubble, physically separated from loved ones, occasionally getting spat upon by coworkers, well, who wouldn't want to tuck into a glass of fermented beets on a starlit Saturday night? The ultimate goal of space settlement isn't surviving, but thriving. That means accommodating human frailty in an environment that isn't always welcoming to human existence. If we are ever going to build permanent space settlements, we need to integrate all this into a science just as urgently as we need to build vehicles that can take us to Mars.

The good news for Astrid is that we'll assume everything is worked out, and that she's raised those kids to not spit on coworkers.

The last thing they need to make their habitat complete is the bubble that keeps space out and the energy to keep everything inside alive.

*In fairness, you would lose a lot more calories going from food to liquor than just going from food to wine. Perhaps there's a compromise position on booze for starving Martians.

10.

There's No Place Like Spome: How to Build Outer-Space Habitats

> It is not hyperbole to make the statement [that] if humans ever reside on the Moon, they will have to live like ants, earthworms or moles. The same is true for all round celestial bodies without a significant atmosphere or magnetic field—Mars included.
>
> —Dr. James Logan, former NASA chief of Flight Medicine and chief of Medical Operations at Johnson Space Center

So, humans are squishy and weak. The real estate options are toxic. And pointy. And cold. You'll be growing vegetables in your own waste, tending your food bugs, and fighting off bark scorpions while drunk on beet wine. This is humanity's new dawn.

Well, that's assuming you can keep the lights on.

The final piece of the puzzle for human survival off-world is to take the humans and their environment and find a way to protect all that from the general horribleness of lunar and Martian climes. Plenty of work has gone into all the little details, but two major problems for us to consider are these: energy and shielding.

Energy

Taking data from 2020, almost 60 percent of worldwide electricity comes from fossil fuels. Unless we discover something *very* surprising about Mars's past, and then decide to set it on fire, anything with the word "fossil" in it is a nonstarter. Most renewables won't work in space either. Hydropower requires flowing water. Wind power requires wind. Although there are, in fact, occasional proposals for Martian wind power, to make use of the ultrathin atmosphere the turbines would have to be huge. Geothermal energy, where heat is drawn from deep underground, won't work on the geologically quiet Moon. It might work on Mars, but would be another enormous on-site construction project, and the best locations for geothermal power may not be the best locations for an early Mars habitat.

This leaves two options: solar power and nuclear power. Most likely you'll use both. If nuclear power makes you nervous, you'll be tempted to go all in on photovoltaics. But consider what that would entail on the Moon or Mars. Typical Mars proposals for tiny crews call for acres of panels to be set up just for an initial foray, never mind running the vast underground greenhouse required to supply local banks with taco sauce. If we assume Martians have power needs comparable to average US citizens, a Martian city of 1 million would require around 130 square kilometers of solar panels.* Martians, however, will almost certainly require far more power given their need to do most of their farming and manufacturing locally, including the need to generate air, clean toxic soil for use, and extract fuel from the atmosphere itself.

On the Moon, unless you're on the Peaks of Pretty-Much-Eternal Light, the panels will simply not work for two weeks at a time. On Mars, they'll work during the day, but remember, the day is about half as bright as what you get on Earth. Plus, the Sun is occasionally blotted out by dust storms. In both settings, you'll want battery banks for the dark periods.

*To be clear, we think of this as an ultralow bound. Getting an actual number would be extraordinarily complex. As Dr. Manfred Ehresmann told us in conversation, you'd even have to think about things like how container ships on Earth are highly efficient because they float on water. The actual number might at least be ten or a hundred times larger.

The total weight of batteries depends on the location and power needs, but at the very least we're talking about thousands of tons of equipment, possibly quite a bit more. Even if you have these huge backup power systems, you can't just plug in solar panels from the Home Depot. You have to have panels that can endure high levels of radiation, meteorite strikes, abrasive regolith, and unearthly hot-cold cycles. On Mars in particular, you'll need to clean off dust—a process that, on Earth, usually requires plenty of water and plenty of workers.

A common response to this concern is to shout "Robots! By God, robots!" We like robots, but we are skeptical. A recent mission called Mars InSight made use of a tool called "the mole," designed to drill five meters into the Martian surface. It managed two to three centimeters before failing, due to a combination of low friction, extremely compacted regolith, and diminishing power supply as dust settled on the robot's solar panels. After months spent trying to recover the experiment, the team was forced to give up. We don't mean this as an insult—your authors can't even build Lego robots, much less ones that operate autonomously in space. And that's not to mention the more successful mechanical inhabitants of Mars—the truck-size robots and the little helicopter bot flying over the

red hills as of this writing. But we're a long way from an autonomous robotic system that can assemble, place, and maintain hundreds of acres of solar power on unfamiliar terrain.

So let's talk about your friend and mine, the atom.

Space Nuclear Power: The Energy Source That Doesn't Usually Scatter Radioactive Debris over Canada

We suspect we are probably a bit more comfortable with nuclear power than a typical reader. We're those type of people who talk at parties about how nuclear waste handling is actually a well-developed technology, and that France doesn't glow green at night despite its electrical grid being 70 percent nuclear power. That said, even if you agree with us, you may still be a bit squeamish about the idea of launching, say, huge hunks of plutonium into space. Would it help you to know we've been doing it already for decades? There are really two ways this happens, which are quite different and often confused.

The First Way: Hot Hot Hunks

The simplest use of nuclear power is for heat, in something called a radioisotope heater unit. If you like, you can think of it as a magic box that never* gets cold. The heat source is radioactive metal—typically plutonium, americium, or polonium.

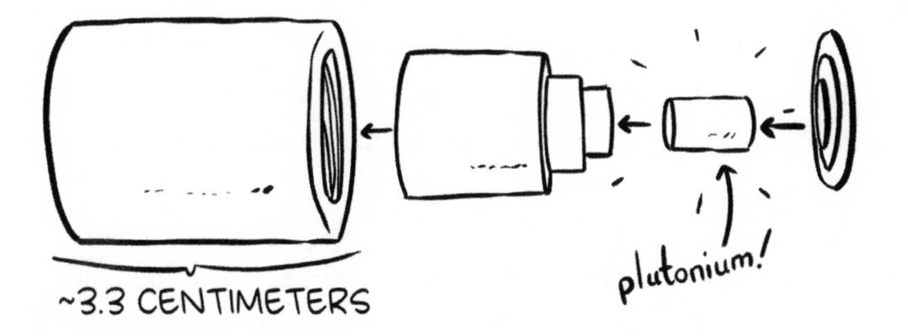

*It's not quite never—these materials decay, but only over a long period. For instance, if you're using plutonium-238, the amount of power generated drops by half every eighty-eight years.

These substances undergo radioactive decay, meaning their atoms break apart over time. In the process, they generate heat, which is something even robots need to keep their bodies running in space. In fact, nuclear space bots have a rich history. The Soviet Lunokhod Moon rovers had solar power, which they used during the daytime, but they were kept at operating temperature thanks to a hunk of polonium-210. As recently as 2018, the Chinese space agency sent the Yutu-2 ("Jade Rabbit") rover to the never-before-roved far side of the Moon using plutonium-238 to stay cozy on its travels.

These systems have not been without incident. Perhaps the worst was the time in 1969 when the very first Lunokhod was launched, but then was destroyed when its rocket exploded, sending polonium over parts of the USSR.

A more advanced system is called a radioisotope thermoelectric generator. Basically you take your nuclear heat system and add what's called a thermocouple. Now, some of the system's heat is used to produce electricity.

The appeal here is you've got a more or less permanent and extremely reliable power source. In the past these have been used as Arctic heat sources and even pacemakers. Plutonium-238 versions powered scientific equipment left by Apollo astronauts on the Moon, and they've also been used on Mars rovers. They are absolutely crucial for very-deep-space probes, from whose vantage point solar power is a nonstarter. Thanks to their radioisotope thermoelectric generators, the Voyager 1 and Voyager 2 probes launched in 1977 have left the solar system and continue to transmit data home.

The cool thing about devices like these is that they're simple and just work. The biggest downside, other than general public opposition, is that you don't get *that* much electricity per unit of mass. Plutonium gets you about 500 watts of heat per kilogram. If your heat-to-energy conversion is really good, you might harvest 10 percent of that. The leftover heat is still useful, especially during the lunar night, but roughly speaking, for every two kilograms of plutonium, you get enough electric power to run a laptop. This is the most intimidating way to do word processing, but not terribly efficient, and it'll cost you somewhere in the $10 million range.

If you could magically transfer an insane quantity of plutonium-238 to Mars, it could be a very convenient way to get power—heat, electricity, and no risk of any of the classic nuclear fears like meltdowns or coolant leaks. But you can't. These systems just don't provide enough energy for the mass. If you use them, it'll likely be for familiar purposes in space, like keeping rovers warm.

The Second Way: Good Ol' Fashioned Nuclear Fission Reactors

When people refer to a "nuclear reactor," they're usually talking about a fission reactor. The systems we talked about above are powered by radioactive material that, if you will, "naturally" decays. Nuclear fission reactors *force* atoms to decay rapidly. The result is a tremendous amount of power per mass.

However, we suspect the words "nuclear reactor" in a book about space travel are making some of you nervous. Rest assured space agencies have done it plenty of times, and as we'll see, 98 percent of these missions didn't accidentally scatter radioactive debris over Canada.

The only fission power source ever orbited by the United States was the SNAP-10A, an experimental reactor launched in 1965, whose electrical components failed on mission day 43. Although a few pieces of it appear to have broken off as part of what is euphemistically called "shedding," SNAP remains in high orbit, where it will most likely circle the Earth the next four thousand years, or at least until technologically advanced descendants of current humanity arrive to sigh, shake their heads, and hurl it into the sun.

The USSR launched over thirty nuclear fission reactors to space aboard satellites. When everything went right, once the mission was over, these nuclear satellites burned a little propellant to transition to a "long-lived storage orbit," sometimes more spookily referred to as a "graveyard orbit," where they will remain for centuries, sort of like Saturn's rings only made of radioactive garbage.

But things didn't always go right. Two of the reactors accidentally reentered Earth's atmosphere, including the infamous Kosmos 954, which, well, accidentally scattered radioactive debris over Canada. We'll hear a bit about this later when we discuss liability law in outer space, but suffice it to say nobody was particularly happy.

No fission reactor has ever gone farther—not in a distant probe or in a rover. But testing has happened on Earth since the 1960s and continues to this day. In 2015, work began on a compact nuclear reactor, called "Kilopower," meant for space projects. The Kilopower reactor is small, providing enough energy to run a few average US households. It was tested in 2018 (under the name Kilopower Reactor Using Stirling TechnologY, or KRUSTY*) and did quite well. Work is now under way on a bigger unit that could provide enough power to run thirty households, and NASA is planning to test a fission power system like this on the Moon eventually.

Fission reactors come with risks. Launching them comes with the danger of what aerospace engineers call "rapid unexpected disassembly," better known as *BOOM!* Once they're on-site, the main issue is that they emit radiation, so you'll need to shield them, most likely by siting them far from your habitat and burying them in regolith. There is, of course, some additional risk of a nuclear meltdown, but let's remember that Mars is already

*A 2019 paper includes other Simpsonian acronyms, for the Demonstration Using Flattop Fissions (DUFF) and the Fission Reactor Integrated Nuclear Kinetics code (FRINK); author D. I. Poston appears to have been doing this for a good twenty years and is still going strong.

covered in toxic regolith so it's not going to devastate the local watershed or whatever.

We get that people are nervous about nuclear reactors in general, and nuclear reactors launched to space in particular. Our view is that these systems can be made about as safe as you like by abiding by some basic rules. Reactors should never be used to power launch. Instead, they should only be fired up far away from Earth, because the really nasty nuclear by-products are only made once the reaction is running. Second, the radioactive material inside the reactor needs to be contained so that in the worst case, where something goes wrong and the equipment comes crashing to Earth, cleanup is confined to a small easy-to-deal-with area.*

That said, we're sure some of you reading this book are still not reassured. We get it, but what we want you to understand is that saying no to nuclear power is basically closing the door on space settlement until some exotic future power sources are available.

A nuclear reactor is a power source in a box, which can be adjusted up and down at will, and which can go years without refueling. It doesn't care about lunar night or Martian storms. It only needs batteries in case of emergencies. It will require maintenance, yes, but its footprint is a tiny percent of the equivalent solar plant on Earth, and let's remember that Mars gets about half the average sunlight we do.

If one day there are permanent outer-space settlements with local manufacturing setups, they may lean on photovoltaics as a power-generating technology that could be developed locally. For a large-scale settlement, nuclear fission will remain the best power source in space.

Shielding

The word "shielding" may cause you to imagine spacecraft from *Star Trek* with invisible protective fields. This will not be your life in space. The

*In principle, one day, nuclear fuel could even be extracted on-site, because both the Moon and Mars have elements that are good for nuclear fuel. But like helium-3, they are at ultralow concentrations. Still, if you want the ultimate safety for Earth, it at least technically exists.

shielding of choice in space will be dirt. That toxic jagged clingy dirt we talked about earlier will blanket your habitat, perhaps as thick as a few meters to thwart meteorites, radiation, and temperature fluctuations while maintaining internal pressure.

There are more or less posh ways to go about this. One option is to simply bury yourself under the dirt. The downside here is that rocket landings and dust storms might knock your shields off. More involved proposals call for sandbagging the regolith, baking it into a solid, or bonding it into bricks. Or, if you're building a rotating space station, loading it into containers and firing it into space.

If you can successfully bury everything in regolith, you now have to figure out how to keep humans from breathing it. Experience on the Moon suggests it will be a serious danger for health and for equipment. This is an old problem for habitat designers, who have suggested a number of solutions over the years.

High-tech ideas include dust-repelling space suits or specialized anterooms that clean you up before you go back inside.

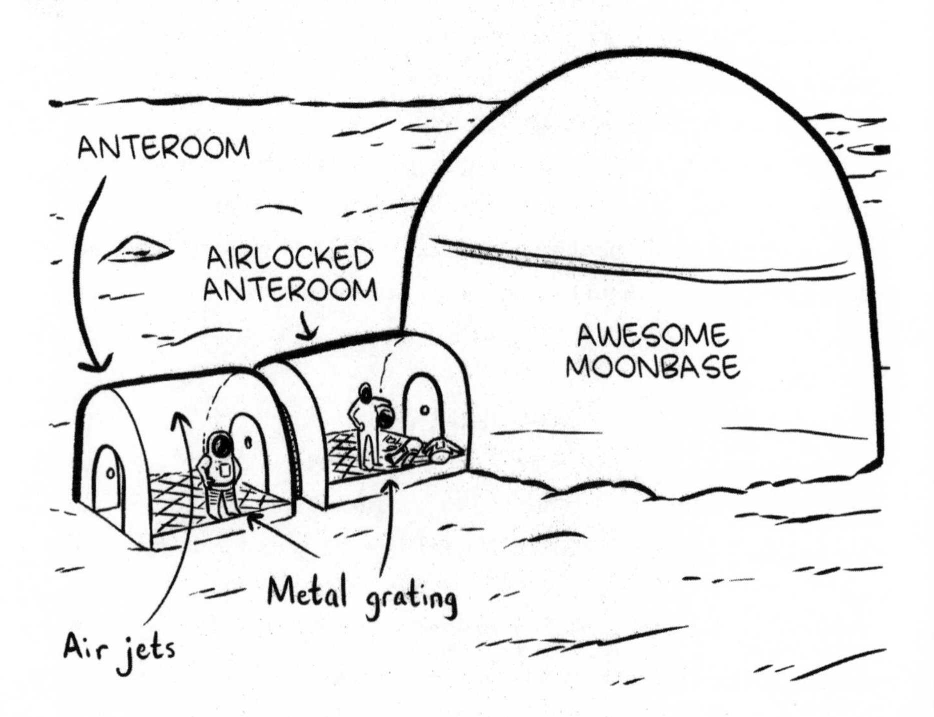

Another idea is to have the suit itself attach to the exterior of your habitat. You crawl inside the suit from the inside of the habitat, and then seal it off before going on walkabout. That way, the habitat itself never has direct exposure to dust. But this idea has some drawbacks. For example, each connection between the space suits and the habitat is essentially another airlock that you need to worry about maintaining. Plus, you're probably going to bring those suits inside from time to time anyway to check them for damage and make repairs, meaning you'll need another solution for dealing with the dust anyway.

The simplest solution may be the best. On Earth after technicians install fiberglass insulation, they don't clean it off in an airlock or with a suit of fiberglass-repelling armor. They put on a disposable jumpsuit and then throw it out after they're done. With this in mind, one design expert proposed something resembling cloaks over the space suits. Cheap, simple, and everyone looks like a Jedi. Using disposable materials is not ideal for your recycling operation, but with a little creativity, maybe you can sew some banana leaves together after you've harvested enough fruit for your wine.

The second most popular shielding option is to use water, which is surprisingly effective at thwarting radiation. Water will be scarce, but as

shielding it's doing double duty. Effectively, the outer part of your habitat would be water storage, likely a more complex setup than simply heaping dirt on top, but hey—you needed to store that water somewhere anyway.

Another solution is to limit the use of regolith by employing more exotic materials. For instance, there's everyone's favorite hydrogenated boron product: hydrogenated boron nitride nanotubes. Visualize a bunch of tiny molecular straws. These are not cheap to make, but they're especially good at neutron absorption and absorbing the cascades of spallation that can be generated when particles strike other types of shielding.

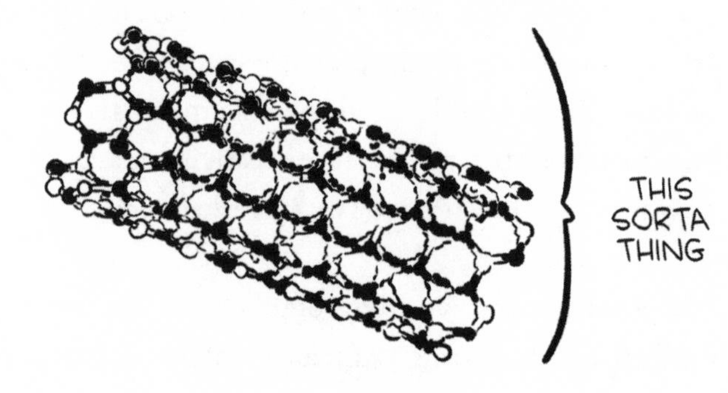

You can also block charged radiation using what's called "active shielding." Visualize a set of towers around the base, generating mighty magnetic fields that deflect charged particles as they arrive from space. Combine this setup with your pal the boron nanotube and you've got a pretty robust radiation defense. The downside is that a perpetual force field doesn't help you with problems like meteors and temperature swings. Plus it takes a lot more energy to operate than living in a heap of dirt sacks.

The big picture here is that shielding will be difficult, and any artistic rendition of a space settlement where everything is under a glass dome is likely wrong. As space architect Brent Sherwood laments, writing about lunar architecture: "The image of miraculous, crystalline pressure domes scattered about planetary surfaces, affording a suburban populace with magnificent views of raw space, is a baseless, albeit persistent, modern icon. Such architecture would bake the inhabitants and their parklands in strong sunlight while poisoning them with space radiation at the same time."

One day in the future, on-site manufacturing may allow for more aesthetically pleasing shielding than dirt piles. For now, if leaving Earth is humanity leaving the cradle, well, humanity is going straight to its neighbor's basement.

Your Home Away from the Homeworld: The Classes of Space Habs

Now that you know what you're up against and where you want to live, you're ready to build. In the field of space architecture, researchers generally divide things into three categories, imaginatively named Class I, Class II, and Class III. You can think of these as the likely stages we'll go through in the approach to settlement.

Class I: The Humanbowl

A Class I habitat is really just a chunk of space station that is transported to a surface in space instead of orbit. It's a sealed container designed to fit in a single rocket payload fairing. In the past, that has meant a diameter limit of about 4.4 meters. A few monster-size rocket systems may double this, if they work, but Class I spomes will always be cozy.

Whatever its size, imagine a cylindrical base that gets plopped on the lunar surface. Top with regolith and voilà, you live on the Moon. You might be able to do this slightly easier by finding an appropriate-size crater and pushing regolith over the top. Dr. Manfred Ehresmann called this "the poor man's lava tube."

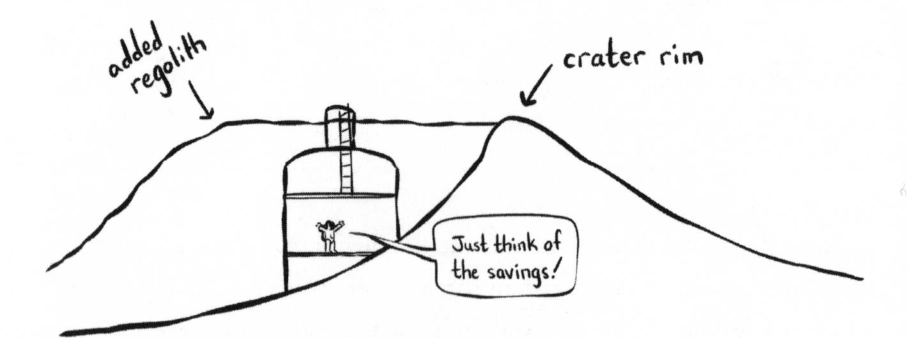

Not exactly *Star Trek*, but the point right now is just to successfully exist. If we decide the Moon is the space-settlement proving ground, a

Class I will be where we build up the Moon Settler's Handbook—how to deal with dust; how one-sixth gravity and space radiation affects the human body; how to handle all the little engineering, safety, medical, and psych needs in this new environment.

We'll also have a chance to do some exploration. One proposal to squeeze a bit more science out of Class I habs is the "Habot"—part habitat, part robot. Basically a lunar trailer park.

The obvious best way to do this is to couple a habitat with a design called the All-Terrain Hex-Limbed ExtraTerrestrial Explorer (ATHLETE), which has six bendable legs that end in wheels. It can walk, it can roll, and most importantly, it looks like a giant Moon bug, which to be honest is the main reason we're mentioning it. Throw in a few Jedi cloaks, and we might even be willing to overlook the lack of robust radiation shielding and the possibility that giant multijointed legs may not be optimal when rolling through regolith in the lunar climate. Or as Robert Wagner pragmatically suggested, "Maybe the giant space-bug needs a pair of giant space-pants to keep the dust out?"

Class II: Some Assembly Required

In a Class II habitat, your home is still built using parts boosted from Earth, but the parts are more complex, requiring on-site construction. This could be somewhat Earth-like construction, but the classic example is an inflatable habitat. Inflatables mean you can cram a lot more habitable volume inside a rocket's cargo fairing.

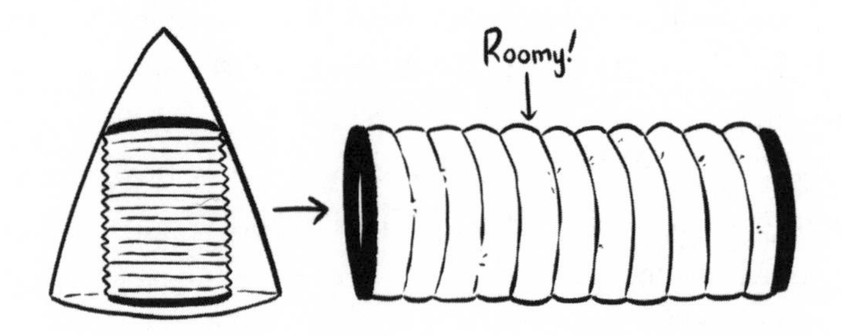

Arrive on the surface, inflate, and enjoy. You can also "rigidize" the habitat by using materials that solidify thanks to heat, UV radiation, or other weirder means.

One highly developed inflatable design was NASA's TransHab. It was three levels high, inflated around a hard central core. You shouldn't imagine a simple balloon here—the exterior was a series of complex layers, such as an outer layer to block meteors, Kevlar to withstand the inner pressure of the habitat against the vacuum outside, and "redundant bladders" that are meant to stop or slow the spread of a fire. They even had an internal "scuff barrier" since just because you're living in a heap of poison doesn't mean you're fine with scuff marks.

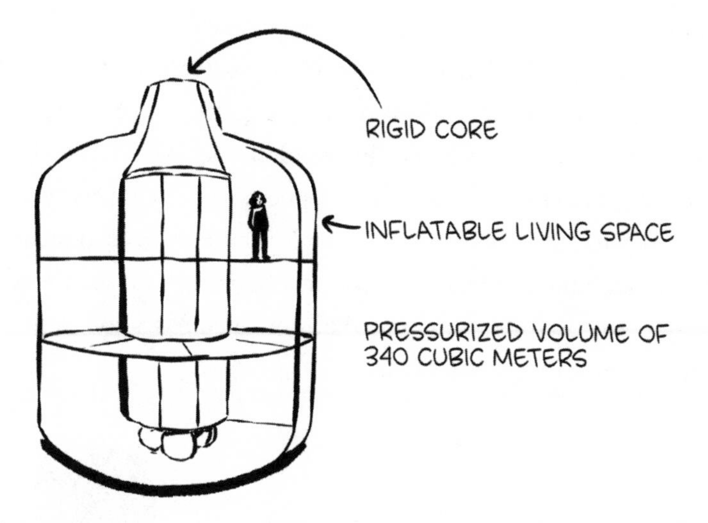

TransHab was actually tested on Earth, sustaining simulated micrometeor strikes while pressurized. But it never flew, and the patent was eventually sold to private firm Bigelow Aerospace. Bigelow created a smaller version called the Bigelow Expandable Activity Module (BEAM), which successfully inflated as an ISS attachment and has been used for storage since 2016.

A big downside right now for inflatables is lack of knowledge. Steel, wood, concrete—this is stuff we know about. Outside of inflatable arena domes and bouncy castles, we don't typically set up these structures on Earth, let alone airless worlds with rocky surfaces that get bombarded with space rocks. If we choose to go inflatable in space, the learning curve may be substantially steeper.

Class III: Do-It-Yourself

Class III is really just everything else beyond Class II. These are the sort of habitats space geeks aspire to, and their defining feature is that they are built from on-site materials.

There are a lot of proposals for how to do this. Lunar regolith is high in silicon, from which glass is made, and so some proposals call for microwave systems that would cook regolith into bricks for construction. Another proposal is a sort of rolling microwave system that zaps the ground under it into roads. There have also been proposals to create "lunarcrete," meaning concrete-like substances made of regolith. Many methods have been suggested, but we'll tell you the weirdest one. As a September 2021 article in *New Scientist* reported, "Alex Roberts at the University of Manchester, UK, and his colleagues extracted a protein from human blood called serum albumin, which is vital in maintaining the balance of fluids in the body, and used it to bind together simulated Mars soil to produce a concrete-like material." Thanks, science!

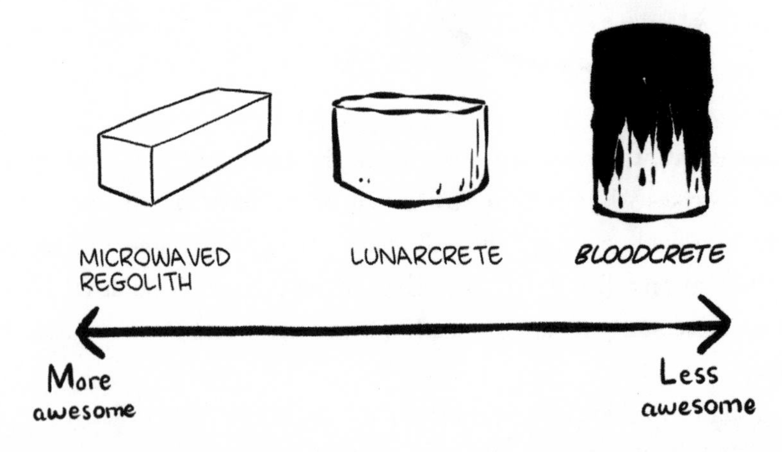

There are also plenty of metals on both the Moon and Mars. But just to emphasize it once more—the bare existence of elements does not mean they're worth using. For instance, one paper in favor of lunar manufacturing suggests using an 1800°C furnace to melt titanium. That may require a few extra KRUSTYs.

For our purposes, the main thing you need to know is that Class III habitats are possible, but will be extremely energy intensive. Earlier we talked about requiring small nuclear reactors or acres of solar panels. Graduating to Class III is going to require power more on the order of a city or very large factory.

Premium Real Estate

By now it should be clear why lava tubes will be so tempting. No need to pile up regolith if there's a huge, probably stable hole just waiting for you. Plus, there shouldn't be much dust, except around the entrance. So you just go in, blimp out an inflatable bouncy castle that only has to protect you from leaks and moderate temperature swings, and voilà—the spomestead of your dreams. In more ambitious proposals, you coat a portion of the tube with a sealant, put an airlock at the entrance, and pressurize the inside, quickly achieving a space habitat larger than anything we could build on the surface this century.

Sealed lava tubes would be an incredible foothold for future construction because within the sealed area you can just build. No fancy habitats with complex shielding and airlock systems. No burying everything in clingy toxic regolith. No connecting little habitats with long tunnels between them. Inside the huge sealed cave, if you have the materials for an English cottage, you can build an English cottage. The ideal setup would be something like this:

If there are ever cities on the Moon or Mars, there's a decent chance this is where we'll put them. If they can be converted to livable space, you effectively get a huge amount of construction and shipping work completed for free. Finding a way to build a town deep inside an airless hole won't be a simple process, but if it can be pulled off, the first group to settle these places may get an advantage that persists for decades.

Or at least that's what the geology might allow. But whether you can settle in a place in space forever isn't just about hard science. It's also a question of international law and geopolitics.

Astrid here thinks all her problems are solved, but in fact her legal troubles are just beginning.

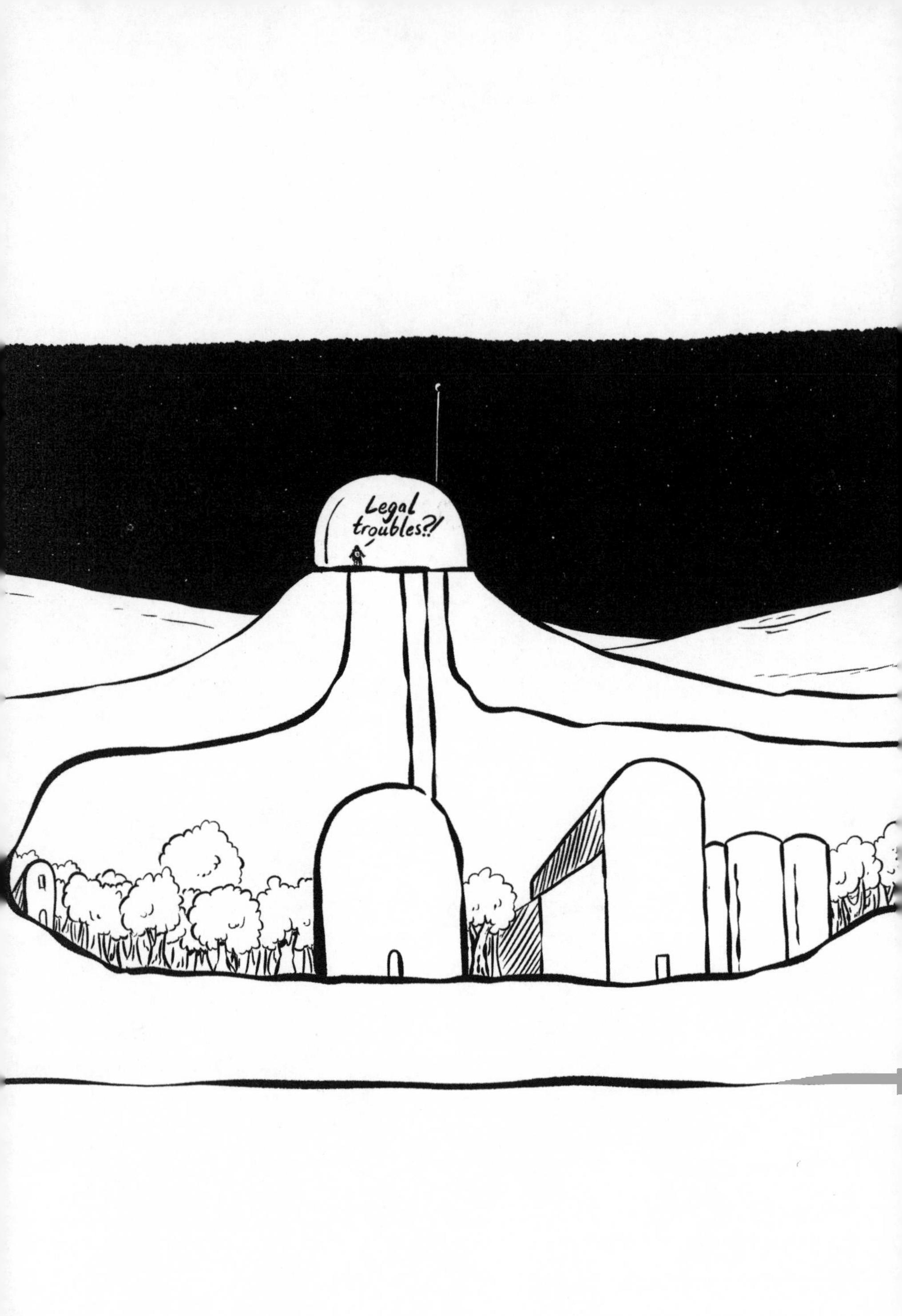
Legal troubles?!

Nota Bene

THE MYSTERY OF THE TAMPON BANDOLIER

You may have heard a story about women astronauts of the 1978 NASA class being given an absurd amount of tampons by clueless techs. It goes like this: Kathy Sullivan and Sally Ride, both members of the '78 class, are asked to check a hygiene kit for women in space. Ride begins pulling out a series of tampons fused together in small sealed packages, sort of like links of sausage. And they just keep coming. And coming. Sullivan later recalled that "it was like a bad stage act. There just seemed to be this endless unfurling of Lord only knows how many tampons." When Ride finally got to the end, the male engineers asked, "Is one hundred the right number?" Sally Ride, with the controlled emotions of a natural astronaut, politely responded, "you can cut that in half with no problem at all."

This is an old tale but was widely circulated online in the late 2010s, at one point featured in a popular musical comedy routine by Marcia Belsky titled "Proof That NASA Doesn't Know Anything About Women." It's a great story, and it isn't wrong exactly, but it may be missing some context that seriously alters its meaning.

Here's the thing: Dr. Rhea Seddon, the only combination medical doctor, astronaut, and period-haver in the class of '78, helped make the decision about how many tampons to include. According to a 2010 interview, the large number of tampons was a safety consideration. As she said, "There was concern about it. It was one of those unknowns. A lot of people predicted retrograde flow of menstrual blood, and it would get out in your abdomen, get peritonitis, and horrible things would happen."

According to Seddon, the women were skeptical of the concerns, and their preference was not to treat it as a problem unless it became a problem. But she was involved with the final decision made with the flight surgeons, and according to her:

> *We had to do worst case. Tampons or pads, how many would you use if you had a heavy flow, five days or seven days of flow. Because we didn't know how it would be different up there. What's the max that you could use?*
>
> *Most of the women said, "I would never, ever use that many."*
>
> *"Yes, but somebody else might. You sure don't want to be worried about do I have enough."*

In other words, the story may have been less about idiot male techs and more about the NASA approach of solving all problems with more equipment. As Seddon remembers it, they decided to take the maximum amount they imagined a woman with a heavy period could need, multiplied that by two, and then added 50 percent more.

This would be typical NASA behavior—if you read the 1,300-page long Human Integration Design Handbook, which we unfortunately have, you will encounter the word "maximum" 257 times, as on page 604, which contains a remarkably detailed treatment of Number 1, including what you might call a peequation,

$$V_U = 3 + 2t,$$

where V_U is the maximum total urine output in liters per crewmember, and t is the number of days of the mission.

In the case of tampons, the excessive concern may have been appropriate. Lynn Sherr, longtime journalist, friend to a number of female astronauts, and also Sally Ride's biographer, said the first woman who ever menstruated in space had problems with "leakage." Remember, space is awful. There is no gravity to pull fluids in a generally downward direction. Blood, through a process called capillary action, tends to climb out.*

*We should note, Seddon had heard a different version of events. According to her: "not totally sure who had the first period in space, but they came back and said, 'Period in space, just like period on the ground.

According to Sherr, that anonymous astronaut elected to wear a tampon as well as a pad.

Women astronauts today mostly favor hormonal birth control. These may have to be reworked a bit for a long trip to deep space, since most Earth women don't require birth control pills that are shelf stable for three years in the presence of space radiation. On a first Mars trip, where the major focus is survival, pregnancy would be a disaster. On any attempt at permanent settlement, pregnancy will be one of the goals.

We apologize for slightly ruining the tampon story, but look—if you want to hear about NASA engineers not understanding female anatomy, better options are available. Take a look at the urination devices they originally proposed for women, of which Seddon once said, "Borrowed from chastity belt designs for sure!"

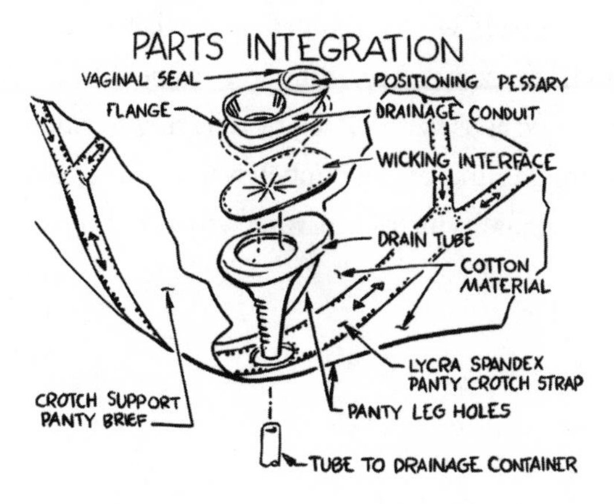

ARTIST'S NOTE:

THIS IS A TRACE OF A DOCUMENT FROM JPL. NO ARTISTIC LIBERTIES WERE TAKEN BECAUSE NO IMPROVEMENT WAS SEEN AS POSSIBLE

In what you might call a literal example of structural sexism, the engineers were trying to duplicate the condom-shaped system used by male astronauts. As Amy Foster wrote in *Integrating Women into the Astronaut Corps*, "it seems that none of the male engineers assigned to this project felt comfortable enough to consult a woman first." The female-anatomy edition never flew, and ultimately women wore a version of what we now

---

Don't worry about it.'" Jennifer Ross-Nazzal, "Edited Oral History Transcript: Margaret Rhea Seddon," NASA Johnson Space Center Oral History Project, Houston, TX, May 21, 2010, https://historycollection.jsc.nasa.gov/JSCHistoryPortal/history/oral_histories/SeddonMR/SeddonMR_5-21-10.htm.

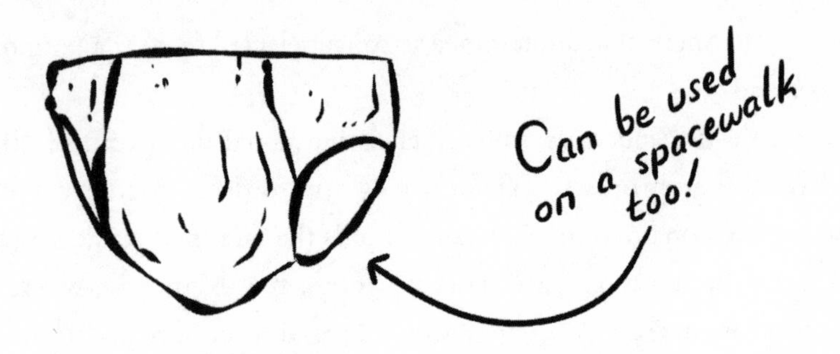

call a MAG: maximum absorbency garment. Basically, adult diapers. MAGs are now the standard clothing for situations like launch and landing, where astronauts can't just get up to use the potty.

This is true for men as well, and it's a blessing. In order to use the old system, men were required to specify whether they needed a small, medium, or large. The choice between being honest with the medical attendant and potentially wetting yourself while strapped in for launch was apparently Scylla and Charybdis for some. According to Michael Collins, among Apollo-era astronauts, male egos were spared by referring to small, medium, and large as "extra large, immense, and unbelievable."

This is perhaps not the best "equality helps everyone" story, but it may be the weirdest.

PART IV

Space Law for Space Settlements: Weird, Vague, and Hard to Change

> No moonlit night will ever be the same to me again if, as I look up at that pale disc, I must think "Yes: up there to the left is the Russian area, and over there to the right is the American bit. And up at the top is the place which is now threatening to produce a crisis." The immemorial Moon—the Moon of the myths, the poets, the lovers—will have been taken from us forever.
>
> —C. S. Lewis, the Narnia guy

THIS IS WHERE OTHER SPACE-SETTLEMENT BOOKS TYPICALLY LEAVE off. Or, well, they would likely provide you a lot more detail about exactly how the rocket ships and chemistry work, possibly along with promises of riches to be had in the near future.

We think skipping over the law is a colossal mistake if you want to have any idea about how space settlement is likely to happen. In our experience talking to enthusiasts, they frequently assume that space law will just melt away due to the amazingness of space settlement. Or will be bowled over because nobody has the right to stand in front of the United States, or just Elon Musk in particular. There's no reason to believe this, and in the meantime the law already affects spacefaring behavior. Chris Lewicki, who was president of Planetary Resources, Inc., told us that the vagueness of the law was a serious barrier to investment and contributed to the fall of his asteroid mining company. Legal scholars we talked to had even bigger concerns—that the law is unprepared for modern developments, and that any kind of race to grab, say, Moon territory, could spark conflict. That's a depressing thought, especially if you agree with our assessment that the Moon doesn't hold a lot of economic value.

What we'll argue is that international space law is currently too broad, leaving room for states to interpret what is allowed in dangerous ways. It needs to be updated. The problem is that while space launch has gotten easier, changes to international law have proven surprisingly difficult. Our ideal change would be toward an internationally managed regime, both for space activity and space property. More *Star Trek*, less *Star Wars*.

Whether you agree with our assessment or not, we think you should at least accept that this stuff matters. The framework we have for human behavior in space will shape settlement every bit as much as the technology available. Understanding how international law works and combining it with your knowledge of what space is like gives you a chance to think about how space ought to be governed. Although things are changing rapidly, humanity is still in the earliest days of space settlement. We can still make choices, and those choices may have consequences that echo for centuries.

Our goal for this section is to explain to you how we got the space law we have, what the space law says, why it's dangerous in its current form, and how we might hope to change it.

11.

A Cynical History of Space

> Perhaps maturity in the Space Age, as in any technological revolution, can be measured by the growing realization that these latest creations of man do not and will not change man himself.
>
> —Dr. Walter McDougall,
> Pulitzer Prize–winning historian

The Apollo program is the closest thing in history to an epic for engineers. In a fight to achieve a technical goal thought impossible just one generation earlier, some of the smartest engineers in the world built some of the most complex machines ever devised and then flew men, who would all become culture heroes, into danger in order to save civilization.

This is why if you browse a bookstore, you'll find most books about space are about the Moon Race, and about Apollo 11 in particular. The problem for our purposes is that if you've only read about the heroic deeds of astronauts and agencies, you miss a lot of the context that led to space law as we have it.

Here, we want to look outside the agencies and focus on the behavior of politicians and militaries. Doing so will give you a sense of both the unique conditions that created today's Outer Space Treaty, and why it's been relatively hard to get major new agreements ever since. In the process, we hope to also show you that however beautiful an aspiration space travel has been, it has also been a cynically wielded tool of militarism.

Rocketry from About the Year Zero to AD 1945

Rockets have been military tools for millennia, going back to the discovery of gunpowder in China around two thousand years ago, and have been used by militaries around the world. However, for most of their history they weren't a prominent feature in warfare. They were small, hard to control, and not very efficient at exploding enemies. Humans riding them above the sky would've seemed about as sensible as trying to ride an arrow.

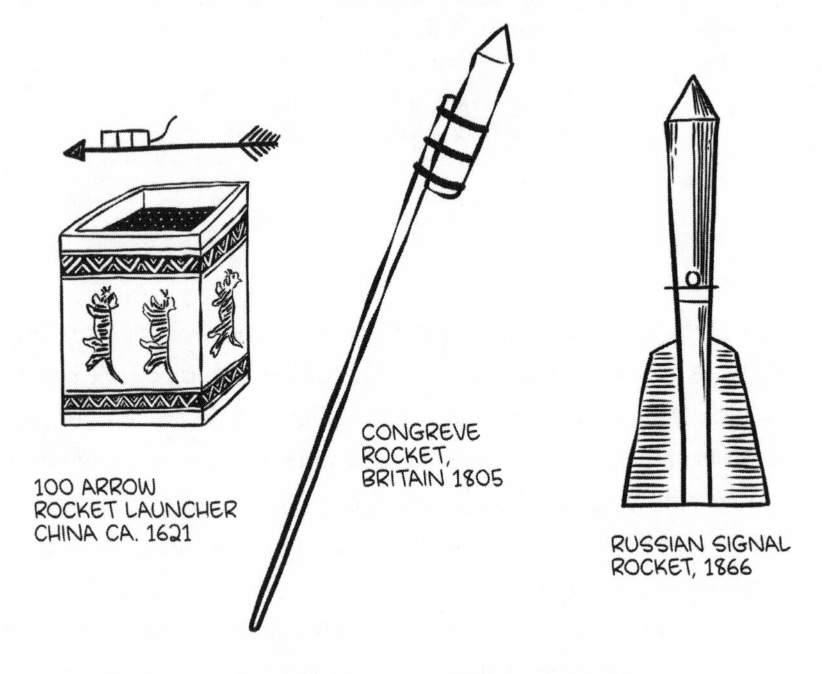

Based on image in Mike Gruntman, Blazing the Trail: The Early History of Spacecraft and Rocketry *(Reston, VA: American Institute of Aeronautics and Astronautics, 2004).*

In the late nineteenth century things started to change. Chemistry advanced, powerful reactant gasses were successfully liquified, and new engineering technology all converged to suggest that space travel was at least possible. Still the kind of thing that'd get you labeled as a crank if you talked about it too much, but anyway the numbers were looking better.

Three unusual men are generally considered to be the founding fathers of modern rocketry: Russia's Konstantin Tsiolkovsky (1857–1935), America's Robert Goddard (1882–1945), and Germany's Hermann Oberth (1894–1989).

Tsiolkovsky worked out much of the basic math of space travel around 1900, and even proposed some still-unrealized ideas like "mansion-greenhouses." But he spent most of his life as an obscure, impoverished schoolteacher. His biggest contribution to history may have been in inspiring a generation of young engineers, including the future master designer of the Soviet space program, Sergei Korolev. But Stalinist USSR proved to be a less than ideal place to do rocket science. Many of the best engineers were purged in the 1930s. Korolev himself got a last-minute reprieve from execution, only to spend many years in the infamous Gulag system, where he lost nearly half his teeth.

The United States had a somewhat better beginning, and Goddard remains famous for fielding the first liquid-fueled rocket in 1926. But he became a recluse after being widely mocked for his ideas about rocket-based Moon travel. He continued to conduct small-scale experiments, but largely cut himself off from America's research community. Thus, in the words of physicist and rocket engineer Dr. Theodore von Kármán of the California Institute of Technology, "There is no direct line from Goddard to present-day rocketry. . . . He is on a branch that died."

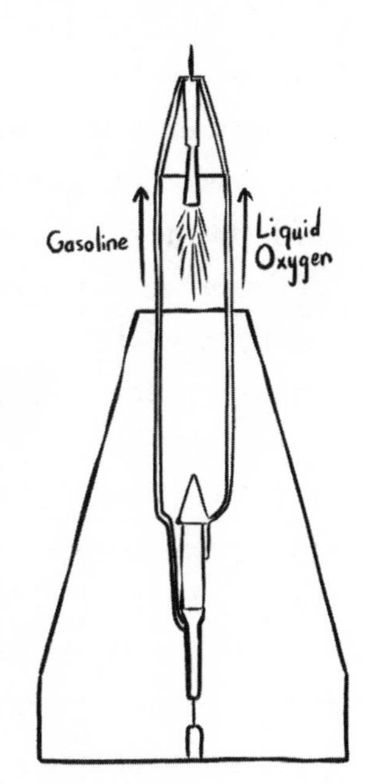

So, with the United States busily mocking its nerds into isolation and the USSR purging them, the study of large rockets only really took off in Germany. This was especially true after Hermann Oberth, the guy who would later explode himself for movie promo, published his 1923 book, *The Rocket into Planetary Space.*

The most advanced rocket group on Earth during the 1920s and early '30s was likely a club of young amateurs, presided over by Oberth, called the Society for Space Travel, or VfR (*Verein für Raumschiffahrt*). They put in long hours for no pay, building rockets out of scrap at the "Raketenflug-plaz," a testing ground on an abandoned munitions depot in the suburbs of Berlin. This was dangerous work. Rockets frequently went off course or exploded. But the VfR did find a permanent berth in history when they were joined by a sharp young man named Wernher von Braun—the blond-haired, square-jawed nineteen-year-old son of Silesian patricians, with the accent and deportment to match.

In the mid-1930s, the VfR got the attention of the German Army. Although the question of whether to work with the military caused a rift in the club, von Braun chose to go where the money was. He would later be showered in resources, building cutting-edge rockets for the Nazi empire.

In our experience, sometimes people think von Braun's weapons were a major part of the Nazi war effort. In fact, for almost the entire war they were still an experimental weapon. One of the weirder stories from the early 1940s involves Hitler going to visit Peenemünde, the small village where the world's most advanced rockets were being constructed. Von Braun, who was known for his good looks and charm, and would later host a popular series of programs on space travel for Disney, failed to impress. Hitler was bored. Bored to the point of irritation. At one point, he angrily asked why the rockets needed two liquid fuels instead of one—something von Braun had just explained. The young man patiently restarted his lecture. Later, as History's Greatest Villain left the facility, he offered an apparently sarcastic "*Es war doch gewaltig!*"—"Well, it was grand!"

Only later, as defeat became ever more likely, did Hitler change his mind. Von Braun was ordered to take experimental rocket designs and convert them to mass-produced weapons. This is the infamous V-2 rocket, where the *V* is for *Vergeltungswaffe*—"retribution weapon."*

The V-2 is remembered as terrifying, but was a relatively primitive, ineffective weapon. In fact, most of the deaths directly related to the V-2 were not from its use, but its construction, which consumed approximately

*And 2 because V-1 was taken by a sort of early cruise missile with wings.

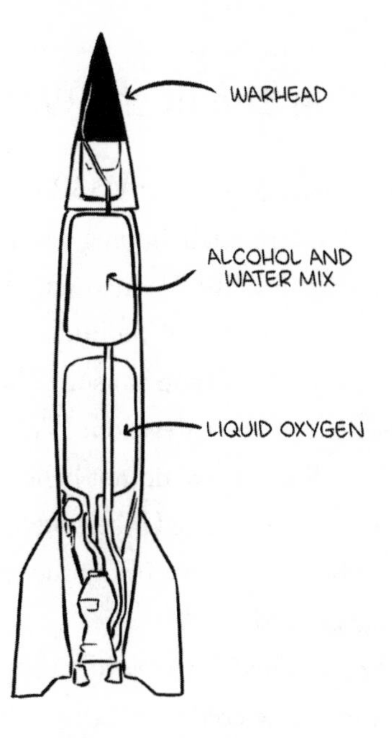

twenty thousand lives. The V-2 factory was built by slaves supplied by the SS, in a new facility near the town of Nordhausen. Most of them were political prisoners, conscripted into a program designed to work them to death. In a scene of Tolkienesque horror, skeletal prisoners dressed in rags were made to blast their way into a mountain in order to create the assembly facility for the new wonder weapons. Frenchman Yves Beon survived the camp to write a memoir called *Planet Dora*, which documents how men died of beatings, of starvation, of disease, of freezing to death as they were made to stand in rags against the German winter. At one point, describing the arrival of new prisoners, he writes: "They know they won't get out of this festering carbuncle on the asshole of hell. . . . Like trapped animals, they fight against death, but death laughs at them. The trap is tightly closed. No one can escape." Although von Braun didn't initiate the slave labor, he was well aware of it. In one of Beon's entries he writes, "The project directors are here. First, there is Wernher von Braun, the same chap who, after the war, will be venerated and shown as a fine example to young generations of the West."

The End of the War

As the armies of the United States and USSR overran the Nazi empire, they wheeled for Nordhausen, each hoping to snap up German rocket technology and talent first. But the US had an unwitting advantage—in January 1945, months before the fall of Berlin, von Braun and his staff and families left Peenemünde in an attempt to surrender to US forces. As one staff member later recalled of their decision: "We despise the French; we are mortally afraid of the Soviets; we do not believe the British can afford us; so that leaves the Americans." The United States not only nabbed most of the German rocket talent, but most of the rocket parts during a frantic week before the Soviets arrived.

When the Red Army rolled into Nordhausen they found the place picked clean of anything that could be easily removed. The factory remained, and they were able to coerce or entice some of the remaining Germans to come to their side, but Stalin was infuriated. The United States not only captured von Braun and his literal tons of documentation, they had absconded back to New Mexico with enough rocket parts to build one hundred V-2s at a secret facility in White Sands. As Stalin reportedly said shortly after, "This is absolutely intolerable. We defeated the Nazi armies; we occupied Berlin and Peenemünde; but the Americans got the rocket engineers. What could be more revolting and more inexcusable?"

After American planes dropped nuclear weapons on Hiroshima and Nagasaki, the United States reigned as the world's sole nuclear power. From a geopolitical security perspective, this was a Soviet nightmare. Where the USSR had seen entire cities swallowed by the war, the United States had the most advanced weaponry, an unrivaled fleet of bombers, and an unscathed national infrastructure. The Soviets, who should have now been enjoying a period of safety after losing tens of millions of people in history's bloodiest conflict, instead found themselves at a severe disadvantage in the emerging postwar conflict. They poured resources into two crash programs: one to develop the atomic bomb, followed by one on methods to deliver these bombs to targets. Human spacefaring wasn't yet a serious consideration, but that would change as the value of Cold War propaganda became apparent.

In the context of that global conflict, it's important to remember that Earth had more than two countries. The hottest part of the Space Race coincides with worldwide decolonization, which largely occurred between 1945 and 1975. The speed at which this happened is about as shocking as the speed of space advances during the same period.

When the League of Nations, the predecessor of the United Nations, was founded in 1919, it had 32 member nations. Even that low number was a bit inflated, because several of these nations were under British control. By 1945, there were 51 states in the UN. By 1957, when the first satellite, Sputnik, beep-beep-beeped its way over the Earth, there were 82. By 1975 it was 144. Today's UN has 193 member states.

As these vulnerable young countries came into existence, they did so in a world dominated by two ideologies and an open question about which mode of governance would rule the future. One of the main psychological tools the United States and USSR used was human rocket travel. Why? Well, you could say that leaders successfully steered away from war toward scientific contests. This is a frequent claim, and it's nice, but isn't really quite true—nuclear arms kept being built en masse and the world came to the brink of apocalypse more than once. A better way to think about human spacefaring may be as "costly signaling."

Here's the idea: if a nation wants to convey to the world that they are the strongest and best, they can of course just announce it at the United Nations. But it won't be convincing. Talk is cheap. Space programs are not. Very few nations can successfully fire a guy around the world at 7.8 kilometers per second, then land him and send him on a goodwill tour.* Human spacefaring has little utility for the price, especially compared to things like military or commercial satellites, but what it does do is dramatically demonstrate wealth, organization, and technical competence. Throw in the fact that early space rockets were often literally the same as military rockets, and

*Alex MacDonald's book concerning this topic, *The Long Space Age*, argues that the tradition of spending on space research as a costly signal predates rockets, having first begun with observatories. The fact that the research is unlikely to create a return on investment is part of why it's such an effective signal.

you have an excellent show of raw power that demands to be taken seriously. You of course never hear a politician say, "We choose to go to the Moon, not because it is easy, but because it'll provide short-term geopolitical advantage," but something like that is a pretty solid explanation.

It also comports with the behavior of leaders at the time.

Sputnik launched in the context of a race to get the first satellite in orbit. Korolev, the Soviet master designer, kept American press clippings about space to goad the Soviet leadership into more space spending. The original design for humanity's first satellite was supposed to be a complex scientific apparatus called Object D. Due to a desire to be first, this was scrapped in favor of Sputnik, which was basically a big ball filled with a power source and radio transmitter. The Soviet leader at the time, Nikita Khrushchev, had been reluctant to even send the first satellite to space. He didn't have that many rockets at his command and didn't want to waste them on a satellite, even though it could produce some public relations benefits. According to his own account, even after the successful launch of Sputnik in 1957, he simply congratulated the team and went to bed. Only when it became clear the worldwide public had gone absolutely bonkers over Sputnik did he lean in, his skepticism almost immediately changing to a demand for more space spectacles, ideally coinciding with days special to Soviet history. This is why Laika the dog went up soon after. Was this a great use of money for science? Probably not, but it was great public relations. It was, as Khrushchev himself admitted many years later, a bluff. And it worked.

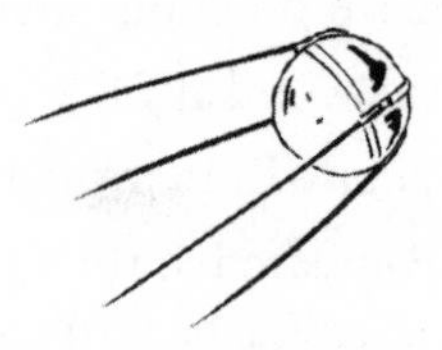

SPUTNIK 1

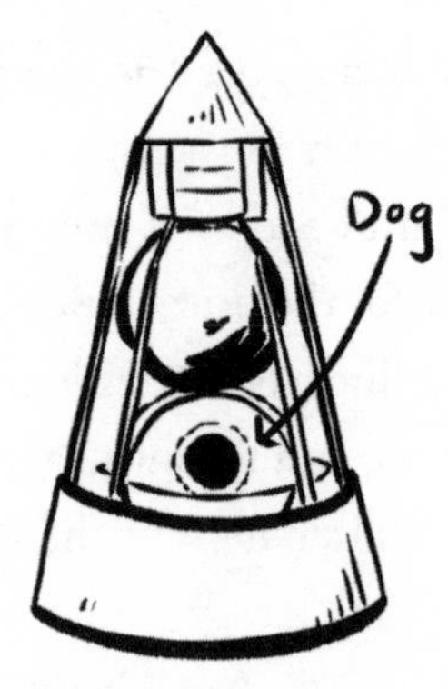

SPUTNIK 2

SPUTNIK 3

President Eisenhower didn't want a race in space, for satellites or anything else. "Hysterical" was the word he used for the people who wanted massive spending on space stations and Moon bases in the wake of Sputnik. Regardless of the geostrategic aspects, this reluctance ended up being a bad look and ultimately part of why John Kennedy won the 1960 election against Eisenhower's vice president, Richard Nixon.

But surely Kennedy was a space enthusiast, right? He did all those speeches about "new frontiers" and this "new sea" that we have to sail on. Here's the thing. If you judge a public figure only by public speeches, you are making a terrible mistake. There is good reason to believe Kennedy wasn't really that interested in space, at least not as a scientific endeavor. In the years after Sputnik 1, but before his presidency, Kennedy and his brother Robert had met fellow Bostonian Charles Stark Draper at a fancy restaurant. Draper was head of MIT's Instrumentation Laboratory, and was looking to convince the two brothers that rocketry was serious business. It didn't go very well. According to one source, they treated him with "good-natured scorn." In an interview years later, Draper recalled that then Senator Kennedy "could not be convinced that all rockets were not a waste of money, and space navigation even worse."

So what changed? Politics. Kennedy had one very very very bad week in the spring of 1961, just three months into his presidency. On April 12, Yuri Gagarin became the first human being to enter space, making a full orbit around the planet before safely returning home. Five days later, the United States embarked on the disastrous Bay of Pigs Invasion, in which fourteen hundred CIA-backed Cuban exiles utterly failed to overthrow Fidel Castro.

Kennedy assigned his advisers to figure out some kind of countermove to Gagarin's flight that would turn things around—something like a space lab or a Moon trip that would produce "dramatic results in which we could win." This would ultimately be supplied by the ex-Nazi team led by von Braun, who believed a Moon landing was possible by the late 1960s.

Although publicly pro-space, Kennedy seems to have hesitated in his Moon plans. At the press conference after Gagarin's launch, Kennedy talked about desalinization, noting his belief that if it could be done cheaply, it would "really dwarf any other scientific accomplishments." And yet, on May 25, barely a month after the Bay of Pigs, Kennedy gave a televised speech on "Urgent National Needs" to Congress, calling for a prodigiously expensive crewed Moon landing.* And lest you think this was a complete change of heart, we have evidence that Kennedy, as late as 1962, was not that interested in space. How do we know? Because during a private meeting with advisers, he told them, "I'm not that interested in space."†

This is pretty late in the process to not be that interested. Nineteen sixty-two is the year Armstrong was selected as an astronaut. It's the year John Glenn puts in his rectal thermometer and turns a full orbit. It's the year of the first "group" space flight as the Soviet Union launches simultaneous crewed orbiters around the Earth. If you only look at the astronauts and engineers, things are heroic and spectacular. Change your gaze to the world stage and you have a Soviet leader wielding space dogs as a bluff and an American leader who, with some reluctance, is targeting a Moon landing as a grand act of propaganda—a profoundly costly signal in a war of perception.

The Early Space Age and the Origin of Space Law

However inspired the public may have been, diplomatic negotiators had good reason to keep their feet on the ground. And good reason to worry.

---

*People sometimes conflate this speech with the famed "not because they are easy but because they are hard" speech, given at Rice University in 1963.

†The full transcript quote is open to interpretation, but it's certainly clear that he's not talking about new frontiers in private. Here's a longer portion: "Because, by God, we've been telling everyone we're preeminent in space for five years and nobody believes it because they [the Soviets] have the booster and the satellite. . . . But I do think we ought to get it, you know, really clear that the policy ought to be that this is the top priority program of the agency, and one of the two things, except for defense, the top priority of the United States government. I think that that's the position we ought to take. No, this may not change anything about that schedule but at least we ought to be clear, otherwise we shouldn't be spending this kind of money, because I'm not that interested in space. I think it's good, I think we ought to know about it, we're ready to spend reasonable amounts of money. But we're talking about these fantastic expenditures which wreck our budget and all these other domestic programs and the only justification for it in my opinion to do it in this time or fashion is because we hope to beat them and demonstrate that starting behind, as we did by a couple of years, by God, we passed them." John Logsdon, *John F. Kennedy and the Race to the Moon* (New York: Palgrave Macmillan, 2010), 155–56.

The race to the Moon may have been costly signaling, but both nations also engaged in a more straightforward mode of signaling involving massive nuclear explosions in space. One such project was Operation Fishbowl, which detonated a 1.4 megaton H-Bomb 400 kilometers high on July 9, 1962, as part of something called "Starfish Prime." It lit up the sky for thousands of kilometers, disrupting radio transmission, knocking out streetlights in Hawaii, causing electric surges on airplanes, and disrupting phone and radio communication. It also affected six satellites—four American, one Russian, and the very first British satellite, Ariel 1. One scholar, Dr. James Clay Moltz, even argues that Starfish Prime was the motivator for the ratification of space law, basically because it scared the crap out of everyone.

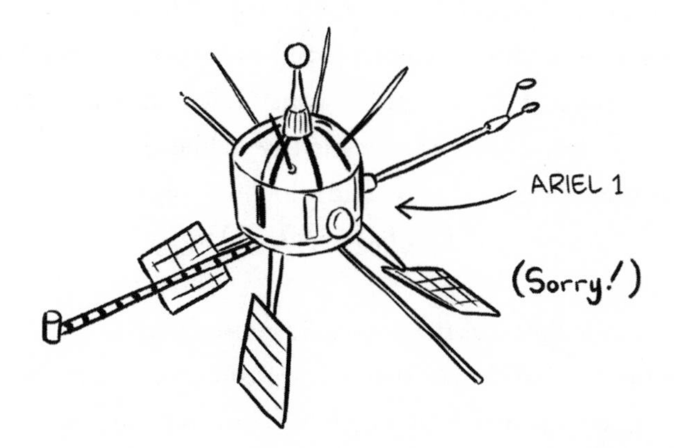

Try to appreciate the cadence of things here. Imagine you're a fifty-year-old diplomat in 1963. You want to avoid another world war, and you want to avoid a nineteenth-century-style scramble for colonies and territory. But the future is coming at you fast. When you were a child in the 1910s, engineers were still working out the kinks in biplanes. In the 1920s, Tsiolkovsky was a little-known old man, Goddard was a recluse, and Oberth was blowing himself up for movie promo. In the 1930s, a few amateurs were doing occasionally fatal work, and rocketry remained an idea mostly of interest to children and the kind of intellectual freaks who purchase books about outer-space travel.

By the mid-1940s, rockets were traveling high into the atmosphere in order to rain death on Britain and Belgium. The year 1957 brings the first intercontinental ballistic missile and the first satellite. By 1961, humans are hurtling through space, and by 1962 nuclear weapons are too. Nothing in your life so far suggests that space travel is going to make human behavior better. Nothing suggests that the use of rockets is fundamentally about science or inspiration or unity. You are someone who's seen almost twenty years of an arms race and now an enormous increase in spending in a worldwide public relations battle. Then, all of a sudden, nuclear weapons are making miniature suns in the sky, and the world sits on the edge of its seat watching the Cuban Missile Crisis.

This then is the view of space most relevant to space law. In this picture, space activity, especially when it involves humans, is tactical. It's about power and survival. It's not a world of heroes—it's a world of dangerous motivations that need to be constrained. But in 1962, space is largely unregulated. The good news is that while the high-flying heroes were doing their thing, scholars and diplomats were piecing together a legal order for space—one that has persisted largely intact into the modern era.

We want to emphasize the two main points you should carry for the rest of the book. First, however much we the nerdy are moved by space travel, and however alien and fascinating its environment is, no social magic happens there. Space is one more place where humans will be humans.

Second, the law we have today largely comes out of a very particular era of history—a time when there are only two players in space, one capitalist and one communist, and when space travel was totally new and moving very fast toward an unknown future. Keep in mind the level of uncertainty here. You the reader know what space was like for the next sixty years. Diplomats and leaders of the 1960s did not. Alternate histories were still possible. Von Braun himself had proposed an orbital battle station that would keep peace on Earth and "space supremacy." As he said in a 1952 speech in Washington, a space-to-ground missile could "stop any opponent cold in his attempt to challenge our fortress in space! The space station can destroy with absolute certainty an enemy space-craft prior to its launching."

Both sides of the Cold War considered militarizing the Moon—an act that would presumably involve something like a scrap over territory. The United States undertook a serious program called Project Orion, designed to propel cruise liner–size spacecrafts off-world via a sequence of nuclear explosions.

So, middle-aged diplomat, it's the early 1960s. Children born during the year when atomic weapons were detonated over Japan are now young adults. The world has managed to put itself back together after a world-spanning conflict, but its most powerful nations are evidently careering toward some vast conflict in space, and in the impending *Götterdämmerung*, it is simply not clear who has the upper hand. This is not an auspicious moment for humanity, but the fact that there are only two space powers in a world that has suddenly gone terrifying is a welcoming environment for new international regulations.

The Creation of Space Law

Nineteen sixty-three was a fine year to be a space lawyer. You get the Partial Nuclear Test Ban Treaty, where the United States and the USSR agree to stop testing nuclear weapons in various places, including outer space. You also get a UN resolution wherein both the United States and the USSR say they will not orbit weapons of mass destruction in space. And finally, after years of negotiation, the UN released a declaration of principles for outer space, which would later form the bulk of what is now called the Outer Space Treaty (OST).* Currently ratified by 112 nations, including the two major spacefaring powers of the period, it came into force in October 1967.

What the OST has meant for humanity remains a subject of debate. Some see the OST as a triumph of peace and liberalism. As one author wrote: "the onset of the most intense period of competition . . . proved to be precisely the moment for reconciliation and progress in space *law*."

Others see something less uplifting. Dr. Everett Dolman, a military theorist who admittedly has the writing style of Darth Vader, once wrote, "The highly touted international cooperation that produced the 1967 Outer Space Treaty was not in truth evidence of a newly emerging universalism; rather, it was a reaffirmation of Cold War realism and national rivalry, a slick diplomatic maneuver that both bought time for the United States and checked Soviet expansion."

The strange thing from today's vantage is how much the world has changed while the law has stayed the same. The Soviet Union is no more. Its successor nation built history's largest space station with its former rival. Meanwhile Europe, China, Japan, and India became serious competitors in space as Russia grew ever weaker and more isolated. Strangest of all, the most commonly launched rocket of 2022 is not one built by a

*The treaty's full name is "Treaty on Principles Governing the Activities of States in the Exploration and Use of Outer Space, including the Moon and Other Celestial Bodies." For obvious reasons, we'll just call it the Outer Space Treaty or the OST from here on out.

nation, but by the private corporation SpaceX.* But while space launch and communication have become modern market-based enterprises, the OST remains the product of that brief moment of rivalry between nuclear-armed powers posturing before a wide-eyed planet. This has weird consequences for the future of space settlement, and any plan for space settlement must account for them. So let's dust off this old document and see what it says. And perhaps just as important, what it doesn't.

*Though one that receives quite a bit of government contract money.

12.

The Outer Space Treaty: Great for Regulating Space Sixty Years Ago

The Outer Space Treaty is the closest thing to a governing document for space, yet space-settlement books rarely deal with it in any detail. This is sort of like having a plan to build a giant mining operation in Antarctica, and being more concerned with your transportation or water-melting operation than the fact that its existence would clearly be illegal in a way that'd antagonize dozens of powerful nations at the same time.

Our particular concern with the Outer Space Treaty is that while it strictly limits the ability of nations to claim land, it can be interpreted as allowing all sorts of behaviors that go right up to the edge of de facto land grabs. In a world where nations are planning trips to the Moon and where powerful rocket corporation owners are promising settlements, the risk is that powerful nations interpret the law in ways that are favorable to them, pushing us to a crisis point, possibly creating conflict on Earth.

International Law: Actually a Thing

But first let us note—and we want to emphasize this because in our experience space-settlement geeks don't tend to believe it—international law matters. It doesn't work like domestic law, it isn't perfect, but it decidedly exists.

International law is sometimes called "anarchic" because there is no higher authority policing nations or deciding whether they've properly interpreted their own treaties and obligations. The closest thing is the United Nations, but you can think of the UN as something like a legislature with a court, but without a lot of enforcement power, and whose laws only bind states that have consented to them. International law isn't direct and functional on nations in the way domestic law is on people, but that doesn't mean it's imaginary. In our experience, people who think international law is a farce tend to point out times when the law has been violated. But, as one scholar we talked to noted, this is a little like believing there are no laws against murder because sometimes murder happens. Earth really does have rules that have developed over the years, and which nations generally obey.

A slightly more nuanced claim sometimes made is that international law is just whatever powerful nations want it to be. But while powerful nations have a disproportionate say in how the law is made and interpreted, they don't have unlimited say. Also, even if you believe the law is just the will of the powerful acting for their own benefit, that doesn't make the law go away. If you get pulled over by the police and given a speeding ticket, you can't say, "But speeding laws are only the will of the powerful!" International law actually constrains the behavior of nations, and so to understand space settlement, we have to understand how international law works in practice.

The two ways international law gets created are agreements and what's called "custom." Agreements are probably what you think of when you hear the words "international law"—written deals between nations that are binding and which nations enforce. "Custom" are the norms we develop through behavior, which are sometimes, but not always, later codified. When Sputnik 1 flew in 1957, the Eisenhower administration considered there to be a serious silver lining: the Soviets had established a precedent for freely orbiting over other nations. The United States quickly took advantage of this standard to field spy satellites. Although space's status as an "open skies" domain was debated as late as 1963, the custom has solidified, and it is now beyond question. It's worth noting that things could have gone otherwise. Consider the rules of Air Law, which since the Chicago Convention of 1944 say that airspace is *not* a zone of free passage.

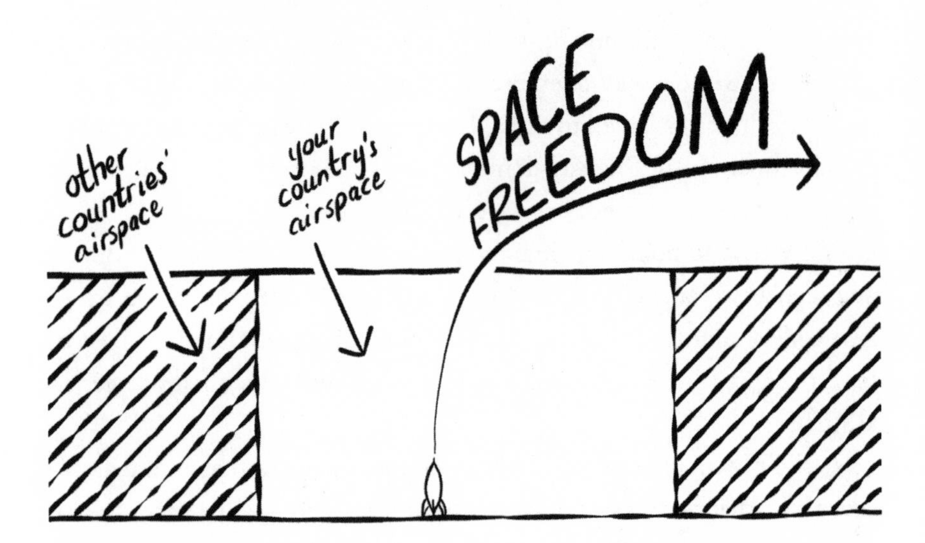

Written-on-paper agreements are important, but they get their oomph from ongoing worldwide custom—from nations doing stuff and other nations tolerating it.

So, how's space custom? In short, it's okay. For the air we have about a century of custom. For the sea, we have many centuries. For near-Earth spaceships, we have about sixty-five years, but for situations pertaining directly to space *settlement*, we have almost nothing. So when we talk about questions like whether you can own an asteroid, we have to work from formal agreements, the most important of which is the Outer Space Treaty.

A Brief Caveat Before We Get into the Nitty Regolithy

Here and in part V we're going to get into the law of space in detail, and we want to be clear: all this is interpretation. While international law has plenty of agreed-upon standards, we found that when we talked to scholars they were not always sympatico. This is entirely reasonable for a field like law, which is ultimately about the opinions and behavior of talking apes wearing clothes. But in order to keep this section a reasonable size, and to elucidate likely paths for space settlement, we've been forced at times either to come down on one side of an issue or to say an issue is unsettled when some scholars believe it *is* in fact settled law. We believe we've got a reasonable interpretation here and have done our best to relate the most agreed-upon views, but as you read, understand that we have our own biases. In particular, we are in the uncomfortable position of arguing that US views are especially important, for power reasons, while being US citizens ourselves. We have done our best to solicit the opinions of scholars and thinkers of many backgrounds, and we believe our overarching points are correct, but the finer points of interpretation here are very much open to debate.

The Outer Space Treaty of 1967

The first thing to know about the OST is that it doesn't say that much. The English-language version is around twenty-five hundred words. It's also notoriously vague. International treaties often come with a section that defines terms to reduce ambiguity. The OST does not.

This can create some serious weirdness, which we'll illustrate here with two examples.

Question 1: What is an "astronaut"? We all feel like we know—it's somebody who's gone to space, right? Well, what about tourists? They're clearly astro, but have they nauted? At least under recent US guidelines, they have not. Clause 5 of the FAA Commercial Space Astronaut Wings Program Order 8800.2 says it's not enough to hit fifty miles high (about eighty kilometers); you also have to qualify as "flight crew" by meeting certain standards and then showing that you were "essential to public safety, or contributed to human space flight safety" during your flight. In other words, you need to be part of the crew, doing crew stuff, not just a rich person with a window seat.

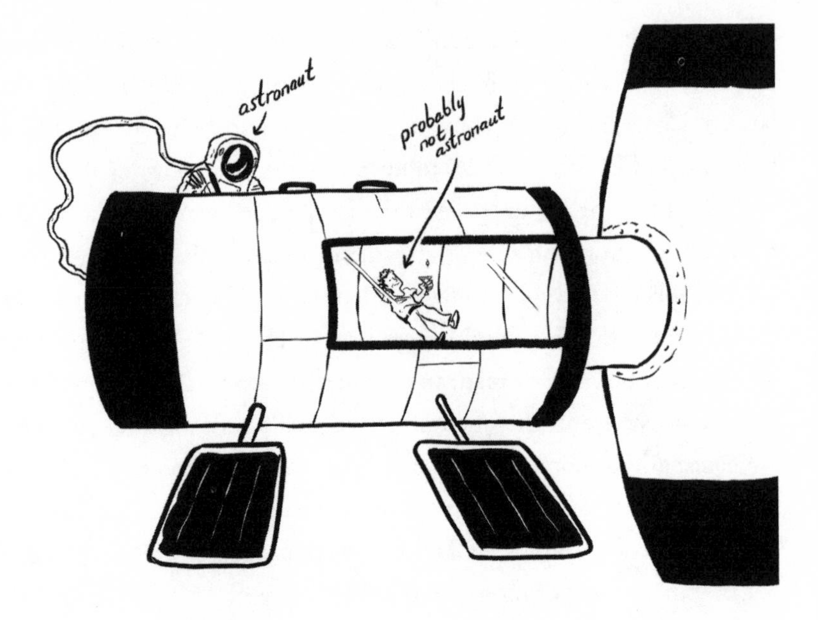

This matters for more than the porcelain egos of space billionaires. Under an elaboration to the OST called the Rescue Agreement of 1968, countries have certain obligations to aid astronauts. If a Russian cosmonaut were forced by circumstance to land in Alberta, Canada, the Canadian government would be obligated to promptly return the ship and cosmonaut home in working order. This was sensible during the 1960s, when all spacefarers were professionals who acted something like ambassadors. The OST literally calls them "envoys of mankind."* If Jeff Bezos takes a tourist orbit and accidentally lands in Russia, he is likely just an envoy of Jeff Bezos, without the customary considerations afforded to astronauts.

We think. Again, it's vague.

Question 2: Where precisely does airspace end and, like, *space* space begin? This is more of a political question than a scientific one because the atmosphere doesn't have an abrupt end point. Custom suggests space begins at the "Kármán Line"—the altitude at which you can't use a plane anymore and have to use a rocket. That's usually said to be about one

*"Envoys of mankind" is also undefined, but generally understood to basically mean nothing. That is, you don't get, for instance, diplomatic immunity.

hundred kilometers up. Other proposals have set the line as close as twenty kilometers, which would confer astronaut status on a few balloonists, or as distant as several thousand kilometers, which would limit the current pool of astronauts to the small cadre of elderly American men who went to the Moon and back. Probably the worst idea, as well as our personal favorite, comes from John C. Cooper, whose proposal we found in a book from 1952. In short: for a given country, space is whatever you can't defend with missiles. This hearkens back to old sea law, which gave nations three miles (about five kilometers) sovereign from their shore, on the basis that artillery could hit targets roughly that far away.

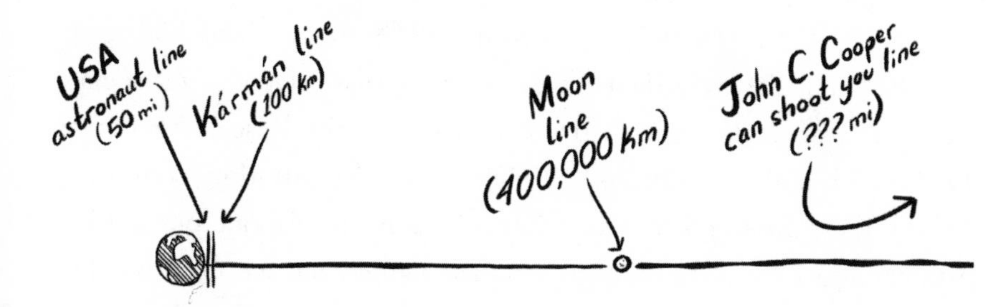

The issue of where air ends and space begins may seem like a legalistic quibble, but it could have real consequences. Suppose SpaceX follows through on a proposal for rocket travel between, let's say, Kiev and Tokyo. The rocket could start in *air*space, but then be in the "open skies" of *space* space during its passage over Russia, only to then return to the airspace of a friendly territory. This might be technically legal, but we suspect Russia would feel iffy about it.

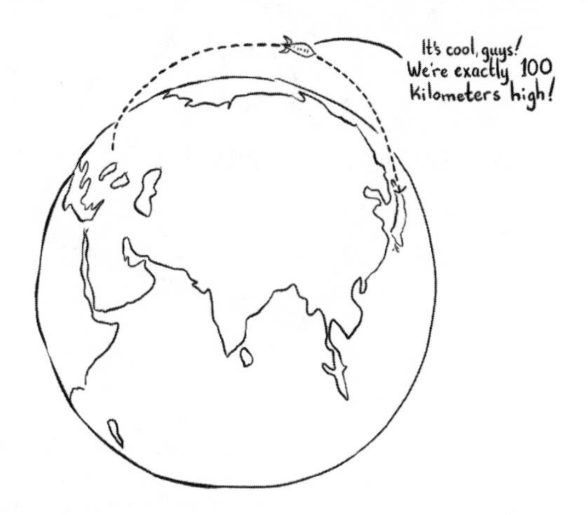

So the OST is vague. In fact, talking to scholars we found there was vagueness around whether vagueness was good or bad. Clear laws are good because they permit certainty—a valuable thing both in an investing and a not-accidentally-starting-war context. Vague laws, on the other hand, have the advantage of allowing wiggle room as things change. In the case of spacefaring during the 1960s, some flexibility was likely desirable, both for dealing with the unpredictable future and for getting everyone to agree to sign the document.

Despite, or perhaps because of, these ambiguities, the OST has been widely adopted. As of this writing, 112 countries are party to the treaty, and that includes all major spacefaring nations. Just as important, the OST has been around a long time and basically abided by, which suggests it's at least nudging its way toward being customary law.

Since the OST was ratified, three more UN treaties have both come into effect and been ratified by the nations that have the technology needed to launch their own citizens into space—the Rescue Agreement (1968), the Liability Convention (1972), and the Registration Convention (1975).* These are important, but they are really just elaborations of OST sections. You have already forgotten their names, but we mention their existence now because they will come up as we discuss what the current law is.

*Try doing these in one breath: The Agreement on the Rescue of Astronauts, the Return of Astronauts and the Return of Objects Launched into Outer Space, the Convention on International Liability for Damage Caused by Space Objects, and the Convention on Registration of Objects Launched into Outer Space.

Because this is a book about settlement, we're mostly interested in the question of what territory and resources you're allowed to use in space. Effectively then, there are two areas of space law we care about—ownership and, well, everything else.

Everything Else

Cooperation

One of the vaguest parts of the OST is a clause saying space activities "shall be carried out for the benefit and in the interests of all countries." The clause has been debated, but the tradition is, in short, that it doesn't mean much. Countries are supposed to act in good faith, but there is no way to enforce cooperation. And in any case, just about any space activity can be pitched as helping everyone. That US flag put on the Moon by Americans for Americans? It inspires *everyone.*

That said, the legal scholars told us not to get too cynical. Conversations about what nations are required to do are ongoing, and given how much stuff is going to space today, the exact norms of cooperation may become more important and more codified by custom.

The OST also says everyone is supposed to play nice, which in the geopolitical setting means be honest and maybe clean up after yourself too, please? Let other countries observe your launches, allow other people to visit your facilities, and pretty please consider not contaminating space.

The noncontamination rules have been openly violated by four of the major space powers for the same reason and in the same way—blowing up their own satellites with missiles. The USSR and the United States were first to explode their old satellites using weapons from Earth. Today, this is considered gauche because explosions in orbit fling bits of metal at high speed in every direction, threatening other objects in orbit. The worst such offense was the destruction of a defunct weather satellite by China in 2007, creating thousands of bits of high-speed shrapnel, much of which remains in orbit as we write. India, a rising power and a rising player in space, took its own turn in boomtown in 2019.

If it's not clear to you why countries test antisatellite weapons this way, well, India's prime minister Narendra Modi said the quiet part out loud after the 2019 test: "India registered its name as a space power." In other words: costly signaling. These incidents tend to cause international outrage, but so far there have been no major repercussions. However, that may simply be because nothing too terrible has happened yet.

Liability

Let's suppose you launch a space object* and it slams into someone else's space object, or airplane, or city. What happens? Between the Outer Space Treaty and the 1972 elaboration called the Liability Convention, *some country* is liable for damages.

Exactly which country can be weird. Article VII of the OST says the liability goes to the state party to the treaty that "launches or procures the launching of an object into outer space" as well as "each State Party from whose territory or facility an object is launched." That is, if you're the state where the launch happened or the homeland of the people who set up the launch, you can be held liable if something goes wrong.

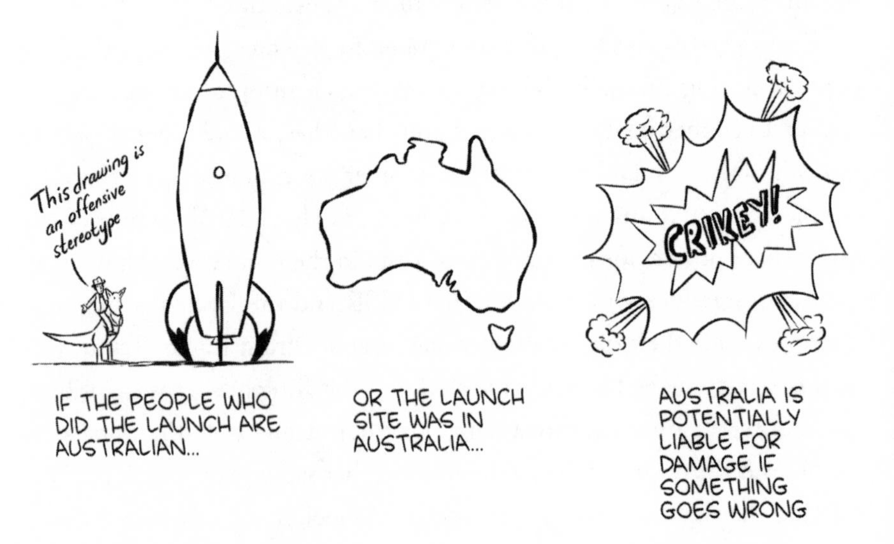

*Also not defined.

A number of nations have passed laws to bring themselves into compliance with this rule, but those national laws don't always agree on what the rule actually says. Some countries prioritize the physical territory from which the object launched. Others believe what matters is the nationality of the people who did the launching.

This creates potential legal gaps. Suppose a group of Canadians, using Canadian money from the Canadian government, launch from Paraguay. Is liability now held by Paraguay or Canada? Under the OST, that's for Canada and Paraguay to work out between themselves. This sort of thing could really matter for space settlements of the future. In our experience, people often have a vague idea that if Elon Musk does something on Mars, that is the business of Elon Musk. But legally, Elon Musk would likely be the US's responsibility, making the US liable for anything he does off-world.

Part of why things are fuzzy is that the terms of the Liability Convention have only been invoked once. Remember when we told you about the time the USSR whoopsied one of their nuclear-powered satellites onto Canada? Operation Morning Light, the joint US-Canada cleanup operation, cost millions. Canada asked the USSR to pay, citing (among other things) the statutes in the Liability Convention.

Registration

In 1975, the Registration Convention came into force, and it basically just says that if you're going to launch something into space, you have to register the fact with the state you're launching from, and the launch state ought to register this fact with the UN. In our experience, people often think space billionaires launch satellites willy-nilly, perhaps while wearing cowboy hats, but only the second part is true, and only of Jeff Bezos. Space is quite a bit more boring, with registration rules similar to those for ships at sea.

But there's still room for shenanigans. In the maritime setting, a common cost-cutting tactic is the "flag of convenience." Ships are obligated to register their home nation so that if there's a legal issue, you know what rules they're under. In practice they often fly the flag of countries with the lowest taxes and weakest regulations for things like safety inspections, labor law, and environmental impact. This sort of thing isn't widespread in space just yet, but as the number of satellites ramps up, that may change. In 2021, for example, a company called AST & Science wanted to launch 243 gigantic (450-square-meter) satellites into a 700-kilometer-high orbit. NASA expressed concerns about so many huge satellites and the possibility of debris or collisions. But the company's behavior was potentially more worrying than their technology. AST & Science, which is a US-based company, got its launch license in Papua New Guinea. Why do this? Quite possibly because that country has not signed either the Liability or Registration Conventions. As James Dunstan pointed out in an article on the topic, the value of the satellites in orbit would be larger than Papua New Guinea's entire governmental budget, raising concerns about their ability to pay for damages even if they *did* agree they were liable.

Weapons

In 1963 the Partial Nuclear Test Ban Treaty prohibited testing nuclear weapons in the atmosphere or outer space. That same year the USSR and the United States agreed to a resolution in the UN agreeing not to orbit nuclear weapons in space either. With the arrival of the OST four years later, "weapons of mass destruction" in general were banned from outer space. But, guess what—"weapons of mass destruction" is *also* not defined. Nuclear weapons are specifically ruled out, and international law in general rules out biological and chemical weapons.

However, at least in principle, a country could orbit enough regular old missiles to enact quite a good amount of destruction, and it'd probably be legal as long as none of those missiles had a nuclear tip. Also, you can't have a Moon fort. No off-world "military bases, installations and fortifications." No military maneuvers in space either, although you are specifically allowed to put military people in space for peaceful reasons—a sensible policy from a time when most spacefarers were military test pilots.

Ownership of Space and Space Stuff

Who Gets to Be Space Sovereign?

For many space-settlement enthusiasts, the underlying hope is to create new societies. New countries with totally new rules in totally new lands. Can they do this? The answer in the OST's Article II can be summarized as "nope":

> *Outer space, including the moon and other celestial bodies, is not subject to national appropriation by claim of sovereignty, by means of use or occupation, or by any other means.*

"Sovereignty" is an important concept for a space settlement, so it's worth sitting on for a minute. Although it's another one of those words the OST doesn't define, it's at least a word with a long tradition, and is pretty close to your commonsense notion of what a modern state is. Canada is sovereign over Canada. This means Canada is the highest and sole authority over a particular territory, that it can regulate people and stuff going in and out of its borders, and that other states generally agree to all this. The OST clearly states you cannot have Canada-oid entities on celestial bodies.

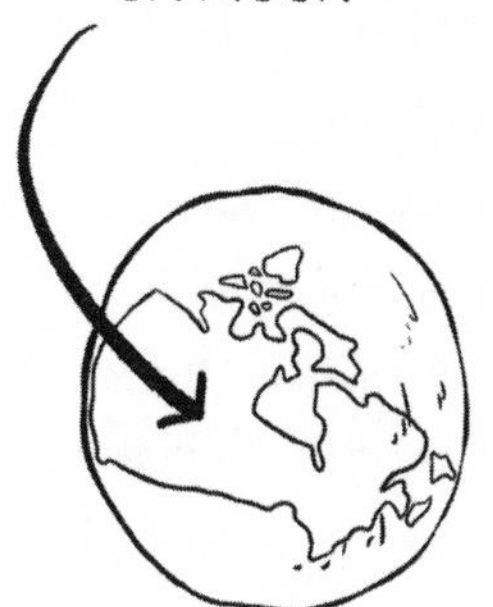

If you're a geek, like us, your first reaction is to try to find the loophole. For instance, hey, it only says *nations* can't claim sovereignty. Can Kentucky Fried Chicken claim the Moon? Sorry, friends, the Moon will have no Colonel. Under international law, nations are responsible for their people, also known as their "nationals." The KFC employees attempting to set up a lunar chicken-ocracy still have to launch from *somewhere* and be from *somewhere*. So if they go to the Moon and claim sovereignty, they're doing it under the jurisdiction of *some nation*. Also, just as a matter of common sense, the law couldn't permit this sort of thing because it would make Article II pointless. Try to imagine how the international community would feel if the United States gave $10 trillion to Kentucky Fried Chicken to claim all of Mars. They'd be finger lickin' pissed. More important, they would refuse to recognize the land title the US government granted to KFC.

What you *can* do is set up a research station where your sovereign law applies, similar to how nations have sovereignty over their ships at sea. That seems pretty reasonable, but remember, premium real estate is limited, especially on the Moon. Dr. Martin Elvis (who you may remember as the space resources expert whose name is literally Elvis) and his coauthors note that the Peaks of Eternal Light account for about "1/100 of a billionth of the lunar area," meaning "a single country or company could, on its own, occupy them all, effectively denying that resource to others." So, although achieving Canada-ness on the Moon is forbidden, you *could* create moon bases with Canada-ish powers inside themselves that also sit on all the best spots. They would be ruled by Canada and naturally exclude others from a region of the Moon. Technically they could even be packed full of Canadian soldiers as long as the soldiers were there for peaceful purposes, like doing research.

Part of why this wouldn't exactly be sovereignty is that Canada would have certain obligations. But they're pretty weak. Under Article XII, Moon-Canada should allow representatives of other parties to the treaty to visit, as long as they provide "reasonable advance notice of a projected visit." The problem is that the needs of space science would pretty much always give you an excuse to say "now is a bad time." As Dr. Elvis and his

coauthors note, the Peaks of Eternal Light are a great spot for solar-observation experiments, which require very sensitive equipment. What's to stop the Moon-Canadians from just putting up a sign that reads: EXPERIMENT IN PROGRESS. VISITING HOURS RESUME IN 10,000 YEARS?

Under the OST, if you really wanna see what the space-Canadians are up to in there, the best you can do is invoke Article IX, which says that if you're worried someone else is doing bad stuff in space, you "may request consultation concerning the activity or experiment."

Take *that*, you Moon-Canadian bastards. We demand that you consider talking things over, please.

For the would-be space settler, the major takeaway here is that although sovereignty is forbidden, you can get a whole lot of the perks of sovereignty within the rules of the OST. No Martian Canada, but possibly a Canada atop Mars. This sort of thing is what we're talking about when we say the current law leaves room for dangerous interpretations. Nobody is supposed to claim territory, but everyone is allowed to just situate bases on the limited premium real estate. If that sounds like chaos, well, just wait until you hear about what they're allowed to do with the nonclaimed territory they're sitting atop.

Exploitation of Space Resources for Fun and Conflict*

You can't be sovereign over a patch of Moon dirt, but as we've seen, you'll likely need that dirt for your radiation shield. Can you use it? And what about more precious resources? Let's recall that, if we're lucky, there's about as much water ice on the Moon as in a small lake. Suppose your authors start a Moon base on Shackleton crater, called Shackleweiner Station. Once set up, we begin converting the water into a gigantic ice sculpture, which can be quite tall thanks to the Moon's low gravity. Humanity's dreams of a lunar way station are now thwarted by the icy countenances of the Weinersmiths of Shackleweiner.

Is this legal?

According to current understanding . . . probably, yeah. It's not sovereignty, so although it's not great, it's not clearly against the rules. The problem is that exploitation without sovereignty can produce some pretty deranged results. Suppose the United States, in its hatred for public trans-

*"Exploitation" has a negative connotation, but in the resource-extraction community it just means something like "extract and use or sell." When we use this term, we are using its latter definition.

portation, decides to pave most of the lunar surface for future parking lots. Other nations might be mad, and possibly even accuse the United States of "contamination," but all the US is obligated to do here is provide a consultation beforehand.

Is this likely? No—30 percent chance tops. But is it legal? Quite possibly. There is at least some precedent for allowing space resource exploitation without interference from international law. The Apollo missions brought back about 400 kilograms of Moon rocks to Earth, which were held by NASA and treated as US government property. American presidents have given out bits of the Moon as a diplomatic gesture from time to time, and on two occasions some of this moondust has ended up at auction. In all this time, there has never been an attempt to treat Moon rocks as unpossessable or as special property that humans must share. This wasn't just true of the capitalists of the United States either. During the 1970s, three of the USSR's Luna series of missions collected a few hundred grams of regolith and returned them to Earth. These were considered Soviet property, and some of them later ended up on the auction block at Sotheby's. Similar rules applied in 2010 and 2020, when the Japanese Aerospace Exploration Agency's Hayabusa missions collected and returned samples from asteroids, and China's Chang'e 5 mission returned lunar regolith to Earth. When it comes to bits of space, legally speaking, finders appear to be keepers. But we also don't have a lot of custom here. Humanity has only ever grabbed a small amount of space objects, and although some of them have ended up in private hands, they were always first collected by government science agencies, not for profit. If some company starts grabbing regolith to sell on Earth, the international community might be less cool about it.

The question, then, is who gets to decide how these rules should be interpreted? Remember, laws can happen by agreements or by behavior. And US behavior, at least, has been in favor of something like an anything-goes approach. NASA recently announced that four companies have been contracted to sell lunar regolith to NASA. In a move that would seem bizarre if you don't account for the power of legal precedent, the plan is for the companies to go to the Moon and formally grab regolith, which they will *not* return to Earth. Rather, having grabbed it, they will transfer ownership to NASA for a token fee, that in one case is precisely one dollar.

What's going on here?

Jim Bridenstine, NASA administrator when the announcement was made, said straight out that the point was to establish a precedent that private companies can extract and sell space resources.

Although the United States appears to be the dominant player in space, they're by no means the only one. Regardless of how the OST is interpreted for exploitation purposes, the United States, Japan, the United Arab Emirates, and Luxembourg have all passed laws specifically stating that under *their* interpretation, private companies have the right to explore, extract, and sell space resources without limit. And to be clear, they're not saying they're going their own way; they're saying their interpretation is consistent with the international law we already have. Russia has registered opposition to this view, but otherwise there hasn't been a strong negative response. A recently established UN Working Group may shed some light on what views other countries hold, but a case could be made that the US interpretation is sliding toward customary. If countries and companies begin landing on the Moon and using its resources in the near future, we will find out pretty quickly.

Stuff You Put into Space

Making things even weirder, the current interpretation is that nations *do* retain sovereignty over stuff they *put* into space. Neil Armstrong's poo bags? Those are Uncle Sam's. But not everything Uncle Sam owns is quite so glamorous—in addition to the literal crap, there's, well, crap. What's sometimes called space junk. Broken bits of satellites from weapons tests, that one dead nuclear reactor, and oh yeah, the millions of tiny needles placed into orbit by something called Project West Ford.

Regardless of which individuals send objects to space, states are responsible for them and retain ownership. Does that mean an *individual* can own things in space? At least one individual thinks he does. Richard Garriott, millionaire video-game developer and son of astronaut Owen Garriott, purchased the Luna 21 spacecraft and Lunokhod 2, the rover it deployed, both of which sit on the Moon's surface. These were sold by Lavochkin Science and Production Association (the manufacturer of the items) and won at a Sotheby's auction in 1993 for $68,500—about the cost of a souped-up Ford

Bronco. In Garriott's view, the OST does not apply to individuals, and so he claims to own the land under the rover. According to one person who spoke to him, not only is he serious about this, he also believes the track traversed by the rover belongs to him. No space lawyer we talked to shared this view.

The Law Is Weird

The various shenanigans we've described above might give you the impression that the OST is what's sometimes called a "parchment barrier"—a piece of paper with no power over behavior.

But generally speaking, the laws passed between 1967 and 1975 are obeyed. Space objects get registered, launches get announced, and for the moment nobody would dare put a nuclear weapon in orbit. Despite the weird loopholes some space geeks contemplate, the OST has succeeded in creating a general norm of hands-offiness in space. But that may have been less about the law and more about how space stayed prohibitively expensive for forty years. It's now been about a decade since the prices began to drop again, and today's major space powers—the United States and China—are both looking to land in the Moon's premium real estate. Legal scholars have noted a tendency for international law to go uncreated until a crisis looms. Personally, we would not like to see a crisis between nuclear powers. So can we get the OST amended or superceded before something frightening happens? Later we'll argue that there's a chance, but you should know that a more clear legal regime really was tried for in the 1970s. It failed to get support from most states, including the only three states capable of launching their own citizens into orbit. The reasons it failed will tell us what we can hope to achieve in the future.

For now, we'll assume Astrid and her family are Americans under American jurisdiction.

We went with whoever flew the most space monkeys. No offense!

13.
Murder in Space: Who Killed the Moon Agreement?

> It is traditional, of course, for explorers to plant the flag, but it struck us, as we watched with awe and admiration and pride, that our two fellows were universal men, not national men, and should have been equipped accordingly. Like every great river and every great sea, the moon belongs to none and belongs to all. . . . What a pity that in our moment of triumph we did not forswear the familiar Iwo Jima scene and plant instead a device acceptable to all: a limp white handkerchief, perhaps, symbol of the common cold, which, like the moon, affects us all, unites us all.
>
> —E. B. White, the *Charlotte's Web* guy

The Moon Agreement*

In 1979, an agreement was struck that would have not just fixed most of the problems described above—it would've created an international regime that strictly regulated human access to space resources. If it had been ratified, perhaps today's would-be asteroid miners would have to first consult with some kind of International Space Authority for exploitation rights. In addition to sounding awesome, having such a framework would likely spare humanity from any current concerns about a conflict-inducing Moon Race Part Two.

However, although it is formally ratified by eighteen countries and is

*Formally, that's the UN Agreement Governing the Activities of States on the Moon and Other Celestial Bodies.

thus formally considered in effect, the Moon Agreement is, to be technical, a great big dud. None of the countries capable of launching humans to space aboard their own rockets signed, nor has the law come to be considered customary. Therefore, it only binds the behavior of those eighteen countries, and soon it will only bind seventeen because in 2023 Saudi Arabia announced its intention to withdraw. Meanwhile, the countries most likely to plant tiny flags on the lunar surface are free to violate it.

For the potential space settler, understanding what killed the Moon Agreement is a way to understand the type of regimes we likely *will not* get in space. In particular, the Moon Agreement would have made space extremely communal, in a way that failed to get buy-in from the biggest space powers. To understand what it was trying to do and why it failed, let's first think about what having a property regime in space even means.

Shall We Socialize the Moon? or, Ways to Organize Space Property

In the regrettable 1953 pulp novel *Space Lawyer*, written by Nat Schachner, himself an attorney, the handsome and intelligent Kerry Dale manages to thwart the villain and win the hand of his space boss's daughter, the lovely Sally Kenton. How does he do this? By, in the climactic scene, saying the immortal words:

> *Earth lawyers from the beginning of Earth law—which is the fundamental basis for Planetary Law—have recognized the status of such alien objects as Comet X; objects that* never *legally belonged to a citizen of the state or the state itself. The Roman legists called them* res nullius*—things which have not or which never had a legal owner. Read the great Pandects of Justinian for the pertinent clauses.*

Not exactly "From Hell's heart I stab at thee," but if there's a lawyer reading over your shoulder they're probably breathing heavily by now.

Rightfully so—*res nullius* is one of the most important concepts in history that most people have never heard of.

As legal hunk Kerry Dale notes, the phrase *res nullius* goes back to Roman law. It's a framework for property, and it means something like "nobody's thing." A classic example would be fish in the ocean that aren't protected by any fishing law.* While those fish swim around, they belong to nobody. They are literally "nobody's thing." The moment someone grabs one, this changes. Like a spell from Harry Potter, "*Res nullius!*" converts the fish into a form of property.

We were told by a space lawyer that it's important to note a difference between the actual way *res nullius* works and the way spells work for fictional teenage wizards. When you *res nullius* a fish, it's not as if the universe itself changes some quality of the fish.† Rather, *res nullius* is a framework that humans can choose to apply to particular types of property. Today we usually talk about objects, like fish, but this hasn't always been the case. Consider the American West after land was forcibly appropriated from Indigenous people. In many cases, US settlers could acquire that land just by working it and paying a small token fee. Under current law, lunar regolith may be *res nullius* while Moon land is not.

This gets even weirder when different entities have different views of what stuff exists in the *res nullius* framework. For example, suppose the Weinersmiths land in a Moon crater, dig up all the water, then claim it for all future Weinersmiths.

---

*Fish today are regulated pretty much everywhere, but imagine an era of greater piscine libertarianism.

†A small number of people do believe something like this, but we'll get to that later.

Plausibly the US government would recognize this transaction while some of the international community would not. Situations like this show why it'd be nice to have a clear international framework, both to avoid international conflict and to make the rules clear for potential investors.

Regardless of what type of property Moon water is, under the OST, Moon territory is very clearly not *res nullius*. It's *res communis*—"common thing"—a framework in which something is *common* property. For example, a city might possess a common pasture, where everyone is allowed to graze their sheep. Or a beach might be open to everyone. Or, on the planetary scale, you can think of the atmosphere, which everyone is allowed to use and from which nobody can be excluded. Commons typically require some sort of oversight to prevent overexploitation. For instance, Cattlemen's Associations were formed in the American West to regulate the overuse of pastures. Similar regulations have been set up for the atmosphere, limiting ozone-depleting substances. We can all make use of the atmosphere to dump our filth, but there are rules so that the atmosphere continues to be useful to everyone.

Whereas *res nullius* works in a pretty straightforward manner, there is a variety of flavors for commons management. This is where the Moon Agreement got snagged. See, the Moon Agreement would have set up the solar system as a particularly communal form of *res communis*, known in international law as "common heritage of mankind" or just "CHM."

Although the term has evolved over time, the modern understanding of CHM is as a commons collectively owned by *all of humanity*. If the Moon were under a CHM framework and you wanted to use Moon water, you would have to compensate *all of humanity* by some means. Seabed law was initially intended to work this way, and a statement by Moragodage Christopher Walter Pinto, a Sri Lankan delegate, gives a good sense of what it means: "If you touch the nodules* at the bottom of the sea, you touch my property. If you take them away, you take away my property." This is far more restrictive than just saying anyone can go do stuff on the bottom of the sea.

We recognize there's been a lot of terminology here, so to illustrate, let's imagine there are alien fish under the surface of Enceladus. The newly formed fish-and-chips chain MuskDonald's got there first and wants to sell a type of fish sandwich entirely foreign to life as we know it. Here's how that plays out under our three frameworks:

*Pinto is referring to valuable mineral deposits on the seabed.

Framework	Description	Amusing Drawing
Res nullius	MuskDonald's grabs an alien fish, which becomes the property of MuskDonald's. They can then transfer the fish to you, with fries and a Muskan Dew™ for just $5.99.	Res nullius!
Res communis	Those alien fish belong to some group of people. MuskDonald's has to work with their framework if it wants those delicious Enceladan trout.	Res Communis!
Common heritage of mankind (CHM)	Those alien fish belong to all of humanity, and if you take any of them, you must recompense all humanity for the alien fish sandwiches that are our birthright.	Humanity is as one, and you gotta write us a check!

That's the gist of things. *Res nullius* and *res communis* are very old concepts, but CHM was a new idea even as the Moon Treaty was being negotiated. This created ambiguity, which created concerns among potential parties to the agreement. A fear among many was that CHM would mean, as one critic said, "socializing the moon."

Whither Lunar Socialism, or, What's in the Moon Agreement?

A lot of the Moon Agreement is just restating what's in the OST, as well as closing loopholes. It also makes clear that countries are definitely allowed to use Moon minerals for mission support, and are allowed to take "moon samples." This may sound vague, but according to *Black's Law Dictionary*, a sample is "a small quantity of any commodity, presented for inspection or examination as evidence of the quality of the whole." This is relatively permissive, but good luck creating your giant orbital station from samples.

The Dreaded Article XI

For our purposes, this is the most important section. Article XI says:

> *The moon and its natural resources are the common heritage of mankind. . . . Neither the surface nor the subsurface of the moon, nor any part thereof or natural resources in place, shall become property of any State, international intergovernmental or non-governmental organization, national organization or non-government entity or of any natural person.*

Translation: "No, you can't exploit it. NO! NO SHENANIGANS JUST HANDS OFF!"

So under this agreement, is the Moon then a sort of wilderness park, never to be tampered with except for science? Not quite. The Moon Agreement calls for an international regime to oversee exploitation. It doesn't say much about how this would work, but in short there'd be a big entity established by states that were parties to the Moon Agreement looking over things, and in particular making sure developing nations got a fair cut.

It can be easy to get lost in the technical stuff here, but this would've been what you might call a huge stinkin' deal. The Moon Agreement is focused on the Moon, but would've regulated *all* celestial bodies. If it were widely ratified, there might now exist a regime regulating resource exploitation throughout the solar system.

What Killed the Moon Agreement?

A lot of stuff. The United States didn't sign it, despite having a major hand in its negotiation. American diplomats liked it, but once it was handed to the US Congress, it was widely seen as a socialist system that would discourage investment. Part of the issue is that the term "common heritage of mankind" evolved over time from a sort of vague aspirational statement to something like a meaningful, and indeed quite redistributive, standard.

But this isn't just about capitalism. The USSR didn't sign it either. In fact, they actively and specifically opposed not just the Moon Agreement, but the CHM framework in particular. No *communis* for the communists, if you will. Even weirder, many nominally capitalist countries like Italy and Argentina *favored* the CHM framework even as the communists opposed it.

So what gives?

There are a few circumstantial reasons the Moon Agreement failed to get the countries capable of self-launched human space flight onboard. As space ambitions in both the USSR and United States drastically tapered during the 1970s, the treaty was seen as premature. Space mining, which looked imminent and important from the vantage of 1970, looked costly and unnecessary by 1980. NASA administrator Robert Frosch, using a phrase normally reserved for teenagers, told Congress he felt "massive indifference" over the Moon Agreement. Why regulate now when we don't even know what to regulate?

Worse, the document basically says "regulation will be figured out later," which was considered especially discouraging to investment. Effectively, you'd be telling entrepreneurs that there will be regulations, possibly quite onerous ones, but we won't tell you what they are until after you've spent time and money to figure out how to make space mining feasible.

On this basis, the L-5 Society,* a group of Dr. Gerard K. O'Neill–style space-station enthusiasts, started a large political campaign, even working with a Washington lobbyist, to kill ratification. Keith Henson, spouse of L-5's then president Carolyn Henson, said at the time, with perhaps a touch of melodrama: "On the Fourth of July 1979 the space colonists went to war with the United Nations of Earth." He also added his feeling that "the treaty makes about as much sense as fish setting the conditions under which amphibians could colonize the land."†

Beyond the specifics of how this all played out, there's a bigger-picture issue. In 1980, as the treaty was debated in various countries, there were precisely two major space powers with any hope of lunar exploitation—the USSR and the United States. If you wonder why the most powerful capitalist country and most powerful communist country would both agree to oppose a resource framework, a pretty good guess would be self-interest. While the OST merely said nobody could claim the Moon, the Moon Agreement would've obligated states to share any benefits derived from the Moon. Neither nation wanted to expend its own resources and effort to get something that would just be handed out to other nations, including rivals and enemies. So they left the law as it was.

As of 2023, the Moon Agreement has not been ratified by most of the big space players: no China, no United States, no Russia, no India. It is generally not considered customary law, though it certainly still has some champions in the space law community. The OST and the three subsequent treaties elaborating it remain the major governing documents for spacefarers—gaps, loopholes, and all.

Thanks in part to the lack of cheap space access during the years after the agreement failed, this hasn't been that big a deal. But things are changing. In the 1960s, as crisis approached, the world got together to

*In 1987, the L-5 Society merges with the National Space Institute to form the National Space Society, which advocates for settlements in space (and rotating space stations in particular) to this day.

†Scholars argue about how much of an actual effect they had, but it has become part of the lore of space advocacy.

change the rules. We appear now to be approaching a second crisis, but for the moment an overarching agreement does not seem likely. If anything, Earth is going in the opposite direction, largely led there by a single nation.

Remember when we warned you parts of this book would be America-centric? Well, grab your cowboy hat and a can of crystal clear beer. It's time to be American for a minute.

Into the Legal Breech! US Space Law and the Artemis Accords

What's happens without a big international framework that everyone's cool with? We don't know about the long term, but what we can say is that right now the most powerful player—the United States—is pushing forward a legal interpretation that sounds a lot like national appropriation in space. China, arguably the second-biggest player in space, appears to be gearing up for a race. Both sides are most interested in the "premium real

NASA-PROPOSED CANDIDATE REGIONS FOR ARTEMIS THREE. EACH IS ABOUT 15X15 KILOMETERS, WITH THE LANDING SITES HAVING A 100 METER RADIUS.

estate" we talked about earlier, and they are particularly interested in the craters and peaks on the Moon.

Earth may be accelerating toward some kind of crisis, and many of the relevant actors seem pretty cool with that. Kelly attended the International Astronautical Congress in 2019, and the view among many law and policy folks from the United States was that routing things through the UN would be a mistake. Scott Pace of the US's National Space Council said UN treaties are too slow, and so we should create nonbinding guidelines instead and then just try to get nations to comply with them via national law.* This is exactly what's happening under what are called the Artemis Accords.

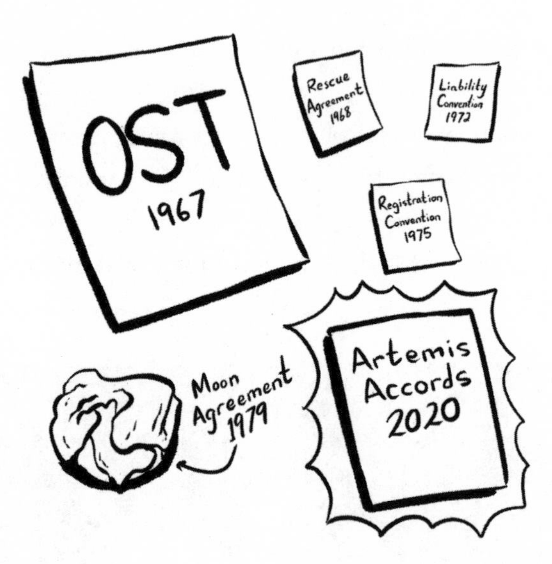

No Limits Nullius: American Space Law Conquers the Moon?

In 2015, President Barack Obama signed the US Commercial Space Launch Competitiveness Act. Regardless of what the OST says humans

*On the plus side, the passage of domestic space law sometimes leads to incredible headlines such as the *National Post*'s 2022 article titled "Canadian Astronauts No Longer Free to Rob and Kill with Abandon in Space or on the Moon." A more accurate but less entertaining headline would've been "Canadian Law also Applies to Canadians When Canadians Are in Space."

can do in space, the 2015 law is very clear about what *Americans* can do: "A United States citizen engaged in commercial recovery of an asteroid resource or space resource . . . shall be entitled to any asteroid resource or space resource obtained, including to possess, own, transport, use, and sell the asteroid resource or space resource obtained in accordance with applicable law."

In other words:

That's a pretty open standard. Note, there are asteroids bigger than the moons of Mars. Under this interpretation, although a space miner can't claim sovereignty over a giant asteroid, they could cut it into pieces and sell them at Walmart. Surely this violates the OST's rule against appropriation, right? Well, that's not the sense of the US Congress. We know this because the bill says, "It is the sense of Congress that by the enactment of this Act, the United States does not thereby assert sovereignty." So there.

In 2020, President Donald Trump, no fan of the previous commander in chief, signed an executive order reiterating the 2015 interpretation, rejecting the Moon Agreement in particular: "the United States does not view [outer space] as a global commons. . . . Accordingly, the Secretary of State shall object to any attempt by any other state or international organization to treat the Moon Agreement as reflecting or otherwise expressing customary international law."

In other words:

NOTE, SPACE-TRUMP CAN ONLY *RES NULLIUS* THAT MOON ROCK, NOT MOON TERRITORY

As it stands, US law allows its *own* citizens to take an extremely loose interpretation of the OST, in which they can exploit space resources without limitation and then personally own the results. Whether the United States is "right" about what the OST allows, the US government is certainly *consistent* about what it says Americans can do. The most recent addition to this approach is the 2020 Artemis Accords. This document pertains to the Artemis program, named for the sister of Apollo, and you can think of it as a US attempt to assert new norms for the Moon.

New Moon Race, New Moon Rules, or, What's in the Artemis Accords?

The Artemis Accords are not entirely about bending norms, and there's a lot of very Outer Space Treaty-ish stuff about peace, helping astronauts, and so on in there. However, there are two parts that are especially of interest to us: First, the Accords reiterate what domestic US law says about extraction of space resources—basically, "finders keepers," but with more syllables. Second, they introduce a concept called "safety zones."

The basic idea of safety zones is that if you're doing something on the Moon, you can designate a radius where it's dangerous to enter. How big a radius? So far, that's undecided. We likely won't know until we get some practice, but there are some decent guesses. Sea law allows 500-meter safety zones for marine operations. NASA's heritage protection rules suggest a two-kilometer exclusion zone around Apollo lander sites. This is probably a decent range of possibilities, though if future landers kick up regolith over much greater distances, the definition of a safety zone might be quite large indeed.

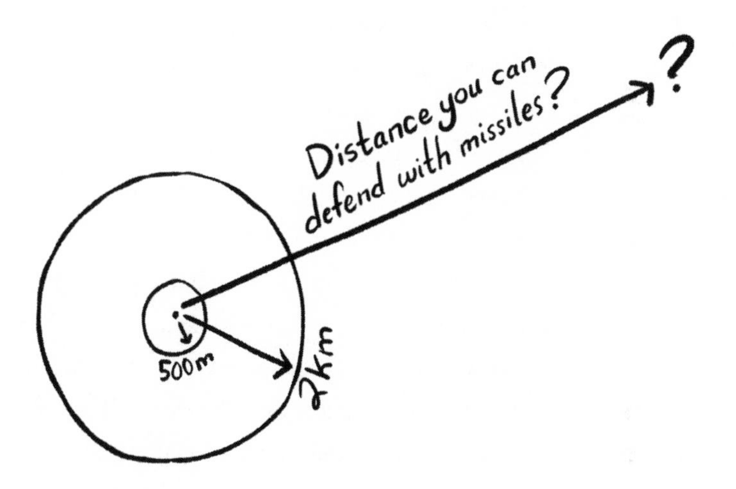

The interesting thing about safety zones is that while they're not precisely a nation claiming a chunk of the lunar surface, they're certainly a nudge in that direction. When countries designate a "safety zone" on the Moon, they'll basically be saying to everyone else, "Look, if you land in this zone, dangerous stuff could happen to you and we can't guarantee your safety. Also, if you land here, you're endangering us. So talk to us before you come into the circle." It's not appropriation of land exactly, and it's certainly not sovereignty . . . but it is something like turf.

This gets especially interesting when we think about the use of "premium real estate" for a settlement. We noted earlier that the Peaks of Eternal Light occupy only a tiny lunar footprint.

To be clear, we don't believe safety zones are a kind of purposeful backdoor sovereignty mechanism. They have real utility. Remember, the Moon is low gravity and its surface is made of jagged charged glass and stone. A rocket landing means all that stuff gets ejected in every direction, and low gravity means it goes quite far. Insisting on some sort of protocol to protect Moon habitats or mining operations is perfectly sensible. Plus, the Artemis Accords do say that safety zones are temporary. But "temporary" is defined as ending when the operation ends.

However well intentioned, the result is that if countries agree to the Artemis Accords they will now be able to lay some kind of claim to larger regions than just a habitat on the surface. To understand why this could get weird, suppose the United States starts a Moon base in a valuable crater and claims a two-kilometer radius circle as a safety zone. They keep it for many years, with generations of researchers coming and going. They've set up equipment, improved parts of the surface for usage, and conducted detailed surveys. They may have places where they traditionally do certain behaviors—not just beloved labs or bars, but places with sacred religious or national significance. Consider, for example, if a comrade died and was buried inside the safety zone. At that point, we're not talking about American *sovereignty*, but we're also not talking about an abstraction—this is a place that has meaning to national identity for deeply human reasons.

Under Artemis, this is all technically temporary. But try to imagine a rival country, especially one like China, which already has a rocky relationship with the United States, telling them to leave. Again, that American safety zone isn't a claim to sovereign territory; it's not a claim that some lunar region is literally America. But it sure gets close. Throw in the limited quantity of good places on the Moon, add the lack of any limits on the total number of bases nations can place, add that the nation pushing the new interpretation is the one with the most space capability, and add that the two countries most interested in the best spots on the Moon are nuclear powers, and you have a recipe for danger. If nations care about who's settling the Moon, well, Artemis would allow behavior that goes right up to the line of appropriating Moon land, and the people most able to claim that land right this second are from the United States.

Is anyone other than the United States cool with this? Surprisingly, yes. As of October 2022, not counting the United States, twenty countries have signed. That includes major powers like Australia, Brazil, Canada, France, Israel, Japan, and the UK. The most notable absences are the European Space Agency, China, Russia, and India.

China's absence is not necessarily a rejection—under US law, NASA is prohibited from signing bilateral agreements with China. But China may have similar goals to the United States, and Russia may end up bandwagoning with them. China has stated plans to access lunar titanium, water, and your friend and mine, helium-3. They have certainly not pushed for a CHM framework, and in fact have stated intentions to set up a profit-making lunar economy.

A Race for the Heavens: Probably a Bad Move

The OST rules space, but the OST is vague. The one major attempt to clarify things through a United Nations agreement failed. Forty years later, the United States is rushing into the gap and possibly dragging the world along with it. This has the making of a conflict whether anyone wants it or not.

Our impression from talking to space geeks is that many are practically salivating for a new Space Race. The reasoning is that back in the 1960s when the USSR and United States duked it out in a race to the Moon, all sorts of spending went into the science and tech they see as valuable for humanity, or anyway *awesome* for humanity. We're not sure this is an unmitigated good: even if we buy that the 1960s Moon Race was a pure net benefit for humanity, that doesn't imply a 2020s Moon Race would be likewise. Neil Armstrong and Buzz Aldrin walking around on the Moon did not prohibit Soviet cosmonauts from doing the same. Under the guidelines proposed in the Artemis Accords, a US base on Shackleton crater could plausibly limit the ability of other players to put a base nearby. You might not have a race so much as a zero-sum scramble. Some say this has already begun.

According to a 2022 report, whose contributing authors hailed from the U.S. Space Force, Defense Innovation Unit, and Air Force Research Laboratory:

> *While the United States space industrial base remains on an upward trajectory, participants expressed concerns that the upward trajectory of the People's Republic of China . . . is even steeper, with a significant rate of overtake, requiring urgent action. The fundamental tonic is to mobilize still greater energies with an enlarged vision and broader set of policy as our nation did in 1962. Specifically, the U.S. lacks a clear and cohesive long term vision, a grand strategy for space that sustains economic, technological, environmental, social and military (defense) leadership for the next half century and beyond. A North Star vision for economic development and human settlement in space should be bi-partisan, multi-generational, and inspirational to all who embrace America's values.*

If you want space settlements to happen someday, this may not be a good development. Why? As we've argued, you cannot have near-term space settlements without a lot of major scientific and technological developments. All of that science and technology development can *already* be done under the OST. Even the highly restrictive Moon Agreement probably would've permitted the relevant knowledge and experience to be acquired.

If you believe, as we do, that there's no obvious economic case for Moon mining, that any settlement science can be done just fine under the old framework, then this new Moon Race is a pointless escalation toward a crisis, possibly even a conflict. If you believe, as we do, that safely and sustainably settling space requires a high level of human harmony, conflict over our nearest celestial neighbor does not advance the goal of populating space.

We believe humanity in general, and possibly the long-term project of space settlement in particular, would be better off with an internationally managed system that regulated both where people are allowed to set up shop and what they're allowed to do with the local resources once they get there. It wouldn't be dynamic, it wouldn't be like a science fiction novel, and frankly it would be very slow and bureaucratic and boring. But it would keep the peace while humanity gets its political and technological act together enough to make space settlement possible.

Is a regime like this likely on our divided Earth? We don't know, and it's worth noting that the United States and China are, as we write, building the rockets and landers needed to ramp up a space scramble. But international regulation is at least possible. There have been two other times since World War II when a massive piece of territory got regulated: Antarctica and the deep seabed. Neither place experienced a major conflict or scramble for territory because both were regulated as commons. Today, both have much clearer, much more restrictive property rules than the Moon under the Artemis Accords. That makes them worth considering as ways to regulate outer space that could actually become agreed-upon international law.

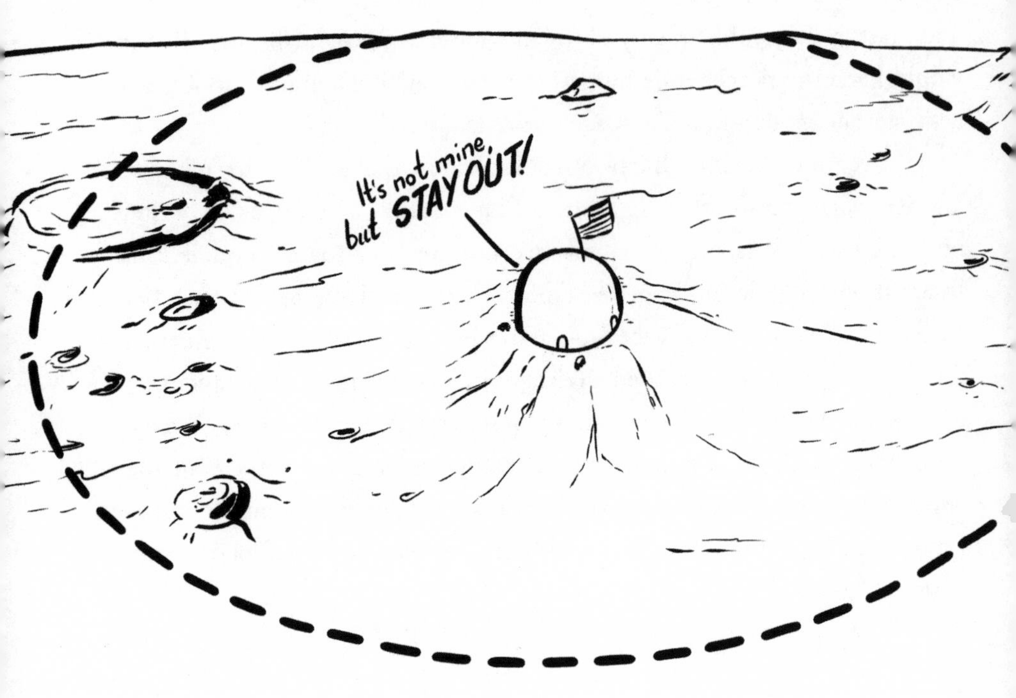
It's not mine,
but STAY OUT!

Nota Bene

SPACE CANNIBALISM FROM A LEGAL AND CULINARY PERSPECTIVE

What do you do when someone dies in space? We'd love to tell you, but there's not really precedent. Death among spacefarers has always been rapid and always consumed whole crews, so there's never been a situation where a few astronauts gather round a lost companion wondering what the appropriate mode of disposal is. The result is that, aside from informing you that if you flushed them out the airlock they should be registered as a satellite (thanks, Registration Convention!),* we can say very little.

So while we had to drop a planned section on death in space, we did get interested in a related question while researching space food: *the forbidden space meal.* The one option on a long Mars trip that is extremely fresh and contains precisely the nutrients a human body needs.

We're just going to be real with you here—the astronautical literature on how to eat the crew is pretty limited. The reader of NASA's Human Integration Design Handbook will search for "human flesh" in vain. Apparently "human integration" is used in a narrow sense.

But we did locate a self-published paper—dare we say an overlooked classic—on the topic: "Survival Homicide in Space" by Robert A. Freitas, 1978, which may provide the needed insight. What's survival homicide? It's a situation where someone or some group can survive only by killing. The classic example is the "Law of the Sea." Not the sea law we'll explain

*Though they probably wouldn't be a satellite for long. In Terry Virts's memoir, he notes that if you had to have this sort of "burial at sea," astronauts would likely direct their deceased crewmate's remains toward the Earth, so they'd burn up in the atmosphere in the not-too-distant future.

in the next chapter. The other law of the sea. The one that says you draw straws and the person who draws the short one gets eaten.

Freitas doesn't mention the dinner aspect of space murder, but he does posit a scenario inspired by the novel *Marooned*, in which three astronauts are trapped in a capsule without enough oxygen. Due to the precise data available to the spacefarers, they know they have enough oxygen for only two to survive until rescue comes. Logically, one should die. And logically, we can extend the scenario to the forbidden meal by imagining your cricket herd have all died and you're running out of delicious food cubes.

Can you kill one of the crew? According to Freitas, the OST actually provides something of an answer. Under Article VIII, jurisdiction goes to the state of registry. Different nations have different rules regarding the "law of necessity,"* so really you just need to know what the relevant legal code permits. If you're in a Canadian ship, get a Canadian judge on the line and explain about the people-meat.

But what if you have options? Writing from 1978, this wasn't in the cards—at that time Skylab was long uncrewed and the Shuttle not yet flown. The only space crews were in Soviet ships with largely Soviet crews launched from Soviet territory. What Freitas couldn't anticipate was the future ISS, which has conjoined sovereign territories from many nations. In principal you could shop around for the nation with the most cannibalism-friendly laws, but your best bet is probably to eat a fellow national inside a module owned by the nation you come from. We don't need space cannibalism *and* an international incident.

Or just do nothing. In a real Mars settlement faced with a situation like in *Marooned*, the likeliest outcome is that you'll just die. Freitas gives a litany of times people could've prolonged their lives by ending others—in sunken submarines, collapsed coal mines, and in boats adrift. Typically, they chose to face the end together. It turns out most of us would rather die than kill.

---

*We elected not to get too in the weeds about exactly what the standards are, but for the future law students in the audience, the two most relevant cases are *Regina v. Dudley and Stevens* as well as *United States v. Holmes*. We've been told these are both used to trick law students into thinking law school will be interesting.

But we did find one exception.

Professor, prolific author, and triathlete, Dr. Erik Seedhouse wrote an analysis of space cannibalism in *Survival and Sacrifice in Mars Exploration.* We don't know Mr. Seedhouse personally, and he didn't respond to our email, but we will note that his book's index contains precisely one entry on "behavioral challenges," a very important topic, but *five* entries on the gustatory mode of crew integration.

Seedhouse asks, "Imagine you're stranded on the Red Planet with three crewmembers. You have plenty of life-support consumables but only sufficient food to last one person until the rescue party arrives. What do you do? . . . One day, while brewing coffee for breakfast, you realize there are three chunks of protein-packed meat living right next to you." He argues that the largest people should sacrifice themselves first, since they both consume and provide the most food. We don't know where Seedhouse would fall in the buffet line because we couldn't find his height and weight online, and honestly we're scared to ask. Mostly because his book includes a weirdly detailed look at how to butcher *Homo sapiens.* Also, on page 144, the reader will find a photo of ten astronauts floating happily in space, with the caption: "In the wrong circumstances, a spacecraft is a platform full of hungry people surrounded by temptation. Is it wrong to waste such a neatly packaged meal?"

That is, as they say, a question for philosophers. But we do have one pragmatic piece of advice for any potential Mars settler: *Leave Erik Seedhouse at home.*

PART V

The Paths Forward: Bound for Moonsylvania?

WE RECOGNIZE THAT THERE'VE BEEN A LOT OF LEGAL TERMS SO FAR. So as we move to the speculative possibilities, let's bottom-line things: the rules that we have are vague. By some nation's interpretations they permit endless exploitation and pseudo-territorial claims, but without sovereignty. Nuclear powers are moving to stake claims in a largely unregulated environment in which they may disagree about what the law actually permits.

Either we wait for a crisis and then see what shakes out or we push for some kind of agreeable framework for space now. But what framework should we choose? Broadly speaking, there are two paths forward. We could create an international regime to manage behavior in space to prevent attempts by nations or alliances to claim all the most desirable places. Or, we could find a managed way to allow national appropriation, so that there would really be parts of the Moon that are China or America or India or even new nations altogether. The latter option would at least provide clarity, but we have trouble imagining a way to get there without conflict.

We'll get into the details of how both of these possibilities might work, but the thing to understand is this: the path we take during this century is likely to affect the human future for a very long time. Much of modern sea law derives from traditions that were already old during the seventeenth century when modern international law was in its infancy. The space law of 2020 may profoundly shape the possibility of space settlements in the year 2200.

14.

Commonsing the Cosmos

Since the mid-twentieth century, technology has unlocked areas of Earth that were previously inaccessible. Most notably for us, the Moon, the deep seabed, and Antarctica. If these areas had possessed valuable, accessible resources three hundred years ago, it's entirely possible that there would've been violent struggle over them.

As far as we can tell, this sort of squabbling is what a lot of people expect, perhaps even desire, in space. However, in the post–World War II era, that's just not how things have typically worked. When we've regulated giant new territories, including space, they've been made into commons. There are no Antarctic cowboys or Marianas Trench claim jumpers. The disputes that exist are handled not by violence, but by negotiation and bureaucracy with speeds that are appropriately glacial. Part of this is due to the difficulty of making use of these areas, but as we'll see they are also bound by agreements on resource use that, compared to the OST, are clear and narrow.

To some this is a disaster. All those potential resources, all those places people might have lived and developed and commercialized have been left alone, to the detriment of human flourishing. Others see unbridled success: the first half of the twentieth century saw claims and counter-claims in the Antarctic that could have led to warfare. Yet Antarctica and the deep seabed have remained peacefully managed commons with reasonably clear, if quite restrictive, property regimes. This has worked on Earth, so it would likely work in space.

It is not the most romantic option or the most awesome option. It's likely not the option that produces the most rapid development. As we'll see, it's not even likely to be especially equitable between nations. But it is an option that could permit widespread space activity while avoiding a needless conflict between nuclear powers. Along the way you'll learn about tactical babies, the time Antarctic Nazis heiled a penguin, and if you're very patient, at the end of this section we've got a Nota Bene on murder and booze in polar bases.

Commons Sense

The first thing to know about commons is that they actually work when they're set up right. We've argued earlier that space will just be another place where humans will be humans, and you may have interpreted that to mean "humans are demonic ogre-creatures and we will inflict our ogreness on space." But we really do share from time to time. We have commons and they work reasonably well. In our experience, a lot of people think commons don't work due to an economic concept called "the tragedy of the commons." The language goes back to Garrett Hardin's 1968 paper on the economics of commons, but the logic is at least as old as Aristotle, who wrote, "For that which is common to the greatest number has the least care bestowed upon it."

Here's the idea: suppose a Martian greenhouse has a nice field set aside for food crickets. Everyone is allowed to use them, including you. Whenever you run your crickets on the field, the field gets worse. You don't care because the benefits go to you while the damages are borne by all cricket ranchers. If everyone else is just as selfish, then pretty soon the commons is, tragically, ruined.

People sometimes see this as an inevitability of commons and tend to suggest privatization as the solution. Divide up the cricket run so each person bears the cost of damaging their plot.

But it turns out not all commons are tragic. In fact, they've existed around the world for ages. Nobel Memorial Prize–winning economist Dr. Elinor Ostrom became famous for her work documenting the ways common resources are effectively managed *despite* the lack of private property or even an overarching authority to enforce regulations.* Cricket ranchers, for instance, could agree to take turns using the commons or to limit the number of crickets per member or to limit membership overall. Or members who wish to run extra food crickets might compensate other members. Possibly with home-brewed beet wine or delicious red cubes.

Commons may lack some of the dynamism that comes with privatizing land. You might, for instance, be disinclined to personally spend time improving the set of plants in the commons because you'll do all the work yourself, while having to share benefits. Or if you want the group to pay

*She did not accomplish this without pushback from economists insisting she couldn't be right, because models at the time didn't support the concept. An "Ostrom's Law" sometimes attributed to her goes: "A resource arrangement that works in practice can work in theory."

for the improvements, you have to achieve cricket-rancher consensus, which may be difficult.

But there are other goals than efficiency. When we think about the future of the Moon or Mars, the goal may not be rapidly extracting maximum value. It may also be things like preserving a pristine environment or preserving humanity from blowing ourselves up in a territory dispute. To the extent *these* things are the goal, regulating space the way we regulate Antarctica is a good option.

The Antarctic Treaty System

A Brief History, with Penguins and Nazis

Antarctica was only glimpsed by human eyes starting in the nineteenth century, but remained forbidding even as Arctic exploration became possible. The South Pole was not reached until Norwegian Roald Amundsen and his crew made a successful dash in 1911. But as late as the 1950s, large unmapped regions remained. Why did it take so long? In short, because Antarctica sucks. This is a trait it has in common with space. In the interior of the continent, winter temperatures hover around -60°C, and sometimes fall below -80°C. It has hurricane-force winds, but is so devoid of precipitation that, ice sheets notwithstanding, it is technically a desert. It was not permanently inhabited until 1957.

Being that Antarctica is such a frost-laden crap bag of a continent, you might expect that nations wouldn't bother with territorial claims. You would be wrong. By World War II, seven* countries had claimed vast pizza slice–shaped hunks of the Southern Continent.

As midcentury approached, it was by no means clear that Antarctica would be a land of peace. At one point, Nazi Germany attempted to claim a chunk and call it New Swabia. According to the account by geographer Ernst Hermann, the Aryan supermen were greeted by a "native" upon

*Argentina, Australia, Britain, Chile, France, New Zealand, and Norway.

arrival—that is, a penguin. “Heil Hitler!” they shouted. The penguin, somehow, “was not impressed.” Despite this setback, both German and South American troops occupied portions of Antarctica during World War II.

More serious was the squabbling over Antarctica’s warm-by-local-standards northern peninsula, simultaneously claimed by Britain, Argentina, and Chile. By 1952, shots had been fired when British ships attempted to land and rebuild an Antarctic base that had burned down. No one was killed, and Argentina apologized, but conditions for peace weren’t exactly sunny.

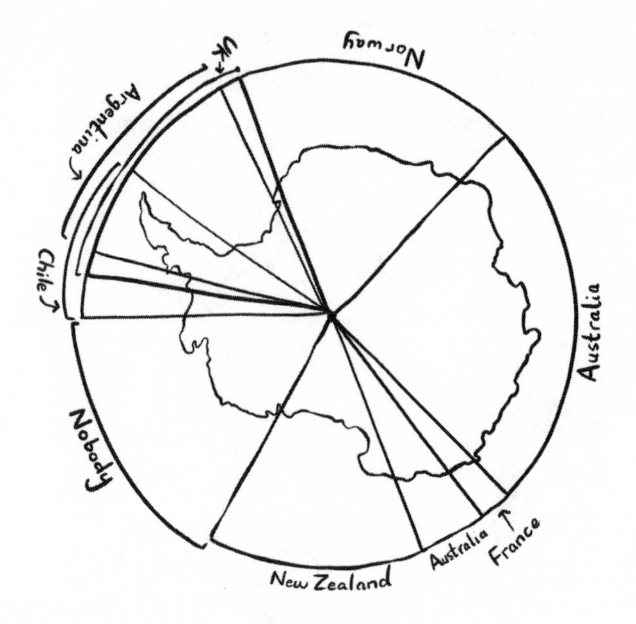

Complicating matters, the titans of midcentury geopolitics, the United States and the USSR, had made *no* claims, despite a full 15 percent of the continent being *unclaimed*. Why this forbearance? In part due to fear of accidentally kicking off World War III over a continent that, let's remember, is a frost-laden crap bag. Sound familiar?

During the 1957–1958 International Geophysical Year, nations poured money into exploring Antarctica.* Twelve participants† made plans to establish Antarctic bases, with the stated goals being science and peace. It was these twelve who negotiated the Antarctic Treaty, which with later agreements became known as the Antarctic Treaty System, generally just called the ATS.

The initial treaty went into force in 1961, and since then seventeen countries have been added as "Consultative Parties" by virtue of paying for their own research setups. Twenty-five additional states are "Non-Consultative Parties," meaning basically they haven't paid for a sweet Antarctic base so they're only allowed to listen, not vote on anything.

*Fun fact: the other big exploration project of that year's IGY was space exploration. Hence, Sputnik 1 going up the same year.

†Argentina, Australia, Belgium, Chile, France, Japan, New Zealand, Norway, South Africa, the Soviet Union, the UK, and the United States.

What does ATS say?

In short: exploitation of minerals is not allowed, at least for the moment. Military bases are not allowed. Nuclear waste can't be dumped and nuclear bombs can't be exploded. Science is cool and nations should cooperate. And those territorial claims? Well . . .

History's Greatest Kludge: Sorta-Kinda-Territory in the Southern Continent

ATS has a territorial sovereignty regime that is unique and bizarre. The basic deal is that by signing the agreement, parties don't renounce any prior claims, nor do they allow any new claims. The claims just . . . sit there. That three-nation land claim between Chile, Argentina, and the UK? It *remains* a three-nation claim. That 15 percent unclaimed land? Still nobody's. Meanwhile, anybody is allowed to build science labs pretty much anywhere.

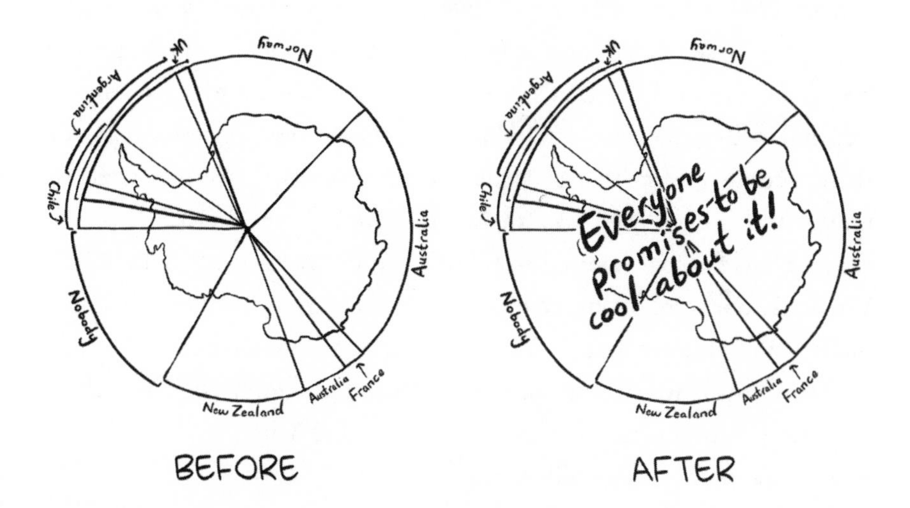

This *feels* like it shouldn't work. It's like having three roommates who all believe they own the same apartment and instead of resolving the disagreement just promise to live together and share some chores. And yet, since 1961, Antarctica has been completely peaceful, with the closest thing to war being occasional posturing in case of future claims fights.

And the posturing is not especially belligerent. The main tactic employed is the construction of more science bases, presumably so that at some future diplomatic meeting someone from Britain can shout "our forefathers hung out in this base taking ice-core samples or whatever and we will never betray their legacy by renouncing our claim!"

There are other methods. In 1953, Argentina built Esperanza Base, and in the '70s converted it to a civilian-friendly facility. Since then, families have been regularly flown in, including in one case a seven-month-pregnant woman. Inside the Argentinian claim, she gave birth to Emilio Marcos Palma in 1978. Why? Presumably so at some future diplomatic meeting someone from Argentina can say, "You British say this is your peninsula, but what of little Emilio?" At least ten other births have happened since. However, the idea of priority via childbirth is somewhat vitiated by the fact that nobody wants to stay in Antarctica over the long term. The motto of Esperanza base—"Permanence, an act of sacrifice."—makes it clear why.

Argentina's move has been taken seriously by at least one counterclaimant. In 1984, Chile constructed *Villa Las Estrellas*, which is bigger than Esperanza Base, including a gym, radio station, school, church, post office, and souvenir shop. That same year, in one of the weirder acts of international one-upmanship, Juan Pablo Camacho was *conceived and* born there, presumably so that at some future diplomatic meeting, well, you understand. In her book, *Antarctica: The Battle for the Seventh Continent*, Doaa Abdel-Motaal referred to Chile's counterstrike as a "counterbirth." We prefer our term: "counterstork." As in the phrase, "Britain has not yet made a similar move, but are believed to possess counterstork capability."

All of this is weird, but note that it's also nice and boring. Yes, there's posturing, but it's prosecuted via science labs and maternity care, which is pretty solid as far as geopolitics goes.

Claims on Ice: The Resource Regime for Antarctica

Peace is nice. Science is nice. Babies are nice. But how about some stuff? There's at least some evidence that Antarctica possesses valuable minerals, and anyway it's about two Australia's worth of land that has never been mined. There's got to be some good stuff in there. Sure, it's under the world's thickest ice sheet, but as technology advances and Earth gets toastier, you might expect a resource-hungry planet to lick its chops and consider a frozen entree.

Well, too bad. They can't. And the reasons they can't inform us about how space exploitation could go in the future.

In the late 1980s, there was a proposal called Convention on the Regulation of Antarctic Mineral Resource Activities, or CRAMRA, which would've clearly opened up Antarctic resource extraction, albeit via careful bureaucratic protocols. CRAMRA failed, largely for three reasons. First, there was a concern that it would reignite fights over sovereignty. It's one thing if Americans set up a peaceful science base in the French claim, but quite another if they start acting all American by drilling for oil there. Second, the UN wasn't super excited about this idea; decisions about how to regulate this commons would be made only by states that were part of the ATS, effectively cutting out poorer states that couldn't afford a research base. Finally, environmentalists led by Jacques Cousteau started a massive international backlash.

In 1998 a fifty-year moratorium on mining went into effect—no exploitation, and in fact you're not even allowed to explore for the existence of valuable minerals. This may not be your favorite option, but hey, it's at least a clear regime. The flow chart looks something like this:

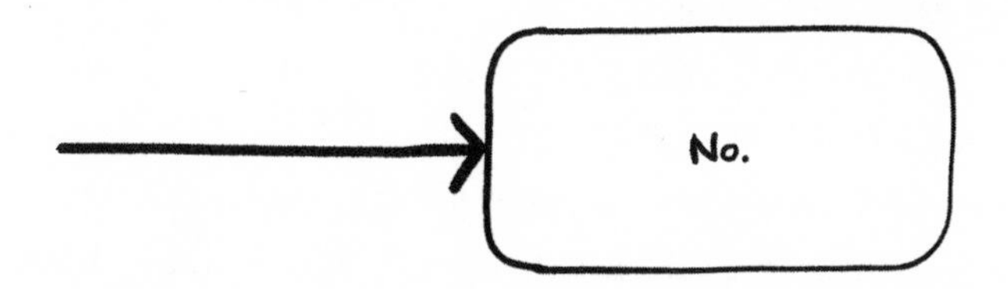

This deal will not be reconsidered until 2048, and even then can only be ended by unanimous consent.

For theories of space settlement, this is important precedent. We cannot tell you how many times a space enthusiast has told us that space law is stupid because all law will go out the window the moment something valuable shows up on the Moon or Mars or wherever. If they believe that, they need to explain why it didn't happen in Antarctica, where there is at least some evidence of valuable metals that would be far easier to extract and sell than anything on the Moon.

What would a similar policy look like for space? Quite restrictive. You could imagine, for example, saying nobody can use the ice deposits on the Moon other than on a small scale for scientific research. You can imagine saying that Moon or Mars bases are only allowed to use local resources needed for base operation. This is not the most exciting possible option, but we think it'd be less negative than it might appear at first glance.

The Case for Space-arctica

If you want space settlements, arguably the two big goals are avoiding conflict and doing the research needed to learn how to settle space. If there were valuable minerals on the Moon, unrestricted mining regimes might produce faster development. However, there's probably nothing worth getting on the Moon to sell on Earth, so trading exploitation potential for medium-term peace is a pretty good deal for humanity.

Clearly, this is not an exciting path forward for your typical space-settlement geek, but in Antarctica it has kept the peace now for two generations. During that time a tremendous amount of research has been conducted, including learning how to live on the frozen bottom of the world. The two biggest problems we've cited for space are babies and greenhouses—both of these have been accomplished, albeit on a small scale, in Antarctica.

One other lesson we can take away from ATS is that Antarctica was successfully converted to a commons *after* countries had laid serious, sometimes overlapping claims. One concern we have for the Moon is that in the next decade claims start getting locked in, making any future treaty difficult. Here, the ATS is somewhat reassuring. Even if some kind of scramble kicks off and nations really come to feel an ownership stake in parts of space, well, ATS shows us the path to a commons remains open. It's a weird, kludgey path, but it's kept the peace for over sixty years now.

If you want space-settlement science, the big concern about ATS is that an ultrarestrictive mineral regime would thwart needed basic research in off-world resource use. Knowing how to use lunar water may be important to a future settlement, but a CRAMRA-style regime might make usage illegal. It's also possible that we Weinersmiths are wrong, and that there are valuable resources worth getting. Maybe those platinum asteroids really are worth trillions. Also, even if we're right, developers seem to want to shell out money, and that money could solve some of the basic engineering difficulties for space settlement. Is there a way to have an ATS-like regime that nevertheless allows *some* exploitation? Of course—that's the Moon Agreement. But it failed, and so we look to one last regime, which may present the best hope for a path forward.

Governing the Deep Seabed

Since 1994, the bottom of the sea has been governed under the UN Convention on the Law of the Sea, or just UNCLOS. Why did humanity wait so long? After all, the deep sea bottom has been there for a few billion years, no doubt patiently waiting for a species of talking apes to shout "MINE!" The basic deal is that like the Moon, until the mid-twentieth century, a proposal to mine the sea bottom would've sounded as plausible as a proposal to harvest pixie dust. By the 1960s, both the Moon and seabed were seen as potential new sources of commodities. Regulation was needed to permit exploitation without conflict. But whereas only about 10 percent of the states in the UN have ratified the Moon Agreement, nearly 90 percent have ratified UNCLOS. This is especially interesting for space, because the way UNCLOS regulates the sea bottom is broadly similar to how the Moon Agreement would have regulated space. It preserves the seabed as a commons while at least in principle allowing humanity to access its mineral wealth.

Putting Holes in the Bottom of the Sea, or, How Does UNCLOS Work?

Under UNCLOS, nations get jurisdiction over an "exclusive economic zone" off their coast. These are big—generally about two hundred nautical miles out, although sometimes substantially longer depending on a bunch of technical stuff at the intersection of oceanography and law. This might seem like a lot, but according to the latest science, oceans are really big. Under UNCLOS, the area beyond these coastal waters is called . . . *the Area*. Here, we should note that UNCLOS weirdly insists on naming a bunch of things as if they're the title of a cheap horror movie. Anyway, the Area is around 50 percent of Earth's surface—substantially more surface than the Moon and Mars put together, and under UNCLOS it is part of the common heritage of mankind. But unlike in Antarctica, you can actually try to exploit minerals in the deep seabed. Similar to what the Moon Agreement would have done, UNCLOS created an international regime to manage resource exploitation.

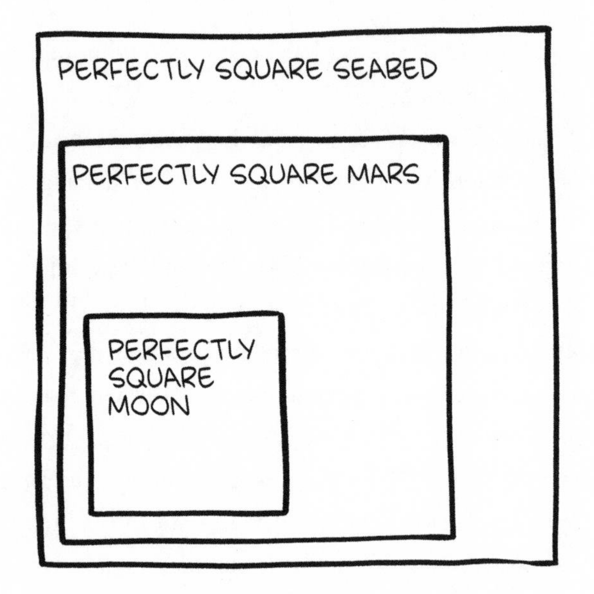

Why were developed nations cool with seabed socialism when they were decidedly opposed to comm-moon-ism? The truth is that at first they were not. The initial framework came out in 1982 and was seen by them as too favorable to developing nations. For example, the original agreement required technology transfer from whomever exploits the deep seabed, with the goal of helping developing countries acquire the technology they needed to mine the seabed themselves. Developed countries thought this would hurt their economic and security interests and pushed back. Similar to some of the problems that killed the Moon Agreement, countries don't generally want to give away their advantages for free.

But by 1994, developed nations were willing to sign, thanks to an "implementation agreement" that made things more favorable to them. Notably, technology transfer became a recommendation rather than a condition. If we want a revised Moon Agreement, we would likely need something similar—a framework that is broadly egalitarian enough to prevent a scramble for territory and get buy-in from developing nations, but which developed nations see as beneficial.

The Race to the Bottom

The way seabed resource exploitation works is complex, but the basic deal is that if you're a company, you can team up with a state signatory to UNCLOS and apply for exploration or exploitation rights. If you would mine . . . *the Area*, you must ask . . . *the Authority*. Better known as the International Seabed Authority. Separate from any theory of whether this would be good for space, we believe people of good faith can agree that having something called the International Space Authority is objectively desirable.

So, how's this worked so far? The Authority has given out over two dozen exploration contracts. When a company starts to exploit, it would go something like this: A company applies to exploit a chunk of seabed. The Authority says yes. The company then gets dibs on *half* of that chunk of seabed. The other half goes to . . . *the Enterprise*. Awesome. The Enterprise is an operation run by the Authority in a way that would ideally benefit developing nations.

So if you want to imagine this for the Moon, it'd go something like this: Weinersmith Incorporated, backed by the good people of Luxembourg, apply for mining rights in Shackleton crater. The International Space Authority says yes and then gives half to us. We begin setting up solar panels and mining the Craters of Eternal Darkness to fill lunar hot tubs. Meanwhile, the Space Authority operationalizes the Space Enterprise. They keep free-market principles in mind, but do their best to involve less-developed nations in the process of exploiting their half of the crater.

If the details sound fuzzy here, it's because they still haven't been worked out. But the Authority has been working on making the rules, and now the clock is ticking. In 2021, the president of Nauru let the Authority know that they're working with a company that is getting ready to exploit the deep seabed, which gives the Authority two years to try to wrap up their deliberations on the laws governing exploitation. However, environmental groups are pushing back, saying two years isn't enough time to learn about the deep seabed to be sure we can mine it without causing environmental devastation. The next few years will be an interesting time as the world watches to see what the rules will be for exploiting the common heritage of mankind.

The good piece of news here for people excited about space resources is that there was ultimately a path to a regulatory framework that still allowed for exploitation. Developing nations signed, and so did many developed countries, including China, Russia, Japan, India, the EU, France, the UK, Germany, and Italy.

Here we should admit that there is one major holdout. A country quite relevant to space and quite opposed to anything that looks like socialism. Have you guessed it? Yep, the United States. The US has signed but not yet ratified UNCLOS, despite occasional internal efforts to try to get us to do so. Why? Well, for one thing, as Donald Rumsfeld said of UNCLOS at a 2012 meeting in the US Senate, it was "conceivable that it could become a precedent for the resources of outer space." That is, once we start sharing everything in the sea bottom, next thing you know we're having to share the whole dang solar system. That said, while the United States is not officially cool with sea socialism, America does claim an exclusive economic zone and broadly abides UNCLOS. How? By saying UNCLOS is customary law. We've heard this referred to both as a diplomatic masterstroke and as a way to get all the responsibility of international law without a seat at the table where decisions are made. The good news anyway is that, speaking pragmatically, UNCLOS more or less works. Peace, cooperation, and managed access to resources. Exactly what we want for space.

Where Do We Go from Here?

If the goal is space settlement, we think something like UNCLOS is the path forward. It would reduce the risk of conflict arising from a pointless space scramble, while still allowing all the most relevant science to be done. Countries could still run labs in space and could learn how to work with on-site resources, and companies could engage in prospecting if they believe it's worth their while. Major space powers would not be pushed into "socializing" their advantages, while developing nations would get a slice of the space pie. It's a deal we can potentially get, which lets humanity take steps forward toward settlement without starting a crisis. Also, and this cannot be overstated, future bureaucrats would eventually get to non-ironically say "operationalize the Space Enterprise."

We suspect most readers find at least some parts of this idea disappointing. Developing nations won't be left out in the cold, but will be operating on terms set by existing powers. Plans for rapid space expansion would be left in the slow hands of international bureaucrats. And do we *have* to replace visions of Mars pioneers with thousands of dreary jurists forever moaning words like "pursuant" and "heretofore" and "as regards the"? It's like a nightmare, only boredom replaces fear.

But is this really so bad?

In his 1952 essay "Who Owns the Universe?" legal scholar Dr. Oscar Schachter worried about repeating the previous centuries of scramble for territory, and the war and colonialism that came with it. Writing five years before Sputnik 1, he said, "first landings on the moon will involve all sorts of acts intended to support claims of sovereignty. Obviously, the flag will be planted and, very likely, names will be given to places on the moon." Governments might "exercise control, perhaps even to issue licenses, and to claim the right to exclude those who are not licensed. All of this would be the old story of territorial rivalry—but this time extended into the heavens themselves." In part thanks to Schachter's work, this scenario was thwarted. In the four decades after his essay, space and space-like environments got new property regimes in a manner that was basically peaceful and oriented around sharing. No Mars pioneers, but then again if we had nuclear-wintered our planet in a scramble for territory, that too would've put a damper on our space plans.

These regimes also have one nontrivial virtue—they exist. You may want a privatized property scheme in space, but in the post–World War II era, whenever humans have been called upon to regulate a gigantic area that was formerly inaccessible, we've gone with a commons. Given the past, and the fact that space itself is *already* a commons, if you were betting on the future of space, "commons" would be a pretty good call.

But there are alternatives. There are ways we could make space much more like the non-Antarctica, non-Area 40 percent of Earth. That is the hope of many space-settlement fans. Here, we present some of those ideas and why we think they probably won't work.

Astrid, however, disagrees.

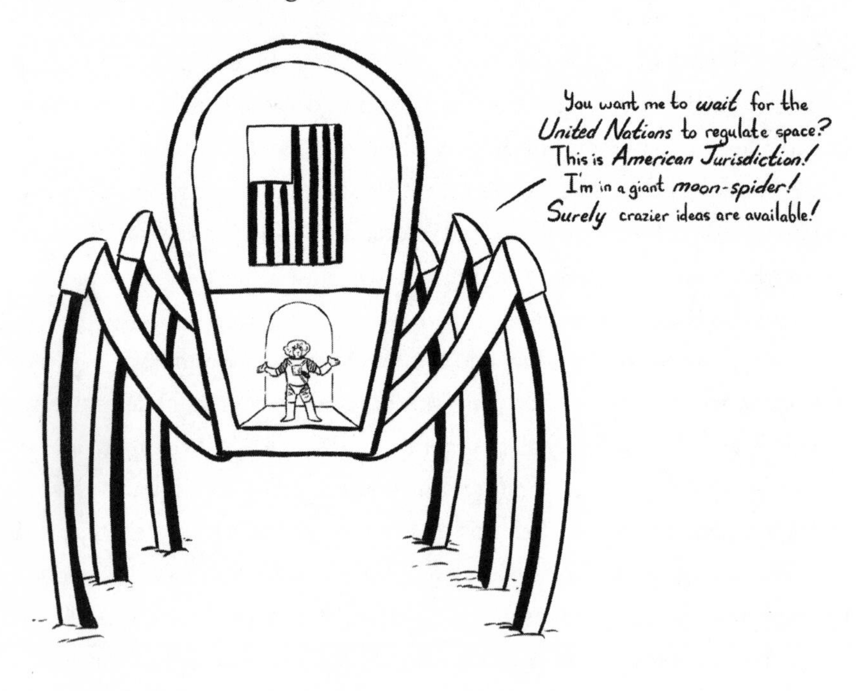

15.
Dividing the Sky

Un-Locke Human Potential!

Space privatization advocates often favor what's called a "Lockean" view of property, after John Locke, a seventeenth-century political philosopher. As Locke wrote in his famed 1690 work, *Two Treatises on Government*:

> *Though the earth, and all inferior creatures, be common to all men, yet every man has a property in his own person; this no body has any right to but himself. The labour of his body, and the work of his hands, we may say, are properly his. Whatsoever then he removes out of the state that nature hath provided, and left it in, he hath mixed his labour with, and joined to it something that is his own, and thereby makes it his property.*

In other words, *res nullius*. In the context of land, that means if you work it, you get it. Or, as Assistant Land Commissioner Joel David Wolfsohn once told a journalist asking about lunar homesteading in 1946: "The best way is to take along the wife and kiddies. A couple of cows, too, would give the undertaking an air of permanence."

Among space-Lockeans, the United States' 1862 Homestead Act is especially popular as precedent. The act, signed by Abraham Lincoln during the Civil War, said that for a small registration fee, white settlers and freed slaves* could grab a 160-acre plot in the West. Work the land for five years and by the magic of *res nullius* they would gain title.

In some cases, the land grab was a literal race, as when Oklahoma was opened to settlers in 1889. At noon on April 22, cannons and signal shots were fired. Men on wagon and horseback, as if parodying the history of settler colonialism, literally took along flags with which to stake claims.

If you want a large human presence in space, such a level of enthusiasm would be welcome. This brings us to the second deep belief that space privatizers hold dear—that opening up a private property regime in space would allow rapid exploitation of space resources and development of space settlements. Well-known space advocate Rand Simberg spoke for many, when he said: "Transferable property rights and free markets are at the heart of how billions have been brought out of poverty over the past two centuries, and they can continue to do so in the rest of the solar system."

In this view, the OST, whatever its merits during the Cold War, has stifled human advancement by acting as a barrier to off-world development. It's hard to know if this is a correct assessment or not—space is a hard place to develop, and it's possible that physical law was the bigger barrier than, you know, *law* law. But as physical law becomes something humanity can deal with better and better, law law becomes a bigger issue.

To privatize space, one must take a posture toward the OST. Pro-privatization parties largely fall into three camps: 1) those looking for OST loopholes; 2) those hoping to find a way to amend the OST to allow private property; and 3) those who propose scrapping the OST altogether. Here, we'll look at a few examples of each.

*The Indigenous peoples dispossessed by the United States were not permitted to use this program.

Approach 1: Only Violating the Spirit of the Law

If you want privatized space, why endlessly debate legal amendments or wait for the plodding governments of the world to finally flush fifty years of legal precedent into the waste containment system. With the right frame of mind, we can claim chunks of space right now!

There are a lot of proposals of this sort, but the basic deal is they don't work unless your goal is to make a legal scholar cry. We don't see these as very serious on the merits, but they illustrate a strain of thought common among space-settlement geeks, possibly including some very powerful ones.

Loophole 1: Martian Liberation

What if we just relied on awe? What if a Martian settlement were so mind-blowingly amazing, that surely we Earthlings would not bind this new branch of humanity with our general earthly crapitude? Consider a proposal from Dr. Jacob Haqq-Misra, who argues that we could route around the OST by agreeing that the OST doesn't bind Martians.

Why would we do this? According to Haqq-Misra, we'd be acting out of a sort of enlightened self-interest. Mars will be a font of "transformative value" for all humanity, benefiting us in the long run. But to get the full benefits, we have to cut the cord, allowing Mars to go its own way.

A priori, this is a nonstarter because it's simply illegal. International law applies to humans and therefore would apply to Martian humans. Even if it didn't, there are several big problems with any idea of this sort:

First, dispensing with international law carries some unwanted side effects. Does this clean break, for instance, involve dropping the Geneva Conventions? Or the rules and norms against wars of aggression? Which human rights is Mars planning to do without? Of course, Martians *could* choose to re-create hard-won international behavioral norms, but then if they're not producing some kind of wild new insight into human morality, why are we suspending international law for them in particular?

Second, relying on awe is probably a bad long-term bet. About six months after the first Apollo landing, network television wouldn't carry

Moon landings live. Why? Because while the fact of space landing is objectively interesting, doing stuff on the Moon is not. It's watching people set up lab apparatuses while talking in technical jargon, only in slow motion due to low gravity and bulky suits. No doubt a Mars settlement would blow some minds, but enough to make leaders around the world agree to forever bind themselves while setting a Mars settlement free from international law? If we're in such a gleamingly altruistic world, it's not really clear that we need a visionary Mars settlement to make improvements. And let's bear in mind that this new nation isn't summoned up from nothing. It's made by people with their own cultural norms and allegiances. They may be a consortium of nations, but suppose they're just Americans or just Chinese people? How long would worldwide awe last then?

Third, and perhaps most important, if Martians can't maintain Earth support after independence, they will probably die. A life-support specialist we spoke to told us that the ISS experiences a breakdown of something once every four or five months that requires new equipment to be sent up for repairs. Examples include the toilet and the Carbon Dioxide Removal Assembly—both things you'd really like to keep in tip-top shape. The number of Martians and Martian industries needed to probably *not* die is a matter we'll discuss later, but suffice it to say that a declaration of independence is a long cry from actual, functional independence. Even if we believed cutting the cord with Earth were desirable or legally allowed, it's not technologically possible, and won't be anytime soon.

Loophole 2: The Moon Agreement Reverse Precedent Switcharoo

Alan Wasser, who was chairman of the National Space Society Executive Committee in the early 1990s, proposed the Space Settlement Prize Act, hoping the US Congress would pick it up. They have not.

The act is deeply Lockean, saying in effect that if a private entity can build a space settlement complete with permanent inhabitants and transportation back and forth to any paying passenger, they should be allowed to own land. He even specifies the size of the estate—if you're on Mars you can claim 3,600,000 square miles (about 9,300,000 square kilometers)—a bit shy of one China of land. On the comparatively easy-to-settle Moon, you are limited to 600,000 square miles (about 1,500,000 square kilometers)—a mere 2.25 Texases.

Under Wasser's framework, the second claimant to a celestial body can only take up to 15 percent as much, and the third claimant up to 15 percent of that, and so on. Pretty quickly, you can barely claim a ranch the size of Spain, so there's a pretty strong first-mover incentive here.

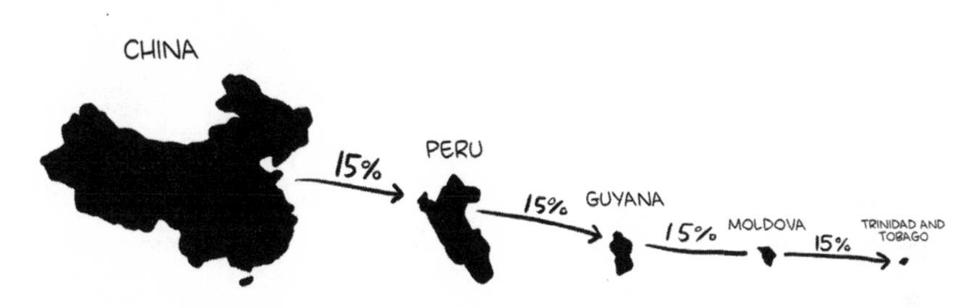

What's the basis for these claims? Natural law, which according to Wasser means that "individuals mix their labor with the soil and create property rights independent of government. Government merely recognizes those rights." In other words, reality is fundamentally Lockean.

We don't intend to take a position in this book on where property rights originate, but we do hold the firm view that property law exists. Whether property rights come from nature, the state, or Alan Wasser's personal blessing, the above interpretation straightforwardly breaks Article II of the OST.

Not so fast, says Wasser. He notes that in the failed Moon Agreement there was language that clearly forbade private ownership of Moon land. This implies that the OST must have allowed private property after all.

To illustrate the logic, here's an analogy. Imagine a man and woman sign a marriage contract, which includes a clause saying "no cheating." Later the man, in a fit of paranoia, asks the woman to sign an addendum stating that there is to be "no cheating in particular with Sexy Dave who lives next door." For various reasons, the second agreement isn't immediately signed. A few years later, the woman concludes that, wait a minute, the original marriage contract must *not* have forbidden a next-door tryst, because otherwise the second agreement wouldn't have brought it up.

Separate from being prima facie ludicrous, there are more technical problems with this "loophole." For starters, much of the Moon Agreement isn't trying to do anything new, but simply clarifying some vague terms. For example, it specifies that nations are allowed to take samples from the Moon. Presumably, Wasser and fellow travelers wouldn't therefore argue that the OST did *not* permit samples.

Second, note that Wasser is not making the clearly bogus claim that *nations* can appropriate land. He's only saying that *people* can. Under Wasser's understanding, the United States can't claim land, but Bob the American, bound by US law, protected by US force, enthusiastic about American cheese, *can*. This is simply not true. Bob is a US citizen and the United States is liable for him. Bob can say all day long that his right to a Moon estate is a matter of natural law, but unless the United States is ready to set the OST on fire, Bob has no title to Moon land. He's just a guy saying stuff.

Even supposing Wasser were offering a plausible interpretation of what the OST actually says, it's a geopolitical nonstarter. Do we really believe that every other nation will stand by while a million Americans claim the

Moon, so long as they insist the claim isn't technically America? About as much as we expect the man in our above analogy to congratulate his wife on her very clever loophole.

Wasser has once again anticipated the problem, noting that hey, it doesn't matter because as long as the world's most powerful nation—the United States—recognizes your land title, you're good. This is, let's say, modestly presumptuous about the continued power status of the United States. But even if we were to grant perpetual American hegemony, well, going back to our marriage analogy, it would be like the wife saying "I don't care if you don't like my loophole, Sexy Dave and I are in love." This would, so to speak, work. But at that point the marriage is in a crisis. Change husband and wife to nuclear powers and you can see why this would concern us.

Loophole 3: The Multilateral Menage

At the 2019 International Astronautical Conference, Rand Simberg gave a paper noting that hey, the OST only forbids *nations* from claiming outer space. It says nothing about a *multilateral agreement* between multiple nations willing to recognize property claims in space. It's not "national" appropriation if the appropriators are *multi*national, right?

This is on its face a bit bizarre. It'd be like a priest noting that the Ten Commandments say "Thou shalt not kill," but remain silent on the question of whether "*Y'all* shall not kill," rendering killing totally cool as long as two people pull the trigger.

It's also a nonstarter for a reason that's familiar now—eventually actual people from somewhere have to go to the actual Moon settlement. Even if the United States and Luxembourg decide to go in on a loophole-based Moon grab by creating the Luxamerican Alliance, eventually Americans or Luxembourgers have to physically arrive at the settlement, at which point the United States and Luxembourg are responsible for them. Again, this violates both the sense and spirit of OST.*

*This attempt to violate the spirit of the law happens in other commons as well. For example, take Travis McHenry, also known as the Grand Duke of Westarctica, who claimed part of Antarctica after cleverly

I Got Holes in Different Areology Codes

It's easy to make fun of these legal theories, which is why we did so, but we hasten to note that we don't think these advocates are stupid. More likely, these theories and similar ones are tactical—they are an attempt to shift the discourse. In 1967, whether a nation could take and keep Moon rocks was an open question. Five years later, two nations had done it and had not made a policy of sharing. Thus precedents are born. A purported loophole, repeated often enough that people in power come to believe it, might meaningfully shape the future, especially if it finds its way to the right politician or businessman.

Consider that, as part of his 2012 US presidential campaign, Newt Gingrich promised a Moon colony within eight years.* This wasn't an offhand remark—Gingrich is a space geek. Back in 1981 as a new congressman, he had proposed a "Northwest Ordinance for space," which said that for Moon settlements, "Whenever any such community shall have as many inhabitants as shall then be in any one of the least numerous of the United States such community shall be admitted as a State into the Congress of the United States on an equal footing with the original States."

This probably wasn't technologically possible in either 1981 or 2012, even with vast spending. The least populous US state is Wyoming, with its six hundred thousand people. It's also hard to know whether a *President* Gingrich would have risked the worldwide backlash even if lunar statehood was a live option. But it's not unimaginable that something like the above "loopholes" could have been invoked to justify a land grab, while the people in authority insisted it was perfectly legal. Long-standing safety zone–style claims might even make it feel reasonable to a lot of Americans. Geopolitically, we don't know what lies beyond that.

---

noting that the Antarctic Treaty System prohibits *nations* from making new claims, but says nothing about *individuals*. As you may have guessed, the international community has not recognized this claim.

*To which Mitt Romney replied at a debate, "I spent twenty-five years in business. If I had a business executive come to me and say they wanted to spend a few hundred billion dollars to put a colony on the Moon, I'd say 'you're fired.'" Stephanie Condon, "Romney Tells Gingrich: I'd Fire You for Your Moon Proposal," CBS News, January 27, 2012, https://www.cbsnews.com/news/romney-tells-gingrich-id-fire-you-for-your-moon-proposal/.

Approach 2: Amending We Will Go

Other advocates are a bit more cautious. They share the view that the OST has probably slowed space settlement, but they don't in general favor a full on Lockean free-for-all. However, they also don't want the large international bureaucracy we favor. What's the alternative? Amending Article II of the OST to allow possession, but simultaneously creating rules and institutions to protect other values, like equity or environmentalism. Here, we offer a few examples.

Amendment 1: First Possession Plus Tax

What if instead of a Lockean free-for-all we had a Lockean fee-for-use? This is a sort of hybrid between the "common heritage of mankind" concept with a Lockean theory of property: you can grab land by working it, but if you benefit from that land, you gotta pay the rest of humanity.

One virtue of this Mars Tax approach is that it would limit settlers' claims to what land they could use constructively—a goal that John Locke himself would have applauded. As settlers claim more land, they compensate humanity by paying higher taxes back to Earth. One downside here is that if you really do get a permanent Martian settlement, you've got a situation where Martians pay a sort of "existence tax" to the home world. We're skeptical of the ability of a Mars settlement to prosecute a revolution against Earth anytime remotely soon, but if we were trying to spark one, this would be a solid strategy.

Amendment 2: Limited Possession

Another proposal would allow the full-on Lockean free-for-all, but would specify that *some* good land must be set aside in reserve for future claimants. This seems like a reasonable enough method, though figuring out exactly how to slice up space will be a fraught political challenge.

History tells us that whatever agreement is reached will likely favor the powerful. The scholars who make these proposals are reasonably generous with the size of domain they would reserve for latecomers, but this may not be what happens in reality. Consider that something like limited possession is already done for the valuable geosynchronous orbit.

Geosynchronous orbit has around eighteen hundred slots available for satellites. Under current international law, each country has precisely one slot reserved to it. If each country gets an equivalent hunk of a place like the Moon, that's something approximately a square of 144 kilometers on each side. Whether that's a good deal for all participants really depends on exactly what areas of the Moon get reserved to them.

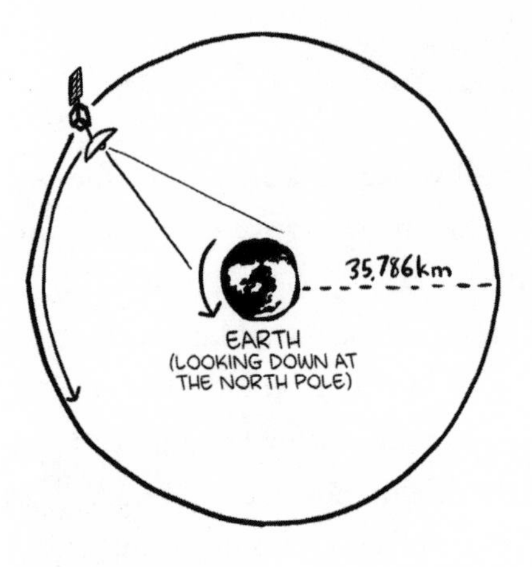

GEOSYNCHRONOUS ORBIT IS 35,786 KILOMETERS HIGH AND IN THE SAME PLANE AS EARTH'S EQUATOR. SATELLITES PLACED THERE MATCH EARTH'S ROTATION, SO THEY KEEP THE SAME POSITION WITH RESPECT TO EARTH SURFACE, EFFECTIVELY ACTING LIKE GIANT OBSERVATION TOWERS.

Amendment 3: Bounded Possession

A related idea is to allow people to just call dibs on land, but within limits, similar to the Homestead Act of 1862. In some proposals, this would mean that a given homesteader or group can claim a circle with a 100-kilometer radius—a rather roomy 7.8 million acres. If at some point their population outgrows the designated circle, they might be allowed to grab more. Thus, possession is incentivized without allowing small groups to claim whole planets at once.

One big issue here is that because not all parts of the Moon or Mars are equally desirable, you've got a pretty huge first-mover advantage, which means this framework would vastly privilege the rich spacefaring nations. Shackleton crater, with its Peaks of Eternal Light and cold-trapped water deposits, is only about 20 kilometers across. Anecdotally, we've seen space geeks note that you could tactically set up lots of small communities, each 100 kilometers apart, thus claiming massive territories despite having only modest populations.

Amendment 4: Possession Plus Parks

For those who see the above frameworks as still a bit too exploitative, there is another option: partially retain the OST, but only to govern certain regions. In a parks framework, some land is held in a way similar to nature reserves, while the rest will be exploitable. Exactly how much land becomes parkland varies by the presenter, with one paper arguing that as much as seven eighths of space should be reserved. Advocates for this direction for space property favor not just setting aside large pieces of outer space, but also making sure those parks include especially beautiful or scientifically interesting regions, such as Olympus Mons, the solar system's tallest mountain.

For a framework like this, the questions are how big the parks are, who decides where they go, and how exploitable is the exploitable part. If for instance, only one eighth of space is Lockean, that's quite restrictive, but still substantially less restrictive than the OST.

Amending We Won't Go?

The biggest hurdle to amendments like these is that you have to get wide international approval for a new framework that likely only benefits a few nations directly. Advocates can argue all day long that the benefits would accrue to humanity in general over the long term, but if you want to get an amendment, you need signatories today.

At the modern UN there are a lot of players and very few of them are rich nations with the ability to launch interplanetary vehicles. They're probably not going to sign on to unleash the small number of space powers for a land grab unless it involves tremendous generosity to the nations that don't go. That being the case, there is one more option.

Approach 3: DIBS!

The other framework is *no framework*. Take the OST, shove it out the airlock, and watch the imperfect creation of thousands of hopeful diplomats burn to cinders in the atmosphere below. Privatize space, straight up, right now.

Wholesale scrapping the OST is not the most popular view, but there are fans of it. However, even if it could be done, it's only the first step in a process. Unless we believe in a natural-law version of Lockean space land, where stateless settlers simply grab, work, and defend land, there needs to be a formal process. There needs to be some entity that lots of humans believe has the power to grant title. This could be some sort of Earth-level Bureau of Land Management, registering land title and telling people to get a few cows to keep up appearances. But the type of people who favor a "DIBS! BY GOD DIBS!" framework don't necessarily want a giant bureaucracy to replace the OST. The likely alternative is nations claiming chunks of space and then handling rights for individuals. But which nation gets which chunk, and how?

One proposal suggests land could be apportioned to each country in proportion to its population or land mass. Thus, Mars becomes a sort of mini-Earth. Kind of like EPCOT Center, only the world is made of poison and if you see a giant talking mouse it's likely because a biomedical experiment went terribly wrong. It's hard to see this sort of thing going down terribly well at a diplomatic conference, where we suspect Russia and Canada would just happen to feel that land mass was really the deciding factor, whereas China and India would find population much more compelling, with the United States insisting on the importance of national wealth.

Even in the unlikely case in which a framework like this got established, quantity isn't the only thing at stake. The United States has about 4.25 percent of Earthlings and about 6 percent of Earth land. Under either framework, that's more than enough to claim most of the premium real estate on the Moon.

Even if nations don't want to do this stuff, could we get dragged along by individuals? One of the most common arguments we hear is that all law is pointless because if Elon Musk has a Mars settlement, *who's going to stop him?* One of your authors has a brother who makes this argument. His name is Marty and *he is wrong*. We call this sort of thing the "treehouse theory of space law." Imagine this: a bunch of kids build a treehouse up a tall tree. The ladder is too light for an adult to climb, so once the kids are up there, they can't be reached. Their father comes out and says, "Kids! I made dinner! Come inside!" The kids smirk confidently and reply, "Never! We do not obey your Dad Law in the treehouse! You will try to oppress us in vain!" In this scenario, what do you expect the dad to do? That's right—stop supplying the kids with food and water, seize all of their stuff, and patiently await their change of heart.

As one of our readers pointed out, this assumes the kids aren't getting fed on the sly by Mom. That is, even if the international community tries to cut Elon Musk off, he can get along fine in this scenario if he's still trading with the United States. This is, of course, true, but under current space law, effectively the United States would just be asserting a bunch of illegal claims, at which point we're in a proper crisis.

Getting from Here to There

If you favor a private-property regime, one strategy might just be to oppose any update to the law and await a crisis. As Dr. Ram Jakhu, one of the world's most prominent space-law scholars, and his coauthors wrote in the 2016 book *Space Mining and Its Regulation*, "The tendency in almost any institution that creates new laws, conventions and especially international treaties is: 'Let's wait until there is a clear problem to be addressed and then we will address it.'" This was meant as a warning, not advice, but we could certainly give it a try. We could wait for SpaceX to declare Martian freedom, or for President Gingrich to welcome the state of Moonsylvania* into the US fold, and suddenly many of the crazy ideas above might look eminently sensible.

Part of why we favor a managed commons is that we think there's at least some chance of achieving it. It happened with Sea Law and plausibly it can happen again. Even if we don't get a perfect regime, simply clarifying the rules so as to avoid a new Moon Race would be desirable.

We see two good arguments against our view. First, if there's really valuable stuff in space, a highly restrictive mineral rights regime would be stealing from humanity's future. That might still be justifiable if it stopped war, but it would at least be debatable. Second, there are philosophical cases. We considered some of these in the introduction, but many people of many political backgrounds see space as providing a chance for a new start. For them, even if international law permits a lot of behaviors, that doesn't answer to the deep need to start new countries and new ways of being.

We think this is unlikely to happen anytime soon because it deeply violates legal custom. However, states do get created on Earth. It's rare. It's often violent. But it does happen. This, then, is the last question we want to know about space law: Are there circumstances where a group of people could violate international law, create a sovereign state in space, and get away with it?

*That was the cleverest state name we could come up with. Alternatives included North Dacrater, South Caroluna, and most regrettably, Regolithithippi.

16.

The Birth of Space-States: Like the Birth of Space Babies, but Messier

In 2016, the space-state of Asgardia was born. Or, at least, that's what the Asgardians say. Named for Asgard, a land of gods in Norse mythology, Asgardia is a remarkably well-organized internet-based entity that at least behaves in a state-ish way. They have a constitution, a parliament, a Court of Asgardia, and they release decrees graced with the level of boringness normally reserved for actual state bureaucrats. For example, "Decree 60 on Acceptance of the Asgardia Government Act," which informs you as to an act of Parliament that occurred on the 8–10 of Capricornus, in the year 0005, better known to you and me as 10–12 December, 2021.*

Are they for real? They've got real money, anyway. Asgardia was started by wealthy Russian-Azeri rocket scientist Dr. Igor Ashurbeyli. In 2017, they rocket-launched what they claimed was a sovereign territory, called Asgardia-1. Space stations tend to be snug, but Asgardia sets a new standard, being an entire country encased in a hard drive about the size of a human head. According to Ashurbeyli, unlike other space-settlement

*The Asgardians have their own calendar because, as they note on their website, the world is already bedeviled by forty different ways to keep time—a problem they seem to think will be solved by creating a forty-first way.

plans, Asgardia is serious business: "[It's] not a fantasy. Going to Mars, the galactics, so on—that's just fake. I intend something more real."

A question presents itself. Well, actually a lot of questions, but the one we'll focus on is why these people would go to so much trouble and expense to be able to claim they're a state and not just an internet club with a microsatellite.

Part of this is philosophical. Earth-independence is, for many, the primary goal of space settlement. As president of the Mars Society, Dr. Robert Zubrin, recently said in a Reddit AMA, "the purpose of going into space is to create new nations." That philosophy explains the desire for space-states, but why does that desire find itself embodied in a three-kilogram piece of hardware in low Earth orbit? It turns out that the word "state" has meaning in International Law, and the Asgardians are attempting, in an admittedly weird way, to meet the relevant standards. Those standards are where our journey into space-state creation begins.

What Is a State and Do Three-Kilogram Satellites Count?

However futuristic the Asgardians see themselves, they are bound by something called the Montevideo Convention of 1933—largely unknown among regular people, but well studied by legal scholars and every crank who ever wanted to start their own country. The basic deal is that in 1933 a meeting was held in Montevideo among states in the Americas, and it resulted in what are now broadly considered customary criteria for what counts as a state. The important part for our purposes is Article 1:

> *The State as a person of international law should possess the following qualifications: (a) a permanent population; (b) a defined territory; (c) government; and (d) capacity to enter into relations with the other States.*

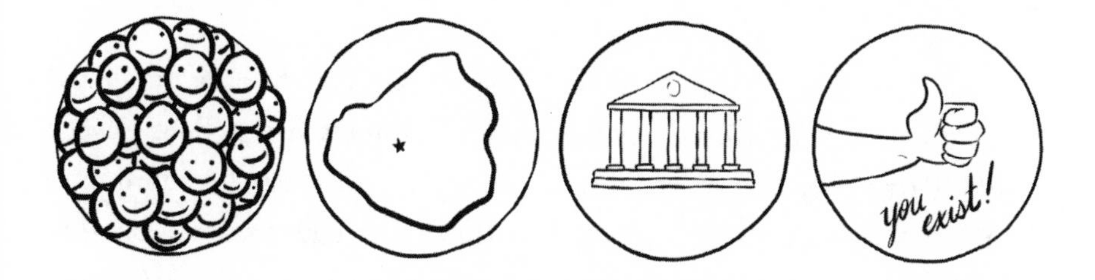

How's Asgardia doing? They in some sense have a permanent "population" in that they claim about three hundred thousand members and over a million followers as of 2022. That said, those people don't live in the shoebox-size territory currently circling the Earth. In international law, to count as population you have to be inside of a country that meets the Montevideo criteria. Otherwise, things could get really weird. If membership alone were the standard for "a permanent population," then the Taylor Swift Fan Club is the China to Asgardia's Liechtenstein.

Criterion (b)—the need for defined territory—explains the orbital hard drive. Admittedly three kilograms is a bit below expectation for space empires, but under international law there appears to be no lower bound on territorial size for legitimate states. Vatican City, the tiniest state, is a mere 0.4 square kilometers, smaller than the parking lot at Disney World's Magic Kingdom. However, "territory" classically refers to a defined piece of Earth, not a hard drive.*

As for criterion (c)—government—the Asgardians do have organized bodies that make decisions, but then again in some sense so does Disney World. Part of why you don't consider Disney World a state (in addition to their lack of sovereign territory) is that they don't really enforce binding rules. If you punch Mickey Mouse in the face, it's the local government you fall afoul of, not the Happiest Police on Earth™. We talked to one of the preeminent space-law scholars, Dr. Frans von der Dunk about this, and he pointed out that Disney World likely has *more* ability to sanction its "population" than Asgardia does for Asgardians.

Although the Montevideo Convention just says you need to have "government," the standard criterion is what's called "effective government." If there's a territory that is actively chaotic and where a nominal government can't enforce its rules, that's not effective government. You can get exceptions when geopolitics comes into play, like Taiwan, which clearly has effective government but is not widely recognized as a state. At other times, states have been recognized before any effective government was assembled. But neither traditional criteria nor any earthly squabbles point to Asgardia being considered a state anytime soon.

*In fairness, the Asgardians are planning to expand eventually.

Which leads us to the severe deficiency of Asgardia on criterion (d)—capacity to enter into relations with the other states. Disney World cannot engage in diplomacy with Germany because Disney World is not widely recognized as a state. Asgardia is in the same unfortunate position and, as we'll see, will likely remain so for a long time.

If your goal is just settling space and declaring the existence of a state, you could argue that all this technical stuff about space is irrelevant. You might say Asgardia can just claim to be a state while the proper Montevidean countries of Earth roll their collective eyes. And you wouldn't be entirely wrong. There's a long-standing debate in the international law literature about whether a state is something that requires recognition to exist, or whether the term "state" is just a description of reality. In other words, if the lost continent of Atlantis were rediscovered and the Taylor Swift Fan Club claimed it before any international agreement could be made, they might not be generally recognized as a legitimate state, but they could still behave like a state as usually conceived. They would have population, territory, governance, and a very popular Instagram account. All they would lack is "capacity to enter into relations with the other states," except maybe Asgardia, whom they could, as Ms. Swift says, "Love you to the Moon and to Saturn."

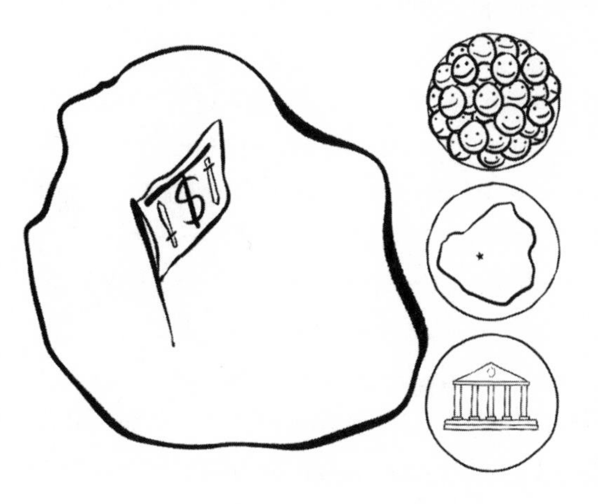

This use of "state" as largely descriptive may have been more useful in the past, but today recognition by other states is pretty valuable. Recognition carries serious perks—like a seat at the UN General Assembly, and the right to conduct diplomacy on an equal footing with other nations. You also get the protection of international norms for states, which includes some pretty swanky stuff, like the norm against having your territory taken by force.

Given how hard it will be just to survive off-world, access to all these things may be crucial for survival. Perhaps this is why Asgardians have a long-term plan to get recognition and a UN seat. There are some obvious issues here, including the unlikelihood of the world setting a new standard where an orbital hard drive attached to a geek club gets a UN vote. But the deeper problem they'll face if they want to create a space-state and get international legitimacy is that, to be blunt, the OST says *no*. Montevideo says you gotta have territory. OST says you can't have territory in space. So here's the math: Montevideo + OST = no space-states.

Probably.

The truth is there are ways to a legitimate space-state, and some are better than others.

First of all, the law could change in the future. During the next century, the nations of Earth may come together to determine criteria to allow people living in space to create their own legally independent states. For reasons discussed already, we think this will be hard, but it's at least possible, especially if widespread space access precipitates some kind of international crisis.

Less plausibly, there are the usual loophole workaround proposals. Asgardia, for example, plans to become a recognized nation, join the UN, *not* sign the OST, then launch their settlement from an Earth-based country that also didn't sign the OST.* However, space stations are not equal to Earth territory legally. Again, under classical international law, they would not meet the relevant criteria.

Even if they somehow surmounted that formal barrier, there's a practical problem here. We sincerely doubt the UN is going to offer a seat at the table to a 0.1-square-meter hard drive owned by an internet-based parliament whose stated goals include routing around widely accepted international treaties.

*Or at least that is our understanding of their plan. Kelly wrote an email to Asgardia's head of administration to try to confirm this, but alas, no response was forthcoming.

Other plans call for a state that already has UN status to just pull out of OST to allow for the creation of independent states. Exactly *why* a state with launch capacity would pull out of long-standing agreements just so someone else could make a rival state is, let's say, unclear to us. Even if they did, if only one country backs out, everyone else could argue the OST is customary law that they are not allowed to break. If a weak state tries it, they'll get slapped down. If a powerful state or consortium of states try it, well, it might work, but you're really just shredding the OST and are now in a proper crisis.

There are more ideas like these, but as we argued earlier, loopholes don't work so well in practice and amendments are hard to come by. Meanwhile, the law we have seems to forbid the creation of states in space, and the difficulty of space settlement suggests that the players most likely to found settlements are preexisting large states.

So, where are we gonna get new space nations? After all of what we've written so far, you might expect us to just say there's no path forward. That's not quite right. States are founded on Earth from time to time, and the way that's happened during the past century or so is our best way to understand how a legitimate new state might be created in outer space.

Let's Have the Talk About Where Baby States Come From

Under modern international law, one of the deepest norms is "territorial integrity." Simply put: states aren't supposed to lose territory by force. This is part of why when Russia recently attempted to take additional territory from Ukraine, there was a worldwide freakout. Territorial integrity is one of the main norms of international law, so just for emphasis, here's a drawing of it as a giant pillar.

Yet during the past century, more than one hundred new states have been born. Legally speaking, where did they come from? The answer, which may one day matter deeply for space, is that territorial integrity isn't the only major norm.

If you go back a hundred years, there weren't that many states. There were, as today, many "peoples," meaning identity groups like Jews, Zulu people, Japanese people, Roma people, Cherokee people, Persians, and so on, but at the top level of governance, much of the world was ruled by a small number of empires possessing vast colonial holdings. This is no longer true.

During the twentieth century there were three major state-creation periods. First, the collapse of the "Central Empires" in the wake of World War I. Second, the rapid decolonization of empires that mostly took place between 1945 and 1975. Third, there was the liberation of states controlled by the Soviet Union, which dissolved between 1988 and 1991. The modern norm that has evolved over those years is that distinct groups of people have a right to "self-determination." Loosely speaking, this means a right of distinct peoples to a say in their government. That's the second giant pillar to concern yourself with in this discussion.

States have a right to stay whole, but distinct peoples have a right not to be oppressed. When these rights are put in tension, sometimes states are born, and in some cases they have gotten full international recognition.

From today's vantage point, it's hard to imagine a major state-creation period in space because there are no empires to dissolve. In a very distant future with long-standing space communities, something like a decolonization event might be conceivable. For now, we want to look at special circumstances when states have been born without the collapse of empires.

So, our first step on the route to a Mars nation is . . . Canada?

Self-Determination and Your Mars Base; *L'Autodétermination et Votre Base Martienne*

The most clear explanation of the modern norm of self-determination comes from *Reference Re Secession of Quebec*, a ruling released by the Canadian Supreme Court in 1998.* The court was dealing with the nature of Canadian federalism, and for our purposes the most important question was whether or not Quebec could unilaterally secede according to the Canadian Constitution or international law.

*If you're wondering why a Canadian Supreme Court case matters for international law, the short version is that they did a really great job. They interviewed a bunch of the relevant experts, and, importantly, they were willing to address a set of questions about the overall legality of secession head on, while major courts usually try to avoid dealing with this question directly. The ruling was later invoked by various international groups, which is how it made its way from a domestic ruling to an international standard.

The judges ended up hearing the arguments of top international lawyers from around the world. There's a lot of nuance to their ruling, but as to the question of whether Quebec could unilaterally secede, the basic finding was that no, no they couldn't. Why? Because although the Quebecois are by general agreement a distinct people with a land, a language, their own customs, and really really good pea soup, they are not persecuted by the Canadian government. They are allowed to run for office, they can vote, and they're not being jailed or murdered or having their culture actively suppressed. There are no doubt Quebecois people who feel persecuted, but their situation is not comparable to, say, European Jews in the 1930s and '40s. The Quebecois right to self-determination exists in their access to an equal say in how they are governed.* If the Quebecois were facing genocide from their government, self-determination could be invoked to achieve what is sometimes called "remedial secession"—the national equivalent of leaving an abusive marriage.

So if your space settlement wants to secede from some Earth nation purely because it feels it has a unique identity, it won't work. By contrast, let's look at a case where secession was broadly accepted, but largely due to conditions you absolutely do not want to duplicate.

Bangladesh and the Power of Self-Determination

The original Pakistani state had East and West portions, separated by thousands of kilometers of Indian territory.

They weren't just separated physically. They were separated by longstanding differences in language and culture. Also in power. The government and military were concentrated in West Pakistan, and things were rocky between the East and West from the beginning.

*We're going for a big picture here, but there really is some serious nuance. For instance, the court noted that Quebec could become a state by first seceding illegally and then somehow gaining broad international recognition. They also said that if a clearly worded referendum showed that a clear majority of Quebec wanted to secede, the Canadian government would be obligated to negotiate a way to amend the Canadian Constitution to create a path for secession. This would be a profoundly messy process. Consider, for example, if Indigenous people in Quebec wanted to remain as part of Canada. Or consider budget issues—how much of the national debt goes with Quebec? And an obligation to negotiate doesn't necessarily mean that the secession amendment will end up in the Constitution. It just means that all sides need to participate in good-faith talks, guided by specific principles such as democracy and respect for minorities.

In the decades that followed, conditions deteriorated, reaching a low point after Cyclone Bhola resulted in the death of over two hundred thousand East Pakistanis in 1970. The West Pakistani government was widely seen as failing to address the crisis. The same year, East Pakistan voted in a political party called the Awami League, which favored greater local autonomy. West Pakistan responded by indefinitely suspending the assembly, then proceeding to what some scholars consider to have been genocide.

Note how, for horrific reasons, the table is set for state creation. You have a clear identity group. They have had their access to government thwarted. They are actively persecuted. Arguably at that point they had the right to remedial secession even though that would violate Pakistan's territorial integrity. The only question remaining was whether they had the power to seize their independence.

This became possible due to aid from India, which violated the norm of territorial integrity by sending their own army into East Pakistan. This is the birth of the nation now known as Bangladesh.

By 1974 Bangladesh was a widely recognized state with a seat at the UN. This was possible largely for two reasons: first, the modern norm of self-determination gave Bangladeshi independence legitimacy. Second, Pakistan was a weak state, meaning it had to accept the outcome.

Bangladesh's route is admittedly not a pretty way to statehood in space, but is a way for which the legal path is relatively clear. The norm of self-determination was so utterly violated that it was widely considered acceptable for the norm of territorial integrity to be violated in order to remedy things.

From a space-settlement perspective this is intriguing: a major norm was violated but the norm remains in place. A new nation is born, no new international law required.

But it wasn't easy either, and you shouldn't get the idea that any Mars settlement can simply decide they're oppressed and be allowed to secede. One of the most important cases of failure to create a recognized state comes from Northern Cyprus.* Despite breaking off from Cyprus and having its own government since 1983, Northern Cyprus has no UN seat, almost no recognition, and almost no ability to conduct relations with other states. They are in a position that would be extremely bad for any type of space settlement we can create anytime soon.

Northern Cyprus and the Limits of Self-Determination

Cyprus, a large island to the south of Turkey, was granted independence from the United Kingdom in 1960. Like other former colonies during the period of widespread decolonization, it was quickly given a UN seat and widely recognized as a state. But internally, things were complex.

At the time, about 80 percent of the island's population were ethnic

*Also known as the Turkish Republic of Northern Cyprus.

Greeks who tended to follow the Greek Orthodox Church, while nearly 20 percent were Turks who tended to be Sunni Muslim. Neither group was especially keen on straight-up independence as Cypriots. Generally speaking, Greeks would've preferred the island to join Greece; Turks would've preferred it to go to Turkey. In the messy way decolonization often went, Britain, Greece, and Turkey created Cyprus as its own state with a constitution that protected minority rights and which had rules potentially allowing the three parties to intercede if necessary.

The constitution had problems from the beginning. At different times, neither the Greeks nor Turkish residents in Cyprus were happy with the constitution they were handed down, and conditions deteriorated over time. In 1974, the ethnically Greek president of Cyprus was ousted by commanding officers of the Greek Cypriot National Guard, who put in place an "enosist"—a person who believes that predominantly Greek parts of the world should be part of the Greek state.

Turkey responded by sending in troops, and to make a very long story short, in 1983 the northern part of the island declared itself an independent state.

But the "Turkish Republic of Northern Cyprus" has not been widely seen as legitimate. In fact, they are recognized only by Turkey, and frankly that was a gimme. Meanwhile, many states recognize a single country called Cyprus, which includes the northern portion. So while

Northern Cyprus may have Montevideo-like qualities—territory, governance, population, and relations with at least one other state—they have no seat at the UN. They can conduct only very limited international diplomacy, and at least in principle, they are also not protected by international norms—all things you'll want if you're creating a vulnerable state off-world.

Why were they considered different from Bangladesh? Likely because their situation was not seen as rising to the same extreme. They were seen as more like Quebec than East Pakistan. Self-determination is a real international norm, but the experience of Northern Cyprus tells us that the standard for invoking it to take territory is quite high.

It may be even harder to invoke in a space settlement.

The Uncommon Problems of State Creation in Space

In order to qualify for self-determination and thus perhaps have a route to statehood, you must first be a people. Being a "people" may sound like a fuzzy concept, but in practice it usually is not. And it can happen pretty fast. Most people today, however they feel about the state of Israel, recognize that there exists an Israeli people, even though Israel was only created in 1948. Documents from at least the early 1950s refer to "the Israeli people" as having distinct cultural characteristics. It seems likely that any actual Mars settlers could pretty quickly be considered culturally distinct.

There'll be the use of taco sauce as currency, and the way they say "Now let evolution take its course" to celebrate newborn children.

Can they get a right of self-determination? History suggests that only very persecuted people get to do that, but early space settlements will likely be backed by, perhaps populated by, the very wealthy. Persecution supplies may be limited.

But let's remember that space is awful. If you live in a tiny survival bubble surrounded by death far from the home planet, any kind of restriction or deprivation could be taken as hostility. If the OST as we know it holds, our Martian settlers might genuinely lack some rights normally afforded to Earthlings. Consider, for example, a fourth-generation Martian in a world where the OST still holds. He grew up in a spome lovingly scrubbed of volatile chemicals by great grandpappy and handed down through the years from father to son. To him, this place is home. Yet under the OST, although the habitat is his, he does not have a right to keep his habitat in place permanently, nor can he own the land beneath it. He would also in principle have a number of weird obligations, such as having to allow representatives of other states to come into his home from time to time, especially if they were in danger. He could also argue that he lacks equal representation. Suppose the settlement is nominally under US governance, but none of the settlers have a serious chance of holding political office on the distant home planet.

It is of course also possible that in larger, longer-lived future settlements, you will get much more unambiguous forms of persecution. Would that lead to a holy grail for space privatization: an acceptable one-time territorial claim in space with the OST left largely intact?

In theory, but not easily. The path to remedial secession is geopolitically and legally fraught. On Earth, modern state creation typically happens when you cut out a chunk of an old country and make it into a new country. But space isn't a country. Under the OST, space is a giant international commons. Arguably, if a space settlement claims statehood, they aren't just taking from one country, they're taking from *every* country. Possibly every *human*.

That's a hell of a new precedent for international law. If you can just declare a state in the space commons, does it follow that you can create a state in the sea commons? Let's recall that Donald Rumsfeld opposed the United States signing onto modern sea law on the precise basis that it would set a precedent for space.

If this seems abstruse or hand-wavy or the kind of thing only diplomats think about, understand that countries really care about this stuff. The best example is the status of Kosovo, which will therefore be our last stop in this magical tour of geopolitical clusterfucks.

Kosovo and Why Self-Determination May Just Not Matter if the Geopolitics Doesn't Work

Starting in 1991, the former Soviet Federal Republic of Yugoslavia began to break into pieces. An often violent process resulted in a number of small independent states, largely drawn along ethnic lines, with familiar names like Croatia and Slovenia. Among them was Serbia, which had an internal, semiautonomous region called Kosovo.

In a setup that will now sound familiar, Kosovo was a mix of two ethnic groups: Serbs and Albanians. But the first president of Serbia was an ethnonationalist Serb named Slobodan Milošević, who promptly stripped Kosovo of its autonomy upon his election in 1990. By 1996, there was a strong Albanian separatist movement called the Kosovo Liberation Army, in active conflict with local authorities. Under Milošević, Serbia initiated a process of ethnic cleansing against Albanian Kosovars that was only stopped by the military forces of the North Atlantic Treaty Organization in 1999. The awkward peace solution was to have the UN itself administrate Kosovo for a while.

In 2008, Kosovo declared its independence and was immediately recognized by the United States and a number of European countries. But it has never received a UN seat because many major international players refuse to recognize it.

Of especial importance for us are the actions of three states: Greece, China, and Spain. Greece opposes Kosovar recognition in part because of their position on Northern Cyprus. If the Kosovars get recognition, might Northern Cyprus be next? China has similar reservations pertaining to the status of Taiwan. Spain's concern is with the Catalonian independence movement. Note well—the motivations for state behaviors are not directly about Kosovo at all.

It is unlikely that the political status of a small group of Albanians has any direct implications for the Chinese economy or military strength, and we suspect it's not a top issue for the average Chinese citizen. Rather, the concern is what the precedent set by recognizing Kosovo would mean for local affairs.

The case of independence for Kosovo then is not simply down to whether Albanian Kosovars were persecuted; they clearly were. But their access to remedial secession and to getting widespread recognition exists in tension with the desires of more powerful states not to have to recognize secessionist movements in their backyards.

The same will likely hold for the relationship between Earth and space. Even in a situation where under regular international norms the Martians have a good case for remedial secession, there will be questions of precedent. Compounding the problem is that, in practice, any space settlement will have particular geopolitical characteristics. Space-settlement enthusiasts often talk about new space-states as starting with a blank slate. There's no reason to expect this, and in fact the specifics will matter quite a bit. If the new country has governance and culture that align with Russia, that's geopolitically different from having a culture that aligns with the United States. Even if all Earth nations were willing to sign off on a space-state in general, many would likely oppose it in particular. Just as with Spain and Kosovo and Catalonia, however much the creation of a new entity is acceptable under international norms, there may be realpolitik reasons why individual powerful nations say no.

The Awful Awful Path to Space Nations

Where does this leave our Asgardians? Not in a great position. If routing around the OST fails, they have a path to statehood, but it comes with obstacles, one of which is, so to speak, the need to be persecuted. Even that may not be enough, because the most likely case is that at least some nations of Earth won't want to set new precedent for the commons. There is a path to space-states of the kind Dr. Robert Zubrin and Elon Musk contemplate, but getting there will require a master politician, a series of unlikely events, or most likely both.

Part of why we want to discuss state creation here is because people often just assume there's a clear path forward. There is not. You either need to destroy the OST, to get unlikely changes made to the OST, or to get through the legal and geopolitical obstacle course we described above.

Dare we say it, this is one more place where wait-and-go-big may be the best way forward. There is no reason to rush into space-statehood. You can do all the science and exploration you want already. With a more fleshed-out international legal regime, this could be accomplished peacefully while allowing some amount of resource exploitation.

We think the ideal scenario if you have to have space nations would be to achieve Earth-wide consent to a new state in space. There are still dangers to this, which we'll explore at the end of the book, but at least you wouldn't have a scramble between existing nations. As we've seen, whether a functional state gets to be a recognized state has a lot to do with the possible effects on every state that already exists. That means if you want an independent space nation, you've got to have something like a harmonious Earth. Given how long it is likely to be until large Mars settlements are possible, pursuing a regime that avoids conflict is probably better than trying to cram through space nations as fast as possible.

We're oppressed by Earth and must be free but please keep sending funding and life support!
ASTRIDALIA

Nota Bene

VIOLENCE IN ANTARCTICA, OR, HAPPY ENDINGS TO STABBY STARTS

In an earlier version of this book we had an extensive section on weird incidents in polar bases. We ended up cutting this, mostly because incidents like this are rare and providing the sometimes literally gory details felt gratuitous. Also, as with space, there's plenty of published work that doesn't seem to be conveying the truth. One story widely shared, including in the scholarly literature, goes something like this: In 1959 at Vostok Station, two Russians had a fight over a game of chess. One attacked the other with an ice ax, resulting in a permanent ban on chess at the station.

There are a few reasons to get your skeptic radar up here, friends. First, the people are unnamed. Second, in some tellings the ax results in death and in some merely in injury. Third, the story is a classic stereotype of Russians as both intellectual and brutal. Fourth, although admittedly we have never had to run an Antarctican base . . . our policy response to an ax attack wouldn't involve the question, "What board game were they playing?"

We confess, we believed this story when we first read it. When we looked for primary sources, however, we found a closed loop of chatter that never seemed to involve actual named ax-wielding Russians. We asked a Russian friend if he could find any reference to this stuff in his native language, and in his reply email, you could practically hear the plaintive sigh over what ends up in the Anglophone media.

But hey, you never know. So Kelly wrote Dr. Vladimir Papitashvili at the National Science Foundation. He very kindly wrote back instead of just turning off his computer and shaking his head for a while. Here's an excerpt from the email:

> *To my knowledge, the story you have referenced is not true. I worked with the Soviet Antarctic Expedition since 1970s . . . and I visited Vostok Station first time in 1983. The only death was registered at the Vostok Station for its entire history since 1957: a mechanic died during the power plant fire on April 1982.*

This doesn't definitively prove the story false, but you'd think someone who spent decades with Soviet polar scientists and hung out at the relevant station would have at least once caught wind of the ax-wielding chess players and the tight board-game restrictions.

To be clear, if you read enough about life at the poles, you can find stories that really are well documented. For instance, the raisin wine killing or that time two Russians really did have a violent altercation, no chess required.

In Vino Violence

Mario Escamilla had had it with Porky. Escamilla had been stuck in the Arctic with Donald Leavitt (aka "Porky") for over a month, and Porky was being a total dick. In particular, Porky was the kind of dick who would steal people's alcohol while wielding a butcher's cleaver. On July 16, 1970, Escamilla's roommate called him to say that a drunken Porky had stolen their wine. Escamilla grabbed a rifle and confronted Porky, whom he found drinking a cocktail of ethyl alcohol, grape juice, and raisin wine with the station manager (a man named Bennie Lightsey). Lightsey and Escamilla ended up having a heated conversation as to Porky's behavior vis-à-vis the raisin wine of other Arctic base residents, during which Escamilla began waving a rifle around. At some point in the process, the rifle went off, inadvertently killing Lightsey.

These events became somewhat famous in the law world, since there was a fair bit of debate over who had jurisdiction over a mobile piece of ice in the Arctic Ocean. The case was ultimately tried in the United States, where both men were citizens, and Escamilla was acquitted.

Stabbily Ever After

In 2018, one Russian overwinterer at Bellingshausen Station stabbed another overwinterer multiple times in the chest. Russian men are apparently superhuman, as the stabbee (who was quickly flown to Chile for medical attention) survived.

You may have heard about this story in the news. The original reports fixated on a weird detail, claiming that Sergei Savitsky stabbed Oleg Beloguzov because Beloguzov kept spoiling the endings to books that Savitsky was reading. Once again we have the Russian stereotype of frat-boy violence combined with intellectualism.

Later versions of this story backed away from the spoiler explanation, and noted that Beloguzov had a long history of insulting Savitsky during their time on the ice. This culminated in the day when Beloguzov told Savitsky that he ought to get on the table and dance for money. And that is when the stab wound to the chest enters the story.

So did the station ban jokes about table dancing? No. Stranger still, Savitsky seems to have pretty quickly realized that this was an overreaction and behaved appropriately from there on out. He admitted what he did, flew home without an escort, and went under house arrest. He was charged with attempted murder committed in a state of passion and cooperated completely. He was forgiven by Beloguzov, and because Savitsky didn't have a prior record and seemed repentant, criminal charges were dropped.

We won't speculate as to what you would, should, or could do in the equivalent situation in space, but we will say this: if we're talking about long-term settlements, *go big.* Consider how much of Earth was needed to handle that stabbing story—not just a transport system, but a legal system, jail system, and trauma medicine. It's hard to imagine a small space settlement having quite such a happy ending.

PART VI

To Plan B or Not to Plan B: Space Society, Expansion, and Existential Risk

SO FAR IN THIS BOOK WE'VE BEEN CONCERNED ABOUT QUESTIONS OF feasibility. Are space settlements biologically feasible? Technically feasible? Legally feasible? As we move into this final section, we're going to assume that by some means, settlements, perhaps even space nations, can exist so that we can ask population-level questions.

First, we'll look at the most proposed and perhaps most likely form of early social organization in space: company towns. There are a lot of ways this might go down, but we'll focus on the literature from historical Earth company towns as a window to the possible future of space company towns. Company Mars towns are, in our experience, the most frequently proposed model when you're chatting up space-settlement geeks. This may worry those of you who see the words "company town" as synonymous with mustache-twirling capitalist villains. While we'll argue the picture is more nuanced, it's not obviously a good one. Thinking carefully about it also provides a window into underconsidered dangers of trying to build an economy 225 million kilometers from home.

Second, we'll look at population issues if the goal is Earth independence. We've said all along that you have to go big, and this is our chance to talk about what that would entail. As we'll see, when it comes to biology we can give decent numbers. When it comes to economics, things get pretty hand-wavy. That said, the hands are decidedly waving at enormous numbers.

Finally, we come to a question we teed up way back in the introduction—war. Is war less likely in space? If not, is it apt to be especially dangerous? If we apply the standard analysis of war theorists rather than trying to reason from pop theories of why wars start, we find that, as in so many areas of human behavior in space, we should expect humans to behave as we do on Earth. If space wars are apt to be especially dangerous, that's a problem.

17.
There's No Labor Pool on Mars: Outer Space as a Company Town

When Elon Musk announced plans to create a new city in south Texas, called Starbase, a number of articles lamented the potential for a "company town." This was a frightening idea because to many people, company towns are synonymous with labor abuse. But in the case of Starbase in particular, the potential for massive corporate exploitation struck us as fairly low. The classic exploitative company town is run by a single company far away from civilization, whereas Starbase is about twenty minutes from Brownsville, a city of 180,000, which possesses at least three Starbuckses. Workers in Starbase wouldn't need to live in company housing or shop at Musk-Mart for everything. They would have other local job opportunities and a relatively easy time leaving the area to pursue better working conditions. While Starbase might formally be a company town, in the sense of being a town where almost everyone is employed by a single entity, the risk of some kind of corporate nightmare scenario struck us as low.

In thinking about company towns for space, details like this really matter. The truth is, company towns aren't inherently evil; often they're just fairly boring small towns oriented around mining or timber. Occasionally the corporate governance is quite well liked. Sometimes, attempts to roll back paternalistic corporate influence are actively resisted by local families. In other cases, like the notorious Battle of Blair Mountain, labor disputes have led to bombs getting dropped on civilians from airplanes.

The danger most relevant to space is that remote company towns tend to put a lot of power in the hands of the company, leaving employees vulnerable. If a company starts a mining operation far from civilization, they have to provide all sorts of services: housing, sanitation, shopping, and in some cases even churches or hospitals. This isn't an act of kindness—it's the basic stuff companies have to offer to get people to relocate to the middle of nowhere. The result, however, is that if workers want to strike or even just negotiate, they have to talk to a boss who is also their landlord, city council, health-care provider, and so on.

The fundamental structure of a company town creates a massive power imbalance between owners and employees that occurs even if we assume the operators aren't mustache-twirling villains. And if they are, things can get frightening indeed. No matter how bad an Appalachian coal town got, there was always a train ride out. In a Mars settlement, this will be far harder, and corporate ownership will control more than any Earth corporation ever did—not just housing and food, but the biosphere itself.

If we're going to have corporate towns somewhere in space, the question is how we steer away from the worst-case scenarios toward a boring town that can prosper, develop, and ultimately transition from corporate to popular governance, ideally without bombing its own citizens during the process. It turns out that there are at least some choices a corporate settlement could make that are likely to improve outcomes. There are also some situations that are unavoidable, and potentially quite bad, which should be planned for.

What Is a Company Town?

For our purposes, a company town is any settlement where the main employer is a single company. The classic examples most relevant to us are towns like Corner Brook in the forests of Newfoundland, which was set up in the early twentieth century. Why build a city more or less from scratch in the more ass-freezing regions of the planet? Because there was plentiful timber, excellent access to cheap hydropower, but no big labor pool available. A town with all the usual town amenities had to be built to get people to come work. Similar setups have been built around the world for activities like copper mining or oil drilling. Space settlements might work this way if some commodity really does turn out to be valuable. Otherwise, they'll have to be oriented around things like tourism, media, and research.

Not all company towns are privately owned. Consider the "monotowns" built by the USSR. Like Corner Brook, they were oriented around resources. One difference, which may end up relevant if space settlement is largely about government posturing, is that monotowns also served political purposes. Stalin originated them in part because he believed cities were getting crowded and run down, and he wanted city people to bring culture to the countryside. He also wanted to get Soviet industry to move eastward in case of a German attack, which was admittedly a pretty good call. Although during the Stalinist era, monotowns were often populated with prisoners of the Gulag system, after reforms in the 1960s these towns had to do the same thing corporate company towns did—supply housing and amenities to attract workers. Typical of these towns was Krasnokamensk. Like Corner Brook, it was founded in a harsh place—local herders literally referred to it as the "Valley of Death." Why set up a whole town in the Valley of Death? Valuable uranium deposits.

Another kind of company town is what you might call utopian. Places with names like Hershey, Pullman, and Fordlandia, named after Mr. Hershey, Mr. Pullman, and Mr. Ford, respectively, were all created to do profitable business while perfecting social relations. Perfection was hard to come by. Pullman and Fordlandia experienced violent strikes and ultimately failed. Hershey had its own issues, including what may be history's only dairyman-on-chocolatier brawl, but it is now a conventional small town, albeit with a lot of candy and a fancy theme park.

These utopian towns tended to inflict the particular idiosyncrasies of their founders on employees. Henry Ford, for example, created weirdly paternalist institutions. His factories had Departments of Sociology that monitored employees for behaviors he disapproved of, like drinking and adultery. They also had ceremonies in which immigrant workers would walk onto a stage wearing clothes from their homeland, then step into a stage melting pot, and emerge wearing American attire. In the overall history of company towns, utopian cities like these are mostly an interesting footnote, but they may have special relevance in space settlement, whose most prominent advocates have ample portions of idiosyncrasy and utopian aspirations.

An interesting element for anyone planning a settlement design is that there may be a strong temptation for social engineering. Recall from chapter 4 on spacefarer psychology the various proposals for managing human harmony by machine. We don't think these are a good way to manage humans, but there is a certain logic to it. A space habitat will be a fragile bubble of life in a hostile world. The more people you have, the more chances there are for someone to do something dangerous. You may think the idea of a Department of Sociology is just weird early-twentieth-century paternalism, but the same impulse for social perfection as a means of corporate efficiency may return in different form.

However, the "classic" company town issues are a bit more pedestrian and have to do with who owns what and why.

On the Care and Feeding of Space Employees

One of the first things to know about company towns is that companies don't appear to *want* to be in charge of housing. In our experience, people often think housing was an actively pursued control tactic, but if you look at the available data and the oral histories, companies often seem downright reluctant to supply housing at all. In Dr. Price Fishback's economic analysis of coal towns in early-twentieth-century Appalachia, *Soft Coal, Hard Choices*, he found that companies able to have a third party supply housing typically did. This is hard to square with the idea that housing was built specifically with sinister intentions.

There are also good theoretical reasons to explain why companies build housing and rent it out to workers. Suppose Elon Musk is building the space city Muskow. Having wisely consulted the nearest available Weinersmith, he decides he shouldn't own employee housing due to something or other about the risks of power imbalance. He looks to hire builders, but immediately runs into a problem: very few companies are available for construction on Mars. Let's consider the simple case where only one company is willing to do it.

Well, guess what. That company now has monopoly power. They can raise home prices or lower home quality, making Muskow less attractive to potential workers. Musk can now only improve the situation by paying workers more, costing him money while lining the pockets of the housing provider.

If he wants to avoid this, Musk's ideal option is to attract more building companies, so they can compete with each other. If that's not possible, as was often the case in remote company towns, then the only alternative is to build the housing himself. This works, but the tradeoff is that he's now managing housing in addition to focusing on his core business. He's also acquired a lot of control over his employees. None of this setup requires Musk to be a power-hungry bastard—all it requires is that he needs to attract workers to a place where there's zero competition for housing construction.

Historically, where things get more worrisome is in rental agreements, which often tied housing to employment. Even these can partially be explained as rational choices a non-evil bastard might non-evilly make. Workers in mines were often temporary. Mines were temporary, too, existing only until the resources were no longer profitable. This made homeownership a less compelling prospect for a worker. Why? Two reasons. First, if a town may suddenly fold in fifteen years because a copper mine stops being profitable, buying a house is a bad investment. Second, if you own a home, it's hard for you to leave. This is a problem because threatening to leave is a classic way to enhance your bargaining position as a worker.

Once you have people whose housing is tied to their job, the potential for abuse is enormous—especially during strikes. Rental agreements were often tied to employment, and so striking or even having an injury could mean the loss of your home. When your boss is also your landlord, their ability to threaten you and your family is tremendous, and indeed narrative accounts refer to eviction of families with children by force. If employees either owned their homes or had more secure rental agreements, power would have run the other way. They could have struck for better wages or conditions and occupied those homes to make it harder for their employer to bring in replacements.

It may be tempting to see this as a purely capitalist problem, but very similar results occurred in Soviet monotown housing. Employees tended to get reasonably nice company-town housing; if they lost their jobs, they had to go to the local Soviet, which provided far worse accommodations. As one author put it, "Thus, housing became the method of controlling workers par excellence." This suggests that there's a deep structural dynamic here—when your employer owns your housing, they're apt to use it against you at some point.

In space, you can't kick people out of their houses unless you're prepared to kill them or pay for a pricey trip home. On Mars, orbital mechanics may preclude the trip even if you're able to afford it. In arguing with space-settlement geeks, housing concerns are often set up as binaries—"Look, they're not going to kill the employees, so they'll have to treat them well." In fact, there's a spectrum of bastardry available. A company-town boss on Mars could provide lower-quality food, reduce floor space, restrict the flow of beet wine, deny you access to the pregnodrome. They could also tune your atmosphere. We found one account by a British submariner, in which he claimed to adjust the balance of oxygen to carbon dioxide depending on whether he wanted people more lethargic or more active. Whether it'll be worth the risk of pissing off employees who cost, at least, millions to deliver to the settlement is harder to say.

This overall logic—companies must supply amenities, therefore companies acquire power—repeats across contexts in company towns. To attract skilled employees who may have families, the company must supply housing, yes, but they also must supply other regular town stuff—shopping, entertainment, festivals, sanitation, roads, bridges, municipal planning, schools, temples, churches. When one company controls shopping, they set the prices and they know what you buy. When they control entertainment and worship, they have power over employee speech and behavior. When they control schools, they have power over what is taught. When they control the hospitals, they control who gets health care, and how much.

Even if the company does a decent job on all these fronts, there may still be resistance, basically because people don't love having so much of their lives controlled by one entity. Fishback argued that company towns, for all their issues, were not as bad as their reputation. In theorizing why, he suggested one problem you might call the omni-antagonist effect. Think about what groups you're most likely to be angry at during any given moment of adult life. Landlord? Home-repair company? Local stores? Utility companies? Your homeowners association? Local governance? Health-care service? Chances are you're mad at *someone* on this list even as you read this book. Now, imagine all are merged into a single entity that is also your boss.

In space, as usual, things are worse: the infrastructure and utility people aren't just keeping the toilet and electricity running; they're deciding how much CO_2 is in your air and controlling transportation in and out of town. Even if the company is not evil, it's going to be hard to keep good relations, even at the best of times.

And it will not always be the best of times.

When Company Towns Go Bad

Unionization Attempts

On September 3, 1921, reporting on the then ongoing miner's strike in West Virginia, the Associated Press released the following bulletin:

> *Sub district President Blizzard of the United Mine Workers . . . says five airplanes sent up from Logan county dropped bombs manufactured of gaspipe*

and high explosives over the miners' land, but that no one was injured. One of the bombs, he reports, fell between two women who were standing in a yard, but it failed to explode.

"Failed to explode" is better than the alternative, but well, it's the thought that counts.

Most strikes were not accompanied by attempted war crimes, but that particular strike, which was part of early-twentieth-century America's aptly named Coal Wars, happened during a situation associated with increased danger—unionization attempts.

Looked at in strictly economic terms, this isn't so surprising. From the company's perspective, beyond unionization lies a huge unknown. Formerly direct decisions will have to run through a new and potentially antagonistic committee. The company will have less flexibility about wages and layoffs in case of an economic downturn. They may become less competitive with a nonunion entity. They may have to renegotiate every single employee contract.

Whether or not a union would be good per se in a space settlement, given how costly and hazardous any kind of strife would be, you may want to *begin* your space settlement with some sort of collective bargaining entity purely to avoid a dangerous transition. A union would also reduce some of the power imbalance by giving workers the ability to act collectively in their own interest. However, this may not happen in reality if the major space capitalists of today are the space company-town bosses of the future—both Elon Musk and Jeff Bezos kept their companies ununionized while CEOs.

Economic Chaos

Another basic problem here is that company towns, being generally oriented around a single good, are extremely vulnerable to economic randomness. Several scholars have noted that company towns tend to be less prone to strife when they have fatter margins. It's no coincidence that the pipe-bomb incident above came about during a serious drop in the price of coal early in the twentieth century. Price drops and general bad economic conditions can mean renegotiations of contracts in an environment where the company fears for its survival. Things can get nasty.

If Muskow makes its money on tourism, it might lose out when Apple opens a slightly cooler Mars resort two lava tubes over. Or there could be another Great Depression on Earth, limiting the desire for costly space vacations. So what's a space CEO to do? In terrestrial company towns, if a Great Depression shows up, one option is for the town to just fold. It's not a fun option, but at least there's a train out of town or a chance to hitch-hike. Mars has a once-every-two-years launch window.* Even a trip to Earth from the Moon requires a 380,000-kilometer shot in a rocket, which will likely never be cheap.

The biggest rockets on the drawing board today could perhaps transport a hundred people at a time. Even for a settlement of only ten thousand people, that's a lot of transport infrastructure in case the town needs to be evacuated. Throw in that, at least right now, we don't even know if people born and raised on the Moon or Mars can physiologically handle coming "back" to Earth, and, well, things get interesting.

The result is that there is a huge ethical onus on whoever's setting this thing up. Not just to have a huge reserve of funding and supplies and transportation, so that people can be saved or evacuated if need be, but also to do the science in advance to determine if it's even possible to bring home people born in partial Earth gravity.

There is some precedent for governments being willing to prop up company towns. Many old Soviet monotowns now receive economic aid from the Russian government. We should note, however, that keeping a small Russian village on life support will be a lot cheaper than maintaining an armada of megarockets for supplies and transportation.

---

*For the astute nerd we should note that the picture is a little more complex. Launch windows could be shortened by wasting a huge amount of fuel and spending a lot of money. At the same time, launch windows could also be missed—for example, due to persistent technical problems or a major storm in the wrong place and time.

When Company Towns Go Not-as-Bad

When Reputation Matters

When we talked to economists about company towns in space, the issue they universally brought up was reputation. Maybe in a first Mars settlement, you can attract employees by just saying: "MARS! LEAVE YOUR FAMILIES FOR MARS!" But if you ultimately want employees who have the option to stay on planets with air, they need to believe they'll be treated well. We think the ideal approach to generating this reputation is to actually treat employees well. But there are other options.

One of America's most famous company towns, Oak Ridge, Tennessee, was originally a secret government project to refine uranium for atomic weapons. A billboard sat within the city with the following injunction:

Ideally, this will not be the approach taken in space, but corporations may have the option. The number of satellite links to a Mars settlement may be pretty limited, allowing for nations or corporations to limit the flow of information. In cases where the success of a space settlement is tied to national prestige, there may be an especially strong desire for the creators of the settlement to control the truth.

When the Law Favors Workers

If the government is more sympathetic to and protective of labor rights, at least some analysts find that labor will be treated better. An interesting question if you want to think about company towns in space is exactly what legal regime you think they'll be under. If we're under current space law, labor law in the settlement will ultimately come from some Earth state. If you're selecting for stability, you may want strong labor laws. However, as we've seen, the laws we have permit a "flag of convenience" approach to space activity. Companies looking to maximize profits could base their operations in countries with weaker regulations.

If somehow you get an independent state off-world, the nature of the labor laws will depend on what local inhabitants decide. The way this goes down could have lingering effects on the development of Moon or Mars culture. An independent state that begins as an international consortium may be quite different from one that starts as a corporate mining operation.

On the Economics of Survival Bubbles Separated from Home by Death Void

We've chosen to examine company towns because they seem like a likely outcome in space, and frankly because space geeks talk about them a lot. They're also well studied enough that we can make actual suggestions for would-be space corporate overlords. But the deep point here that we wish to emphasize is that problems arise when you take a familiar economic structure and then situate it in a poisonous hellscape. A major determinant for how bad all of the above problems become will be "labor mobility," meaning the ability of employees to choose other employers. If you can choose to work elsewhere, you not only benefit from getting a new job from time to time, you improve your bargaining position with your current employer. Although labor mobility can vary depending on where you live, it is rarely *zero*.

On Mars, the next available employer may be on the other side of the Sun. Even if there are other employers on Mars, the situation might remain quite bad.

Suppose Elon Musk, having once again consulted the Weinersmiths, decides to invite four other companies to build towns on Mars, on the assumption that greater labor mobility will lead to happier workers, or that anyway he can kick bad employees to Bezos's Mars company. This would likely be better than zero labor mobility, but not by much.

On Earth, if you are working at Taco Bell and you switch to a job in an Amazon warehouse, you probably don't worry about whether Amazon has adequate supplies of atmosphere and edible crickets. In space, a company can't hire you unless they can provide on-site food and air while recycling or scrubbing all your bodily emissions. Actual settlements will only be rated for a certain population.

This is why Starbase Earth doesn't frighten us—it is near a large city with a large population. People have the ability to leave, and many of the workers are going to be highly skilled, with the option of getting work elsewhere. Even in a coal mine town of the early twentieth century, there were necessarily trains leaving on a regular basis. None of this will be true in space unless transportation and local habitats are extremely developed. The result is that employees will have only limited ability to switch jobs, and employers will know this.

The one piece of good news is that companies should be reluctant to make their employees disgruntled, given the cost of sending them out there. Refusing to work can be pretty effective if your time is worth millions of dollars per day. Also, historically, company towns tend to treat highly skilled employees better than blue-collar workers. That's not exactly a beautiful fact, but it suggests good things for the elite inhabitants of early settlements. The downside is it's not exactly harmony, but agreement via both parties having a pistol to each other's head.

We understand it's weird for a discussion on space settlements to ask what the cost of rent is going to be or whether the local store is charging too much for canned goods, but this is the stuff regular lives are made of. Space-settlement proposals tend to skip this, assuming that our awesome spacefarers will be focused on the task at hand and perpetually oriented around the glory of settlement itself. We should always remember that these settlers will be regular people, and regular people's daily concerns are much less about grand narratives than about their homes, their jobs, and their groceries.

Ultimately, we don't know what the outcome will be for a space company town, or any other economic structure. It's easy enough for authors to suggest corporate governance or to say corporate governance would be awful, but the truth is that the details matter a lot. A large Mars settlement with multiple employers who each have enough life-support resources to take on new employees, and with a well-developed transport infrastructure to get people off planet, isn't just a bigger version of a small settlement. Such a settlement would have the economic diversity to weather downturns. It would provide at least some level of economic mobility. It also means workers would have more negotiating power with bosses and local stores.

The lack of these things is clearly bad for workers and for the survival of the settlement, but it's potentially also bad for those of us who stayed home. Earlier, we worried that a society founded on letting evolution take its course against children might not pair wonderfully with a future in which societies can ever more easily sling large objects around space. Any society with this capability is similar to a nuclear-armed society on Earth. That counsels a high degree of concern over the governance systems employed in early settlements. If having a single corporation run a space settlement is more likely to produce an abusive or autocratic settlement, then it represents a long-term danger to everyone. If we choose to settle space and we care about economic structures, once again we should favor a wait-and-go-big approach. Wait until we can build a society that from the very start can supply the basic economic stuff that developed Earth states take for granted.

SPACE-MAYOR
There. Astridalia now has four general stores, two hospitals, five construction companies, and 8,426 Starbuckses. We good?

18.

How Big Is Big? Plan B Settlements Without Genetic or Economic Calamities

In the introduction, we argued that space settlement isn't the solution to any short-term problem. However, if you're willing to look at the long-term, space may offer a genuine chance at a second holdout for the human species—one that could survive loss of contact with Earth. But can a space society endure for long without trade and immigration from the home planet?

When it comes to scaling your settlement toward the possibility of independence, there are two basic questions: how many humans are needed to avoid inbreeding problems and how many humans are needed to sustain the high-tech space-settlement life without any help from Earth. Here, we get into the weird details, but the short version is that the number of humans you'll need is *a lot*.

How to Have Space Babies Without Marrying Your Space Cousin

It turns out an Adam and an Eve aren't quite enough to restart humanity. Pretty quickly your family tree starts to look like a Celtic knot. This is artistically pleasing, but makes family reunions awkward. Or maybe not awkward enough. So how many humans do we need for the biology not to go haywire?

This isn't an easy question to answer. For starters, do not Google "human breeding population" in a café. You get weird looks. In any case, it turns out there's not a simple number like "4,268 humans." How do we know this? From two fields of study. The first is conservation biology, the science that deals with things like how to save endangered species. The second is what we're going to call Ark-eology—the field of study that concerns how to make an interstellar spaceship that could carry many generations of humanity.

The Sustainable Growth Scenario: Lessons from Conservation Biology

Assuming the space-baby-making business is working fine, from a population perspective there are three things you need: diversity, quantity, and immigration.

In terms of diversity and quantity, the usual metric is "effective population." This is distinct from regular old population, which is just a head count. A few stupid examples will illustrate:

Suppose for instance that you're starting a Moon settlement. In canvassing for settlers, you accidentally only reach out to college fraternities. You soon have a hundred thousand volunteers all of prime reproductive age. Then, to your dismay, after moving everyone to Shackleton crater, a biologist informs you that your *effective* population is precisely zero. They're bingeing on beet wine and chasing feral chickens around Biosphere 3 as part of a weirdly specific hazing ritual, but they are not getting down to the important business of making on-site humans. Because they can't. Actual population: 100,000. Effective population: 0.

You can still get problems even if people are the right age and right balance of sexes. This is where diversity comes in. Suppose that after the Moon-fraternity incident you try to start over on Mars. Having been burned trying to get humans the usual way, you decide to just start with a good Adam and a good Eve and just make fifty thousand clones of each. The effective population here depends on how you calculate it, but we can at least say you're better off than the frat boy–o-sphere scenario. You can now successfully mix genes, but your gene pool is about the size of a footbath. Every member of the next generation will be full siblings genetically. This lack of genetic diversity can be very dangerous for the health of offspring, especially if our clones carry any inheritable genetic diseases. Again, your *effective* population is far smaller than the head count.

These examples are stupid extremes, but the main thing to realize is that a sustainable Mars population likely requires an initially large and diverse set of individuals. In the long run, your best bet is frequent new gene flow thanks to immigration from the homeworld. As long as you maintain a connection to Earth and encourage a variety of people to come, the risk of inbreeding problems will remain more or less Earth-like.

However, for many space enthusiasts the goal is the ability to survive the death of Earth forever. This is literally possible, but not easy.

A Plan B Population

If the goal is a safe reservoir of humanity, what you want to know is something conservation biologists call "minimum viable population," or MVP. Generally speaking, if a population gets down to Adam and Eve, it's in trouble. You may be able to breed for a few generations, but inbreeding is going to catch up with you, negatively impacting health and fertility.

So what's the human MVP? We don't know. "Too few" is a problem humans haven't had in a good long while. A number you sometimes see in space-settlement books is five hundred, because that's a rule-of-thumb number in conservation biology. But in a realistic scenario for humans in space, it's probably not accurate. How do we know this? Thanks to a tiny group of Ark-eologists who've spent a few decades looking at it.

Some early work done in the 1990s seemed to show that you really only need about 80 to 150 humans to have a sustainable population. This assessment was based on anthropological observations that tribal bands are often small and homogenous. If they can survive for thousands of years with tiny populations, it stands to reason space tribes can too.

Unfortunately for any space geeks excited about the low numbers, the analysis turned out to be incorrect. Although there are indeed small tribes of people all over the world, outbreeding still happens. Preindustrial cultures may have had small tribes, but they would have connected through mating with members of nearby groups, effectively creating a total population on the order of thousands.

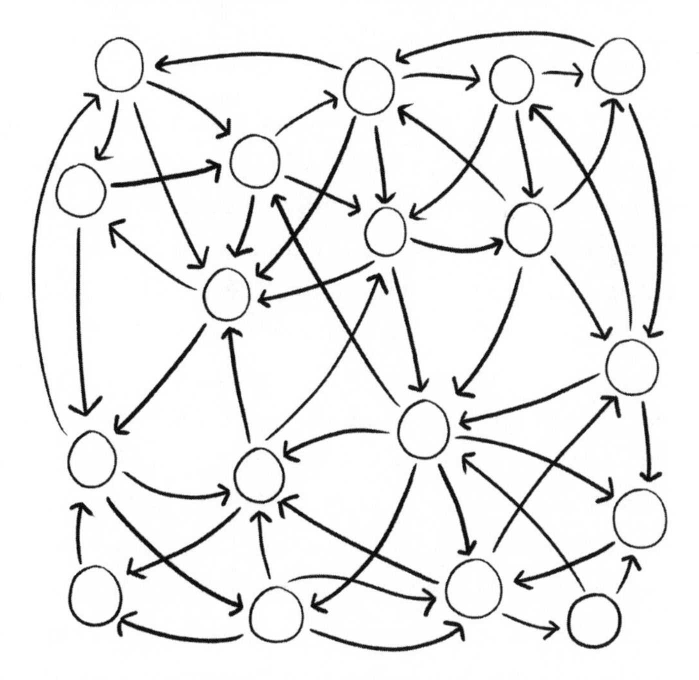

Second, more advanced computer models, although they don't always agree, have tended to find much larger estimates for the needed MVP.

This is especially true if you account for the danger of space. For example, Dr. Cameron Smith, perhaps the preeminent scholar of space anthropology, makes the reasonable assumption that you'll have at least one catastrophe. Inspired by the Black Death, which wiped out around 30 percent of Europe in the 1300s, Smith includes in his model the chance that at least once a random 30 percent of your population gets eliminated by a disease or collision or war inside the spacecraft. Based on his model, he suggests that if you're planning for just five generations of isolation, you want an *effective* population of ten thousand, which he says requires an *actual* population of about thirty thousand. Other models find more modest numbers, with many merely in the four-digit range. That's a pretty wide spread, but it's important to note that the models disagree mostly because they make different assumptions about what you need. But the basic picture is that if you want full independence, at the very least you're talking about a population in the thousands or tens of thousands.

Tech Fixes

Rather than going big, maybe we could just go weird. One idea is to freeze-dry gametes, or even zygotes. Why bother taking whole males when you can just bring the most evolutionary relevant one gram of them? Freeze-dried gametes are small, light, and allow you to bank a huge effective population inside your habitat. And guess what! One recent study sent freeze-dried mouse sperm to the ISS, and then nearly three years later used it to fertilize oocytes on Earth. They were able to successfully make mice babies, though the babies from the freeze-dried space sperm didn't live as long as mice from that particular strain typically live. That said, these mice were able to go on and have babies on Earth, and their babies had babies, too, with no obvious problems. Freeze-dried male gametes would be sort of like a backup reserve of immigrants, albeit who arrive by atypical means. Combine your neighborhood sperm library with your local radiation-proof artificial womb and you barely even need adults.

Your best bet, though, may be relinquishing all bodily autonomy to a very smart piece of software. This was explored in a model called HERI-

TAGE, which looked at ideal population sizes on a ship heading to interstellar space. Just to be clear, these modelers weren't making a serious proposal—rather they were asking what the absolute minimum population would be. They found you could have a human population survive for thousands of years starting with a population of just ninety-eight. This is quite similar to that early optimistic population from the 1990s, but it comes with a minor catch: Mating choices are made by a computer, which has a complex system of rules to preserve genetic diversity. The program also determines things like how many babies women will have, upping the number when the population starts to get too low and reducing it when the number gets higher. Admittedly, obeying a robot probably results in better life choices than most of us typically make, but it'll likely be unpopular.

Our main concern about tech-based biology solutions, which are regular features in the space discourse, is that even if they work, they would probably get a lot of pushback if they were mandated in space. Things like sperm banks and genetic testing are helpful family planning tools here on Earth, and when these resources expand our options they're great. But making them mandatory for everyone? Even if a first generation of settlers are cool signing the waiver, what are the odds that all, or even most, of their children will feel likewise? An ethical space settlement shouldn't involve group-mandated reproductive choices. Once again we're back to wanting a wait-and-go-big approach. Don't install weird gadgets to try to get a rapid start. Rather, wait until you can send a huge, diverse population, with the resources to permit personal choices to be personal.

Studying Ark-eology is interesting because it can help us understand the minimum number of humans needed for biological survival. However, it's unlikely that the minimum viable population is enough to survive the death of Earth without incredibly exotic technology. To see why, we consider the question of autarky.

Let's Talk Autarkic Arks

In the introduction to this book we mentioned Elon Musk's statement that an independent settlement could be achieved by midcentury. Our source for this is a Twitter interaction in which a user named Pranay Pathole asked: "Elon, What do you think is the estimated timeframe for creating a self-sustaining civilization on Mars? 20 years? Self-sustaining meaning not relying/dependent on Earth for supplies." Musk responded, "20 to 30 years from first human landing if launch rate growth is exponential. Assumes transferring ~100k each rendezvous and ~1M total people needed."

Even by Elon Musk standards this is pretty ambitious. In an earlier statement, Musk estimated the first human on Mars would arrive in 2029. As we write, the relevant vehicle is still in the testing phase, never mind the part where you land in Mars's thin atmosphere and stay not-dead for a few years. More important for this section, the "~" in "~1M" is doing a lot of work.

The technical term for complete economic independence is "autarky," from the words meaning "self" and "rule." Physicist Dr. Casey Handmer did an admittedly very rough, and we would say very generous, assessment of what would be required to take Mars to autarky. He noted that the closest things on Earth are likely Cuba (population approximately 11M), which has still had to import advanced industrial products, and North Korea (population approximately 26M), which regularly has shortages of petroleum and food. That is, the most autarkic countries on Earth have much more than 1 million people, are not the most economically desirable places on Earth, and incidentally would both like to be less autarkic. Autarky is hard enough in these places, but it's going to be a hell of a lot harder off-world, where you will be utterly dependent on technology.

One important example is computer chip fabrication. On Earth, this is a very expensive business that requires large amounts of water. It also requires a lot of specialists. As Dr. Martin Elvis notes, the nearly 8 billion people on Earth get their ultra-high-tech chips from just three manufacturers. Even if you had the people and technological level to make these chips on Mars, the economics would still be against you. Chips weigh very little. You can pack a lot in a rocket. They are likely the last thing you'll want to build space locally unless you are absolutely positive you're going to lose contact with Earth soon.

So what number does Handmer come up with? According to him, you need about 1 million, just like Elon Musk said, only with the caveat that he could be off by an order of magnitude in either direction, depending on how ultra-advanced you want to believe robots will be in the future. We consider this a pretty low estimate. Another, from hard science fiction author Charles Stross, puts the number closer to 100 million, possibly even a billion. This seems more plausible to us. North Korea isn't autarkic at 26 million, and whatever else you might say about that country, they have access to air, an ocean, and ground that isn't made of toxic chemicals.

It'll be even harder if you've set up a settlement somewhere other than Mars. Elvis points out that large space stations will never achieve full recycling. Even if you have 99 percent reuse of something like water, that loss of 1 percent mass adds up to 40 percent of your mass over fifty years. You'll want, and need, to trade.

We are aware that giving numbers that range from one hundred thousand to a billion doesn't make us seem like geniuses. The point we want to make clear is simply that it will take *a lot* of people. Given the amount of unsolved problems for settlement, and the fact that the needed giant spaceships haven't even gone to orbit yet, and that there isn't a good economic reason to go independent, we can't expect to have a Mars base that can survive the death of Earth anytime remotely soon.

Also, it's worth noting that the lower you take the number, the more you've got to assume some sort of ultra-advanced robotics. That's fine, but speaking just for ourselves it seems to create a bit of a tradeoff. If it's possible for every human to get forty androids worth of assistance, we'd rather have them on Earth to carry us on a throne, fan us, and pour us glasses of wine that isn't made of beets.

Whither Plan B?

We said at the beginning of this book that there is no "short-term Plan B" in space. There are certainly lots of scientific, technological, and ethical barriers in the way, but the difficulty of autarky may be the last and largest nail in the coffin. If you think Earth is dying and you want to save humanity, you either have to transfer a huge population to Mars—possibly on the order of hundreds of millions—in a short period, or you have to have unimaginably developed robotic technology. Although it's hard to predict the future, it seems to us that if we're so good at robotics and ecology that we can build a permanent bubble world for 1 million people on a distant oceanless planet, well, surely we can clean some carbon dioxide out of the air on Earth.

The other thing is that the attempt at Plan B may be the cause of earthly calamity. There may be good reasons never to build a backup copy—not because we can't do it, but because having a multiplanetary species may not actually render humanity's existence more secure.

"It's cramped, but my children will value the job and mating opportunities."

19.

Space Politics by Other Means: On the Possibility of Space War

> Who controls low-Earth orbit controls near-Earth space. Who controls near-Earth space dominates Terra. Who dominates Terra determines the destiny of humankind.
>
> —Dr. Everett C. Dolman, professor of Military Strategy

In 1999, US Congressman Roscoe Bartlett (R-Maryland) was negotiating an agreement to end the war in Kosovo. At the table was Russia's Vladimir Lukin, former ambassador to the United States and deputy chairman of the Duma. He wasn't happy. Russia, not even a decade past the collapse of the USSR and the loss of its superpower status, had only been added to negotiations late in the process. At a particularly tense moment, Lukin turned to Bartlett and, according to Bartlett, said, "If we really wanted to hurt you, with no fear of retaliation, we would launch an SLBM. That's a submarine-launched ballistic missile. . . . We would detonate a nuclear weapon high above your country, and we would shut down your power grid for 6 months or so." Another Russian present added that they had spares, in case the first one didn't do the job. According to Bartlett, the quantity of spares was about seven thousand.

Bartlett had a peculiar response to this information, which we'll get to later. Whether or not Russia ever intended to use weapons like these, they certainly could have. Both the United States and the USSR tested nuclear weapons in space, and both learned the hard way that these weapons can mess with important infrastructure. The Soviets temporarily knocked out some of the electric grid in Kazakhstan, while the United States, as we've seen, caused streetlights to fail in Hawaii and damaged on-orbit satellites.

Satellites today perform essential services in navigation and communication, but they are not typically hardened against physical attack. Whether or not space is a likely site of war, electronic objects in space are extremely vulnerable. And just in case you've forgotten, any humans in space will be dependent on functional electronics.

At the beginning of this book, we said the best arguments for space settlement were, one, long-term species preservation, and two, because it'd be super awesome and maybe there's no reason not to. Both of these arguments fail about as fast as a power grid near a nuclear explosion if space activity makes warfare more likely.

War in Space in the Near Term

The Peace so Far

Happily for humanity, and unhappily for anyone with fantasies of space battleships and death beams, outer space has remained largely peaceful. This might have surprised a space expert living in the 1960s, who would have seen the rapid evolution from small-payload V-2s to space-based nukes. However, it quickly became clear that giant space explosions are bad for everyone. High-speed orbital metal and radiation endanger civilian and military satellites, while also being decidedly unwelcome among humans in spacecraft. This is a major part of why treaties were implemented in the '60s to restrain the weaponization of space.

Those treaties have basically worked. Sure, Americans sent up space machetes and the Soviets sent up space pistols, but those were technically for the survival kit in case you landed back on Earth in the middle of nowhere, not for awesome swords-vs.-guns space battles. And sure, the Soviets once tested a cannon on Salyut-3 in 1974, but they only fired it once, and they did so while the station was uncrewed just in case something went wrong.

Because the main use cases for orbit are military and commercial, and because satellites aren't designed to withstand attacks, and because almost nobody lives in space, space war, in the sense of spacecraft having awesome dogfights with each other, is unlikely anytime soon. The bigger risk is that escalations in space activity produce regular old war on Earth. While satellites with kinetic weapons have never been used against Earth, military satellites have been in space since the late 1950s. Although there was a steady advance of military space assets afterward, things really got going with the First Space War.

The First Space War? Doesn't ring a space bell? That's because you know it as the First Gulf War, 1990–1991, but to people who study military space, it was a pivot point in history, when space assets were absolutely crucial to rapidly crushing the Iraqi Army. The number of military satellites has grown ever since, and although the United States still has the most, Russia has plenty and China is gaining fast.

Measuring exactly how militarized space has become is tricky, however, because satellites are "dual use." Almost any object orbiting the Earth can be repurposed for tactical reasons. This was apparent from the earliest days of space-travel theory, when Hermann Oberth proposed a giant mirror in space, for combination agricultural and death-beam purposes. The latest example is SpaceX's thousands of Starlink satellites. Launched originally for the purpose of transmitting internet stuff, they became integral to the Ukrainian resistance against Russian invasion after Elon Musk had several thousand terminals shipped over. Soon after, the Russian Federation felt compelled to announce that "quasi-civilian infrastructure may become a legitimate target for retaliation."

Although nobody has yet blown up an enemy spacecraft, several nations have demonstrated that they can shoot down satellites from Earth. If you count the ability to just mess with satellites via cyberattack, the number of combatants is quite large, and likely growing. Managing and protecting satellites was previously done by the US Air Force, but is now one of the main tasks undertaken by the recently created Space Force. Countries like Russia, China, and France have similar organizations to protect their space assets. As the world gets more developed and more nations and corporations field satellites, this particular sort of warfare is going to become more common.

So while there are reasons to believe space itself will remain peaceful over the short term, you should understand with perfect clarity that humans don't experience any sort of change of heart when we conduct business in space. Part of why explosions in space are so rare is that Earth's militaries need to keep space pacific in order to conduct war on Earth.

That's not the best reason for peace, but on the whole we're glad the Earth isn't encircled with missile-toting battle stations. Will this situation hold as space access becomes cheaper? We don't know, and scholars don't agree.

The good news is that much of our forbearance to put weapons in space comes from a highly persistent human quality: self-interest. Space is crucial for commerce, for spying, for environmental data gathering, for surveillance to confirm treaty compliance, and plenty of other stuff people and nations care about. The bad news is that another part of why space has not become a theater of war is that space weapons have, historically, sucked.

On the General Suckiness of Space Guns

Setting aside how everyone would *feel* about it, dangling a gun above Earth seems like it ought to be a strong military move. Analysts aren't so sure.

We'll illustrate the problem by imagining you're at war with Antarctica. You being the United States, because who else is going to shoot missiles at a nearly empty frozen desert?

After the Mars secession of 2063, an army of South Polar graduate students decide it's open season on seizing the global commons and claim the Northern Peninsula despite all the cute babies tactically birthed there by Chilean and Argentinean women. Fortunately, you happen to be in possession of a fleet of ground-based missiles. Here's what it looks like when you launch a first strike:

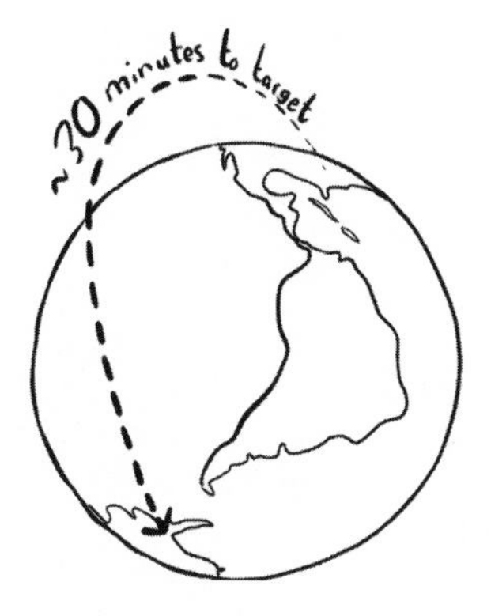

Not bad. The missiles arrive quickly and on target, minimizing collateral penguin death. But suppose you want to launch the missile from an awesome space satellite. Well, that looks like this:

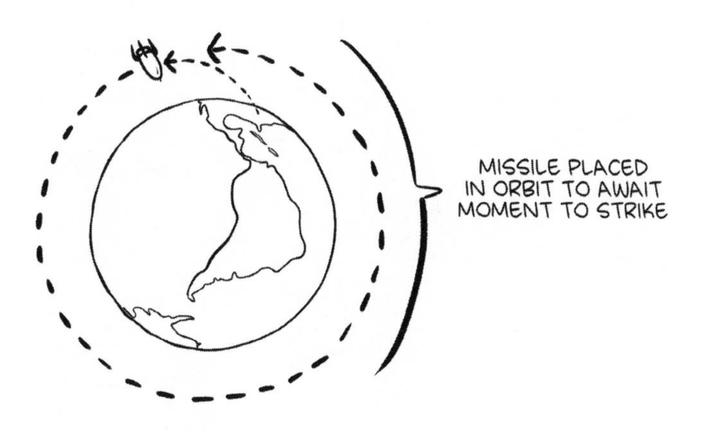

It's pretty clear that this is not a *cheaper* option. Note that you have already had to go through the expense of flying a missile above the atmosphere, but now have the added burden of getting that missile to orbital velocity and maintaining it against space's huge temperature swings and radiation doses, so that hopefully it works when you finally decide to eliminate the Antarctican menace.

And most of the time you can't even do that. Satellites aren't like bombers that can always just head toward the target. Sure, you can park them in geostationary orbit over an equatorial enemy, but now your missiles have to do a long-distance commute to work—35,000 kilometers uphill, then all the way back before even a single graduate student gets atomized.

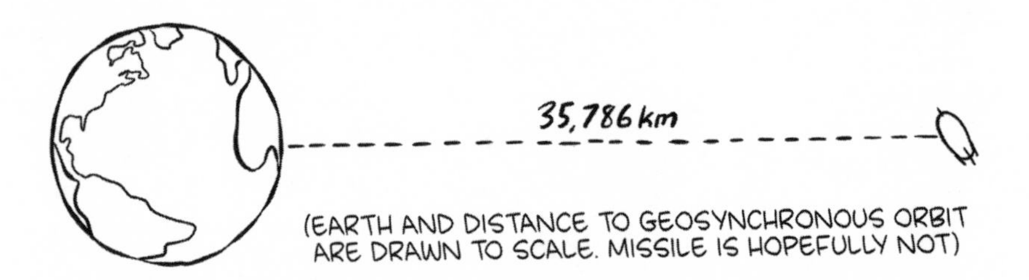

So you put your missile in low Earth orbit, but now it's almost never over the Antarctican menace. A decent amount of time the only thing you can rain death on is the open ocean. What's the solution? Have *lots* of space guns.

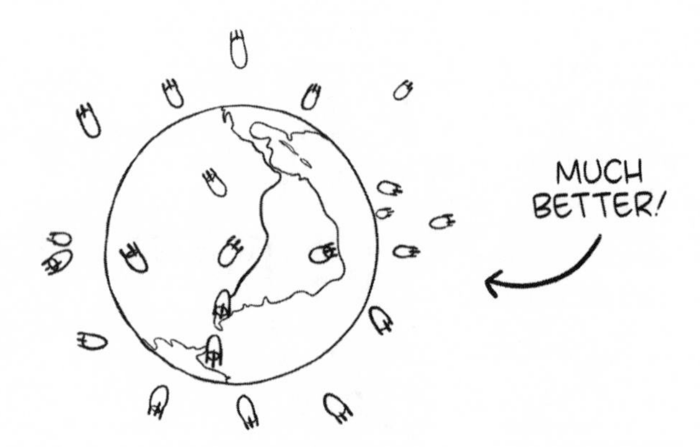

It's like GPS coverage, only for death. But making sure the answer to "can we kill them now?" is always "yes" clearly requires a whole fleet of expensive weapon satellites. Have you made a smart move here?

Well, you've probably angered a significant portion of the planet. Until you've actually released your in-orbit missile, it is true that nothing here is obviously illegal: the Outer Space Treaty clearly prohibits military bases on celestial bodies as well as space nukes, but parking a regular old missile or death ray in orbit isn't explicitly against the rules. Plus, the UN's General Charter allows for self-defense, so you could argue these are there just in case you're in the position of needing them.

Setting aside worldwide anger, what about bang for the buck? From what we've read by military strategists, it's not clear why you'd want orbital kinetic weapons pointed at Earth even if they do work. They largely duplicate your preexisting ability to shoot rockets at the enemy, but are far more expensive—not just to field but to maintain in working order. They're also easy to track and target by your enemies because they follow a Newtonian trajectory. While they're orbiting, if your enemy has the right tech, your weapons are easy to explode or just knock out of the sky. "Expensive and easy to attack" is not a promising set of qualities for an attack system.

That's the *theory* against space weapons, but we also have something like experience. During the Reagan era, as part of a program called the Strategic Defense Initiative—SDI, or sometimes "Star Wars"—the United States looked into a variety of space weapon ideas, many of which were, to be generous, howling-at-the-Moon bonkers. The weirdest was perhaps Project Excalibur, which would have used a nuclear explosion in space to power X-ray lasers, which would be used to blast Earth-based nuclear missiles heading for the United States. Why do all this in space? Speed. The ideal time to stop enemy missiles is in the "boost phase"—the period right after takeoff when they are still building up momentum. If you're worried about a huge nuclear strike from an enemy, having a fleet of orbital weapons that reach the target at light speed is a pretty strong move.

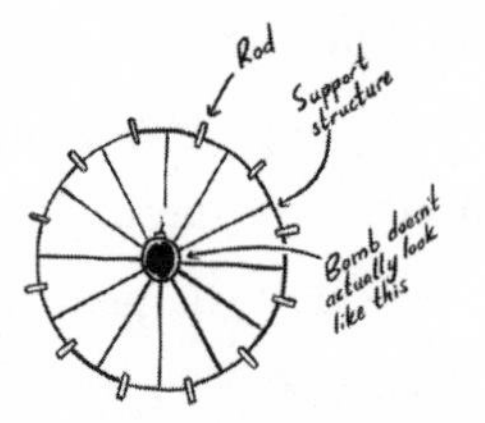

STEP 1: PUT NUCLEAR BOMB IN ORBIT, SURROUNDED BY SPECIALIZED METAL RODS.

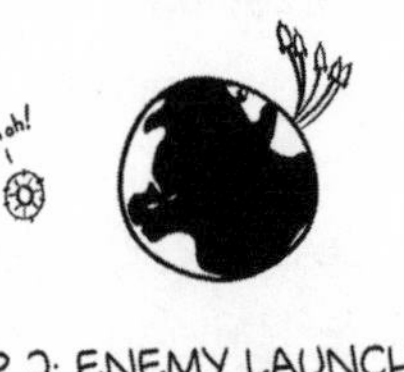

STEP 2: ENEMY LAUNCHES ARMADA OF MISSILES.

STEP 3: METAL RODS MOVE TO PRECISE LOCATION AROUND NUCLEAR BOMB, ALIGNING WITH TARGETS.

STEP 4: DETONATE NUKE. RODS ABSORB BLAST ENERGY, EMIT POWERFUL X-RAY BEAMS. MISSILES DESTROYED. EVERYONE LIVES HAPPILY EVER AFTER?

One flaw with Excalibur and similar sci-fi ideas from the time is that they didn't work. They were too hard to build and field. Controlling a nuclear explosion in space to create a laser beam is enormously technologically complicated, never mind the potential awkwardness at the UN when it becomes clear you've violated both the OST and the Partial Nuclear Test Ban Treaty and never mind the destabilizing effects of having only one country that can both wield and shield against nukes.

In short: humans have had good political, economic, military, and technological reasons not to weaponize space. The one question looming at our particular moment is whether everything will change in this new age of cheap space access. We don't know, but we do know some theorists are worried.

In the modern-day military literature we read, there were plenty of US folks arguing that constraint will be dangerous to the United States in the long run. Their arguments are varied, but in general they believe space will become more militarized and weapons *will* be placed in orbit, given enough time. It follows, they say, that the United States should try to take the lead now to avoid a situation in which a country like China controls the ultimate higher ground. We don't know if this is true or if it's good strategy, but we hope it's clear by now that it's a prophecy that could quite easily lead to its own explosive fulfillment. The future is more multipolar, has more space activity generally, and hey, maybe there's a zero-sum race to the Moon on the horizon. What we can say as we project further into the future is that nothing about the space environment so far appears to imbue the human heart with a desire for peace. This is all the more reason for the near-term creation of a space regulatory body.

War in Space in the Medium Term

So let's assume that no progress is made on the international law front, and things keep moving forward as they appear to be: a small number of powerful nations head to space to set up outposts, maybe even mining operations and lunar gas stations with families living there full time. A number of nations or multinational groups have some kind of presence on the Moon or Mars, and perhaps, *perhaps*, some of them start itching for a war of independence.

What do we think will happen? Probably not much. Conflict between space powers would certainly be possible, but conflict between Earth and space will remain extremely unlikely. If we're talking about space war, in the sense of a settlement on the Moon, Mars, or a rotating space station making war with a nation (or all nations) on Earth, we think there's almost no chance anytime this century. Why? Because a space settlement would get crushed. As we said earlier, autarky is going to be extremely difficult in space anytime soon. The Moon rebels will have to maintain some sort of trade with Earth simply to survive. Also, remember those nuclear weapon attacks Bartlett was worried about? Well, losing your electric grid on Earth is no fun, but losing it on the Moon will be downright embarrassing. Weapons can be specially engineered to knock out power grids via massive electromagnetic flux. Power grids can also be attacked using cyber methods. Try to imagine your electrical system going out on the Moon just as two weeks of night begins. Anytime in the short term, expecting a space settlement to declare war on Earth would be like expecting Malta to declare war on Europe.

The one serious additional risk in the era of early space settlements would be terrorism. If you really do get a lunar mass driver built, well that mass can be driven into Earth's gravity well. That said, given how easy it would be to knock out a giant electrical structure on the Moon, any terrorists going in this direction would be well advised to make peace with their Creator before pressing the big red button.

In short, even in an era of expanded space possibilities—with occupied lunar caves, with giant orbital cylinders, with tourist attractions on Olympus Mons . . . war between celestial bodies will remain unlikely.

War in Space in the Long Term

In this book, we've been reluctant to speculate about the distant future, in part because if you believe speculative space books from the past, you would have gotten your space Chevy van all waxed up with nowhere to go. That said, thinking about the long-term future of space war matters specifically because space-settlement advocates often claim that a vast human presence in space will lead to peace for humanity. While this is a claim about the long term, it is often used as a justification for large-scale spending right now, and in general for not worrying too hard about what massive space access means for the survival of our species.

There are a number of versions of this argument. Most commonly, theorists claim that war is fundamentally about scarcity, so if space access solves scarcity, it solves war. For example, in Dr. Avis Lang and Dr. Neil deGrasse Tyson's book *Accessory to War* they argue that "what's contested on Earth because of scarcity is typically common in space. . . . Even if control of [access to space resources] rests in the hands of people you'd hate to be in control of anything, the resources themselves will not be scarce—and it's scarcity that breeds conflict." Similar claims from other thinkers say that war is about land, so space's new land will end war. An even more circuitous claim says that war isn't about resources, but rather about people thinking that war is about resources—a problem that will be solved by the abundance of space. Still other theories say that the ability to go do your own thing in space, away from civilization, will create peace.

One pretty clear argument against all these ideas is that similar notions were once advanced regarding white settlers entering the American West. We're skeptical about whether space will provide abundance for anyone, but the genocidal appropriation of the West from Indigenous people pretty clearly provided much for the settlers. And yet the conquest of that land, and the question of whether the resulting states would be slave states or free states, precipitated the American Civil War, which left about 2 percent of the nation's population dead.

This should give us a clue that war isn't just about scarcity. So what is it about? Well, we can tell you what Dr. Chris Blattman, a scholar on the causes of war, refers to as "false causes" of war in his book *Why We Fight*: "poverty, scarcity, natural resources, climate change, ethnic fragmentation, polarization, injustice, and arms." These things, he notes, are "terrible for other reasons. And they add fuel to a raging fire. But they probably didn't ignite fighting in the first place."

The truth is that war is a quite complex human behavior, and people who study it don't even agree on the most fundamental causes—about whether war is inherent to humans, or cultural, or ecological in the sense of depending on the environment. We won't try to convince you of any particular theory—rather, we want to convince you that war can come about for reasons that have nothing to do with space, or abundance, or any kind of objective assessment of the well-being of nations. To do this, we'll look at two ways war can start.

Scenario 1: Commitment Problems and the Thucydides Trap

The "Thucydides Trap" is a term popularized by political scientist Dr. Graham Allison, referring to early Greek historian Thucydides's assessment of the Peloponnesian War. In short, after the Persian Wars in which they had been allies, the city-states of Sparta and Athens suffered an increasingly bad relationship, largely because Athens was suddenly flourishing in a way that scared the hell out of Sparta.

Why is war more likely in these circumstances? War theorists often talk of war in terms of bargaining. Nations bargain over all sorts of things—resources, weapons development, trade, treatment of minority groups, and so on. Exactly what kind of deal nations are able to bargain for depends on several factors. For example, who's more powerful economically and militarily? How certain are you of your own country's relative power? How willing would your opponent be to fight if push comes to shove?

The breakdown of peace is, so to speak, another bargaining phase, only it's one where you get very clear information about your opponent's capabilities. Most of us would rather we stay in the bargaining phases that don't involve bullets. How do we do this? Ideally, we each honestly commit to a bargain that guarantees peace. But in the case where the balance of power is shifting rapidly, commitments can be hard.

To see why, imagine two countries on the Moon—Bezostralia and Muskow.

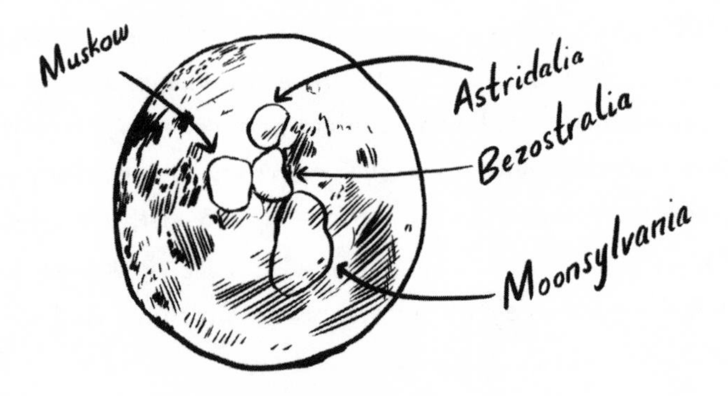

They have lived in relative harmony since the Moon rebellion of 2144, despite a small territorial dispute over Shackleton crater. They have about equal economies and equal militaries. However wary they are of each other, they are both committed to peacefully working out their differences.

But then something changes. For generations, nobody had bothered mining helium-3 because the Weinersmiths said it was a bad idea. Suddenly it turns out to be very valuable, and Bezostralia is sitting on the richest reserves. Their economy grows rapidly. People are moving there from Muskow. They have more money to invest in facilities, and possibly in their military. Bezostralian politicians begin to talk openly about using their new power to seize that long-disputed bit of crater. If Muskow wishes to respond logically, what should they do?

What war scholars would say is that Muskow now has a "window of opportunity." Right this second, they could potentially make war successfully on Bezostralia. If they wait, Bezostralia will become so powerful that Muskow will have to do whatever Bezostralia wants.

Can they make a deal? This is where the trap comes. Suppose they agree to be peaceful forever. That deal obviously advantages Bezostralia because all they have to do is sign it, then bide their time until they're powerful enough to break it. The scary thing here is that *even if both sides want peace*, the inability to guarantee commitment suggests that the smart move for Muskow is to kneecap Bezostralia's rise and lock in their advantage before it's too late. This doesn't *have* to lead to war. It could lead simply to oppression, or even to enlightened peace. War scholars are often at pains to note that, cable news coverage notwithstanding, the default state of most places at most times is peace. What we do want to note, however, is that the increased risk of war in this scenario comes not from some sort of objective assessment of well-being by nations or individuals, but from the fact that one power was experiencing a large, fast, relative change in power that a potential rival could see coming. This happens on Earth and there is no reason for it not to happen in space.

Scenario Two: Leader-Nation Alignment, or Space Bastards in Charge

If you mentally model nations as rational actors obsessed with survival and advantage, you won't be completely off the mark. But, we should remember that nations are led by actual people, and most of us don't feel our leaders are always perfectly reasonable 100 percent of the time. Or, to the extent they're being rational, it may be in their own interest and not that of their nation. James Madison, one of the framers of the US Constitution, wrote about the nature of this problem in 1793:

> *In war a physical force is to be created, and it is the executive will which is to direct it. In war the public treasures are to be unlocked, and it is the executive hand which is to dispense them. In war the honors and emoluments of office are to be multiplied; and it is the executive patronage under which they are to be enjoyed. It is in war, finally, that laurels are to be gathered, and it is the executive brow they are to encircle. The strongest passions, and most dangerous weaknesses of the human breast; ambition, avarice, vanity, the honorable or venial love of fame, are all in conspiracy against the desire and duty of peace.*

Madison for some reason forgot to add: "this is also true on the Moon probably," but we think the logic applies. Unless you see the era of Bezos and Musk opening space as eliminating ambition or vanity, the fears of 1793 will likely apply in 2093 and beyond.

Bad behavior by leaders can be especially likely in cases where leaders are disconnected from the costs of war, as in nondemocratic nations. The recent invasion of Ukraine by one of Earth's great historical space powers is an especially vivid example.

The thing to note is that war isn't the simple result of people or nations deciding they need more stuff. Wars can happen simply because leaders will benefit from the conflict. There is no reason to assume anything about space will alter this, even if it delivers on a series of unlikely economic benefits.

The terrifying upshot? Going to space will not end war because war isn't caused by anything that space travel is apt to change, even in the most optimistic scenarios. That takes us to the final issue: humanity going to space en masse probably won't reduce the likelihood of war, but we should consider that it might increase the chance of war being horrific.

Interplanetary War: Probably Not Super Great

In the very long-term future, if interplanetary war happens, it could be a uniquely awful event in human history. That's not just because of the human ability to fling asteroids in this scenario. An underappreciated concern here is the fact that there's never been a war between two powers down different gravity wells.

Part of why we no longer do regular nuclear weapons tests on Earth is that scientists in the 1950s collected baby teeth and found them to contain dangerous products of nuclear explosions, notably strontium-90, which the body treats like calcium. That is, part of human forbearance to use nuclear weapons in war is that it contaminates this fishbowl called Earth. Mars is in a separate fishbowl. There is no shared atmosphere to be filled with dangerous isotopes. For that matter, there's no wind to blow chemical weapons in your face, no globalized interactions to transmit diseases back to the source.

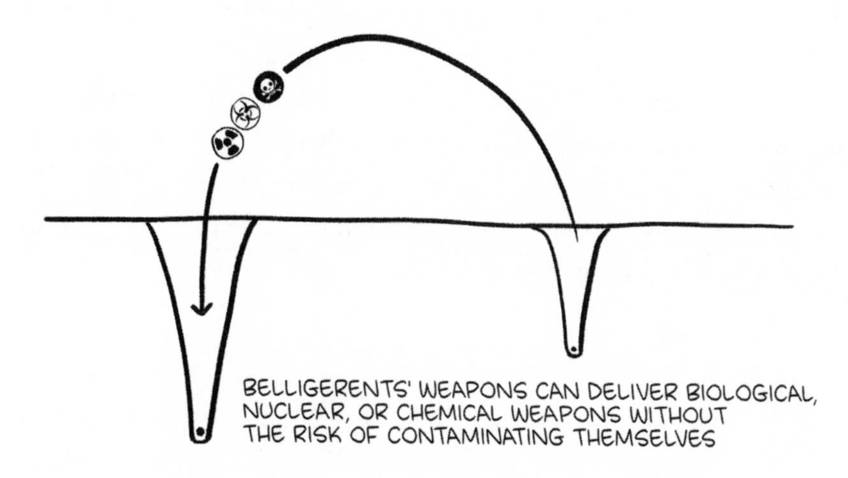

There are reasons beyond tactics not to use these weapons, of course—fear of reprisal, or just a basic sense of fellow feeling for humanity. While we don't know if they'll ever be employed in some distant future space war, what we can say is that at least one of the reasons for not using humanity's most horrifying weapons will be taken off the table. This is a simple fact of physics in our universe.

Space settlement will not end war, but war in a world of space settlements could be especially deadly.

Make Star Peace, Not Star Wars!

Although they are often called war theorists, people who study human conflict are also deeply concerned with keeping peace. The jury is still out on whether war is a permanent feature of our species or something we can reason our way out of. The problems of space war that exist at every time scale lie at the intersection of Newton's laws and human nature. Physical laws are not apt to change anytime soon, so if we want to have any hope of avoiding conflict in space, or due to space, we need to act to constrain human nature and shape the future for peace.

Speaking of which, you may wonder how Congressman Bartlett shaped his own future after hearing an offhand threat of space attack from Russia. He became an advocate, spending years fighting for grid defenses against nuclear attacks. After a long career in politics, he noped out of civilization. Today, he lives on an off-the-grid farm. It sounds nice. There's a lake with two swans, and according to Bartlett the only confirmation he gets that the world outside continues to turn are the contrails of passing planes. This may be the correct solution for Bartlett, but it will not be one widely available to humanity, on- or off-world.

Now you tell me!
Chateau de Beets

20.

A Brief Coda on a Rarely Considered Alternative: Wait-and-Go-Nowhere

Earlier, we mentioned Dr. Daniel Deudney, the international relations scholar infamous among many a space geek for his proposal that the existentially safest move for humanity is just never to create a major human presence in space.

Deudney isn't opposed to space activity per se—he just thinks it should be used for nondangerous stuff, like science, environmental monitoring, and communication. It's not the sprawling space-exploitation regime of many a geek fantasy, but then again, a dead planet has a decidedly limpid economy.

This is a lonely view in the space-settlement community. Deudney recently appeared at a virtual session on his own ideas put on by the National Space Society, and it was essentially him versus everyone else. After a debate between Deudney and NSS's Mark Hopkins, there were two panels in which speaker after speaker attacked his ideas. He was not invited to join the panels to respond. At one point, a panelist basically called him stupid.

Deudney is a more extreme can-we-just-not-ist, but there are other voices suggesting a high level of prudence. For instance, Dr. Linda Billings, a consultant specializing in astrobiology and planetary defense, wants "collective peaceful existence on Spaceship Earth" before we seriously contemplate space expansion. Drs. Carl Sagan and Steven J. Ostro published a paper in 1994 titled "Dangers of Asteroid Deflection," noting that a humanity advanced enough to save itself from asteroid impact is a humanity advanced enough to deflect an asteroid into itself.

We would add to all this the concerns we've expressed about creating an immoral society, especially in the context of human reproduction and economics. Deudney himself points out that in a world of easy access to doomsday weapons, it takes only one society gone autocratically evil to create a nightmare scenario. If some aspects of space-settlement actively increase the likelihood of this outcome, the point becomes even more worrisome.

Space geeks often cite a quote by science fiction author Larry Niven: "The dinosaurs became extinct because they didn't have a space program. And if we become extinct because we don't have a space program, it'll serve us right!" But as Deudney notes, giant asteroids are rare. Humans haven't been around that long, while the dinosaurs had a good long run. "Given these possibilities, perhaps the reason the dinosaurs lasted for nearly two hundred million years is because they did *not* have a space program."

We have come to believe that unless major legal reforms are undertaken, expansion followed by crisis is a likely path. If that expansion is informed by mistaken ideas about human reproduction or sociology or economics, dangers of the sort Deudney fears become more likely.

Things are now moving very quickly, and we don't know to what end. But we do believe that the naysayers should get a little more room to say nay.

Nota Bene

AMUSING ASTRONAUT NAMES AND THE SOVIET TENDENCY TO FUSS OVER WEIRD DETAILS

Well, that was depressing. The good news is we saved the funniest bonus story for last.

In an earlier draft of this book we spent a lot of time on intercultural issues in isolated and confined environments for the space psychology chapter. We ended up cutting the section because the overall picture isn't superinformative: cross-cultural interactions are typically fine when you have elite professionals with cultural training who are on their best behavior. It's more of a human-resources issue than a humans-conquer-space issue. There are a few stories, but they are about as interesting as that one about the guy who thought he had a toothache but didn't.

However, there are a few oft-forgotten tales of intercultural weirdness that occurred during the USSR's Interkosmos program, which was an effort starting in the 1970s to make cosmonauts of non-Russian members of the Soviet bloc. Best of all is the tale of Kakalov.

You probably haven't heard of Bulgaria's Georgi Kakalov. Possibly this is because you aren't a Bulgarian cosmonautics enthusiast. But perhaps also because he didn't fly under his birth name. In Russian, Kakalov sounds something like "Doodoolov" does to an English speaker. More literally, you might render it as "George Poopson." Or, with a Celtic flare, George McCrap. Or, using the traditional Anglo-Norman construction, George FitzDookie. Anyway, prior to flight, at the insistence of the Soviet space agency, he was rechristened Georgi Ivanov, after his father, Ivan. Incidentally, less than ten years later, cosmonaut training would be run by one Commander Vladimir Shatalov.

A very similar situation happened with a Mongolian cosmonaut candidate, but the story is little known because the guy whose name was changed didn't end up flying. However, as Weinersmiths, it is dear to our hearts, so we'll share it with you here: According to one source on the Interkosmos program, "There was one problem of a rather unusual nature. Maidardzavyn Aleksandr Gankhuyag was told that he would have to change his surname to avoid offending or amusing Russian speakers. The surname Gankhuyag is a Mongolian word meaning "armor," but the last two syllables resemble a Russian slang word for the male sexual organ. At the insistence of the Soviets, he changed his name to Ganzorig." Those authors didn't get more specific so we consulted your friend and mine, Wikipedia, and it turns out that the word "khuy" in Russian "means cock, penis, or for an equivalent colloquial register: dick."

We want you to know that out of a profound sense of fairness we tried very hard to find an American astronaut with a funny name, but we came up empty. The best was the more-awesome-than-funny-sounding Bill McCool. As a consolation, we offer the name of the first Dutch astronaut. It isn't amusingly prurient, but does have the virtue of sounding like an alias invented by a six-year-old trying to get away with something: Dr. Wubbo J. Ockels.

Conclusion

Of Hot Tubs and Human Destiny

So, should we do this?

That's not a question we imagined asking when we started this project. Our original assumption was that space settlement was coming soon and that a question of governance was looming. We now believe the timeline is substantially longer and the project wildly more difficult and that the governance work to do is more about regulating the behavior of Earthlings than designing a Martian democracy.

But hang on a minute. You, valiant reader, have come this far: you've worried over Mars babies, considered life in a lunar cave, pored over the work of space lawyers, imagined ways of building society off-world, and contemplated a dark future. With all that in mind, let's revisit the two plausible arguments for space settlement we set out at the beginning of this book. How have they fared?

Argument 1: The Survival Cathedral

Should humanity begin the process of settling space purely because it makes us less vulnerable to species extinction?

Dr. Stephen Hawking liked this argument. Elon Musk likes this argument. We're no longer sure it's a good one. If humanity were governed so well that war and terrorism hadn't existed for ages, then we should build the cathedral. If humanity had some kind of *Star Warsy* force-field technology that allowed us to perfectly control what objects fall through our atmosphere, then we should build the cathedral. If we could build a settlement on a faraway star, making interplanetary conflict all but impossible, then we should build the cathedral. But we have not moved beyond conflict. Our ability to harm ourselves vastly outweighs our ability to protect ourselves. Settling the solar system will likely increase the danger, and we will not be leaving for distant suns anytime soon. During the last century, humanity has stacked up a good half-dozen brand-new modes of self-annihilation. Do we really want to add one more?

The only case for the survival cathedral would be if we could prove that although space settlement creates new risks, the benefits outweigh them. But at least for now, the benefits look slim and the risks look substantial.

Without the confidence that space expansion produces net benefits for humanity on a short or long time scale, the creation of the survival cathedral becomes, appropriately, an act of faith.

Argument 2: The Hot Tub

Do humans have any right to bind their fellow humans to Earth?

We said at the beginning of this book that the question here is whether the choice to settle space is more like the personal choice to buy a hot tub or more like the highly regulated choice to buy a nuclear missile. Given that space settlement is unlikely to deliver huge benefits anytime soon, given that the path forward might lead to legal chaos, given that space activity won't reduce the likelihood of war and might increase it under some scenarios, well, it's hard to argue that this is just a matter of personal preference. Is this merely a personal choice? We think, at the end of everything, the short answer here is no.

Many of the most persistent voices for Mars colonization are philosophically libertarian—they think Earth is too bureaucratic, too rule bound, too oppressive. But even ardent libertarians don't favor an increase in private possession of nuclear weapons. If you agree with us that a large-scale human space presence poses a nuclear-weapon-*ish* threat to humanity, almost any political or philosophical posture should favor a collective right to say what goes on above our gravity well.

Look, we're nerds. What we were supposed to write here in this book is that yes you *will* live in a giant space wheel or a Martian biodome or a lunar cave *with* a lovely space family and anyone saying otherwise is in the same category as the people who said airplanes wouldn't work or that you can't make fire by rubbing sticks together.

We would like to be saying that here. Really. Kelly once attended a space-settlement workshop, and when she announced that her next book was about space settlement, the room broke out into applause. It was a weird moment, because she'd held a more optimistic view of the near-term opportunities when she'd originally signed up for the workshop. By the time she was being praised as part of the movement, she wasn't so sure anymore.

We like all these people. We like applause. We like solidarity among fellow dorkwads. We're science geeks. We're tech geeks. We stay up late to watch rocket launches with our kids and we break out the telescope on clear winter nights. We believe that technological advancement has a major role in creating a wonderful future for humanity.

But we just cannot convince ourselves that the usual arguments for space settlements are good. Space settlement will be much harder than it is usually portrayed, without obvious economic benefits. Attempting space settlement now may increase the likelihood of conflict on Earth in the short term and ultimately increase human existential risk. Even the awesomeness of the view out of a Martian habitat is not worth all this.

Russian cosmonauts have an anthem called "The Grass by Our Home." It is a profoundly 1980s ballad, renditions of which typically involve a keytar and at least three mullets. But it contains some surprisingly moving lyrics, told from the perspective of a man in space contemplating Earth. The chorus goes like this:

We dream not about the roar of the cosmodrome,
Not about this icy blue space.
We dream of grass, grass by our home,
*Green, green grass.**

Having devoted all this time to trying to see a more realistic vision of space settlement, one of the few things we can say with certainty is that we have more appreciation of this grass by *our* home. In fact, not just the grass, but the home itself, and the people inside it.

Fantasies about going to space usually involve escape. Sometimes it's the personal escape of a single character in a story, but just as often it's about humanity escaping institutions and traditions seen as repressive or ugly or dying or dull. But here's the thing: You can't leave. Not really. Not in time to stop any calamity on the horizon or any social decay you see as imminent. And if you could leave and build a new civilization, do you know what you'd do first? You'd start re-creating Earth as we know it. Not just our biosphere, but social institutions we've had to wrench away from the darker side of our nature—things like the rule of law, human rights, and norms of behavior between societies.

Our biggest fear in writing this book is that the very people who helped us learn about space settlement will be disappointed or angry at our treatment of it. Well, dammit, they shouldn't be. We love this community, and we've done our best to further their discourse. We believe space settlements are possible, and perhaps one day could be done in safety. But doing something big requires us to assess the scale of the challenge. In healthy communities of thought, the people in the corner shaking their heads and wagging their fingers aren't barriers on the road to progress, but guardrails.

Earth isn't perfect, but as planets go it's a pretty good one. We aren't saying you should give up on the hope for a life off-world—that's too pretty a dream to part with. What we are saying is that if you *do* want that dream, you have to see the challenges as they are—real, profound, and present at

*The song is written by Zemylane ("Earthlings"). Lyric translation by Artem Kalyanov and Kelly Weinersmith.

every level from molecules to sociology. We hope, at the very least, when you read an article or hear a conversation about space settlement as an idea, you'll be able to see it as a very rich problem, which won't be solved simply by ambitious fantasies or giant rockets.

OKAY OKAY, BUT . . . or, a Moderately Hopeful Coda

If you hate our conclusions here, we have excellent news: we are not powerful people. Zach is a guy who draws funny pictures for the internet. Kelly is the president of a regional society for the study of parasitic worms. The world does not tremble before us.

Although we do not think humanity should buy the proverbial hot tub just yet, space agencies and very very rich people are already figuring out how many proverbial water jets should massage humanity's proverbial butt. So as we close this book, rather than considering what humanity should do overall, we want to consider what humanity should do if space expansion is a foregone conclusion. Imagine then that through some unfortunate sequence of events, a pair of Weinersmiths have been tasked with running a $20 billion agency for space settlement. Here's where we'd allocate funds:

Track 1: Biology, Reproduction, and Ecology

The life sciences pose hard problems for long-term survival in space, and they are also problems about which we know very little. The big answers we need are about the very long-term effects of life in space—not just in space stations, but in partial gravity. We also need answers on how to safely reproduce off-world. Maybe it's broadly similar to how it goes on Earth, or maybe we need a giant rotating baby-drome above Mars. Very little progress has been made in this direction, though as we were writing this conclusion, news came out that China is planning to send monkeys up for some intimate time in their space station. We wish them luck getting primates to have adult time in a space station the size of an apartment and with a crew of three taikonauts.

As silly as this sounds, it is broadly the right direction. Animal research on long-term biology and reproduction in space is important, not just as a way to produce more humans on Mars, but as a way to do so ethically. We'd love to see satellites out past the magnetosphere, rotating at partial gravity, home to multiple generations of rodents who seem as happy and healthy as can be. Or, if we get a research station on the Moon, we'd like that work done there instead. That sounds like a good start. Part of why this sort of science hasn't advanced is that the space agencies generally shy away from anything involving the word "sex." This is a chance for private enterprise to shine! Jeff Bezos's Amazon owns a huge fleet of vans—they aren't Chevys but they look pretty roomy. Elon Musk is responsible for a high percentage of all children born on Earth. We can work together. We can do this.

The other piece of the puzzle is the ecology. Biosphere 2 nearly showed that you can sustain eight people for two years on about three acres. For the price of today's International Space Station, about five hundred such facilities could've been constructed. Better still, five thousand of them, with a lot of variance in size, in order to try out new combinations and figure out the absolute minimum setup that doesn't involve the crew eating only algae for years. Incidentally, on this warming planet, a deep systems-level knowledge of ecosystems and how to produce functional ones in weird environments might just come in handy.

In the fullness of time, a proper ecosystem should be attempted somewhere off-world—perhaps in a lunar lava tube or near the perpetually lit poles. Once we have a self-sustaining life-support system that doesn't break for years at a time, when KRUSTYs keep us warm through the long lunar nights, when human adults can spend years on the Moon without losing their eyesight or losing their minds, and once babies are safely born and grow up on Earth's closest neighbor—then it'll be time to scale up to Mars. We can and should visit Mars in the meantime, to figure out things like how to remove perchlorates from the soil and make sure of local resources, but until we understand how to ensure long-term human survival, these should not go past the research outpost stage. Ultimately, when the right technology is finally ready, all these threads should be pulled together to make one massive Mars settlement during a very brief period.

That's assuming, of course, you can get everyone to agree on a peaceful and legal way to start a settlement.

Track 2: International Law

Human legal norms are substantially more confusing than space monkeys in love. We believe the right law more or less already exists—it should be something like UNCLOS or the Moon Agreement, but calibrated to be signable by the major space powers. There is no doubt research that can be done here, but a big part of the job is advocacy—getting people to understand the actual deal on space settlement. Why it matters, how it should be done, what time scales are reasonable, and where we can go from here. That is part of what we've attempted to do in this book. Most nerds who are interested in space settlement want to study rockets and spacecraft. But this is an equally important research track, and frankly it's one that should be of interest to any young person looking to change the world. International law is one of the few fields where by understanding history and law and writing about it you can have a vast long-term effect on the future of humanity. If you want the sort of space settlements that sci-fi geeks dream about, we need a legal regime that regulates away the possibility of territorial scrambles in the short term, allows peaceful scientific work together in the medium term, and perhaps, *perhaps*, one distant day, will permit the peaceful creation of independent nations off-Earth. Just as Dr. Oscar Schachter writing in the 1950s helped shape the law that governs today's solar system, scholars writing here in the 2020s can shape the law that governs the 2050s and beyond.

Track 3: Geopolitics, Sociology, and Economics

Because this book is designed for a general audience who doesn't want to read texts that weigh more than they do, we were forced to skip some topics that might one day be useful to know more about: the design of constitutions, the effects of religion, how culture is influenced by geography, and detailed treatments of economic factors like the effect of six-month transit times on the value of commodities. Economist friends of ours, it seems, would've preferred exhaustive analyses of every possible scenario, which we suspect would've tried the patience of even the nerdiest of pop-science consumers. However, if humanity must settle space, these details not only matter, but they can be adjudicated rather than speculated about. There really are scholars who study the features of constitutions that tend to survive for a long time, and they are rarely, if ever, consulted by space geeks planning to start space nations. Part of why we didn't discuss this stuff much is simply that we see it as being extremely dependent on hard-to-predict specifics. But if we were in charge of a well-funded agency for space settlement, we would happily devote some of our budget to people prepared to understand both the technical details of space-settlement creation *and* the way the humans inside might successfully coexist. Scientists like Dr. Charles Cockell are trying to get discussions like this started now, and we'd love to see more people getting involved in the conversation. Over the long term, this knowledge too could be combined with experiments in lunar ecology, to try to find the best ways to run small, distant societies with economies partially disconnected from the home planet.

The Good News for Nerds

Can we just say, this is all fun? Our pessimism about the possibility for short-term space settlement has in no way interfered with our fascination for the topic. What other endeavor requires you to understand everything from orbital mechanics to ecology to history, law, and war? Space-settlement studies are often siloed, with technological people having only passing knowledge of the law, and with law people having limited knowledge of technological feasibility. There is so much wonderful work that could be done by young scholars willing to just read a stack of books about the size of Asgardia-1, write up their findings, and submit them to one of the relevant journals. Some of the best papers we came across while researching this book were written by collaborators from distant fields—like the papers by astrophysicst Dr. Martin Elvis, space philosopher Dr. Tony Milligan, and political scientist Dr. Alanna Krolikowski.

Why aren't there more people like this? We're talking about a field where by reading a bunch of fascinating obscure literature you can contribute to the human future, garner public interest, increase scientific knowledge in dozens of fields, and meet some of the most amazing weirdos the planet has to offer. If Dr. Daniel Deudney is right that we are facing *Dark Skies*, you have a chance to show us some light in the clouds—and to have a damn good time doing it. This is especially true *if you think we're wrong.* If you think our species must inhabit other worlds, and not remain, as anthropologist and philosopher Dr. Joseph Campbell wrote, "a kind of recently developed scurf on the epidermis of one of the lesser satellites of a minor star in the outer arm of an average galaxy," then you should think of this book not as reasons to stay home but as challenges on the road outbound.

We don't know how to do it yet, but we still believe that someday, with enough knowledge, we can have Mars. And one very faraway day, other solar systems. But we have to earn it, both by gaining in knowledge and by becoming a more responsible, more peaceful species. Going to the stars will not make us wise. We have to become wise if we want to go to the stars.

Acknowledgments

It is traditional in acknowledgments to say that none of those who helped are responsible for mistakes in the text. We would like to go a bit further. Because this book is somewhat controversial, at least among people nerdy enough to get worked up over international law pertaining to the Moon, we want to emphatically state that none of the people below are necessarily responsible for our views.

We believe a functioning scientific community should welcome dissent that shows up with citations, but in the space-settlement community, people with negative views about aspects of space settlement are sometimes called idiots or "anti-human" or worse. So we especially want to applaud those readers who exercised the proper spirit of academic research and helped us with this book even when our conclusions ran contrary to theirs. In other fields, this would be called "normal." Here, we think it warrants the name "bravery."

Because space research is bottomless in its depth and technicality, we wish to first thank those people who answered questions, helped us track down or clarify difficult details, or get more accurate information: Bryan Caplan, Winchell Chung, Francis Cucinotta, Jason Derleth, Daniel Deudney, Dmitrii Komarov, David Livingston, Alan Manning, Beyond NERVA, Vladimir Papitashvili, Rod Pyle, Joshua M. Rice, Asif Siddiqi, Doug Plata, Larissa Starukhina, Mark Takata, John Vasquez, Lena Yakovleva, Matt Zefferman, and Linyi Zhang. Also, just, the endless helpers on Twitter. Thank you for proving social media can, occasionally, be a net positive.

ACKNOWLEDGMENTS

We wish to thank those people who read chapters, sections, and in some cases the entire book. We are especially grateful to those of you who read the early draft, which somehow had ten thousand words on the history of space simulations and analogs. History will not acknowledge your suffering, but we do: Ran Abramitzky, Timiebi Aganaba, Ruth Nic Aibhne, Grant Anderson, Joe Batwinis, Samuel Bazzi, Chris Blattman, Haym Benaroya, Callan Bentley, Michael W. Busch, Dan Cassino, Yuk Chi Chan, Charles Cockell, Simon Commander, Giacomo Delledonne, Scott P. Egan, Manfred Ehresmann, Martin Elvis, Annie Handmer, Casey Handmer, Robin Hanson, James Hendricks, Robert Gooding-Townsend, Matt Fantastic, Chad Jones, Malik K., Ram Jakhu, Charles Kenny, J. R. H. Lawless, John Lehr, Chris Lewicki, David Luebke, Jonathan McDowell, Timothy Miller, Clay Moltz, Michael Munger, Ryan North, Paul Novitski, Oscar Ojeda, Sinéad O'Sullivan, Frédéric Padilla, Phil Plait, Kevin Ringelman, Alexander Roederer, James Schwartz, Daniel Shaw, Scott Solomon, Louie Terrill, Andrew Thaler, Ben Tolkin, Gary Tyrrell, Tomer Ullman, Frans von der Dunk, Robert Wagner, Dan Warren, Martin and Phyllis Weiner, Chris White, Daniel Whiteson, and members of the UK's Defence Science and Technology Laboratory.

We wish to thank our editors, Virginia Smith-Younce and Caroline Sydney, for gently yet insistently turning a gigantic dossier on literally everything about space settlement into a readable, pop-science book. Thanks to our copy editor, Jane Cavolina, for making us look literate.

We thank our literary agent, Seth Fishman, and our managers, Mark Saffian and Josh Morris, for their continued efforts at making our weird careers work and keeping reality at bay so we can think deeply about toilets on Mars.

And we thank our children, Ada and Ben, for enduring countless conversations (and sometimes monologues) on obscure space-travel topics, when they really wanted to talk about cartoons and pie. We are sorry you are stuck with us, and very glad about it.

Many people contributed to the development of this book, but ultimately any mistakes that remain in the text are entirely the fault of Phil Plait.

Notes

These notes contain only citations associated with quotes presented in the text and manuscripts we refer to directly.

INTRODUCTION: A HOMESTEADER'S GUIDE TO THE RED PLANET?

1 **"It is no longer":** Tim Peake, *Ask an Astronaut* (London: Arrow Books, 2018), 240.

1 **In 2020, for example:** Antonino Salmeri, "No, Mars Is Not a Free Planet, No Matter What SpaceX Says," SpaceNews, December 5, 2020, https://spacenews.com/op-ed-no-mars-is-not-a-free-planet-no-matter-what-spacex-says/.

2 **Consider the 2015 *Newsweek*:** Kevin Maney, "'Star Wars' Class Wars: Is Mars the Escape Hatch for the 1 Percent?," *Newsweek*, December 14, 2015, https://www.newsweek.com/2015/12/25/mars-colonies-rich-people-404681.html.

10 **In 2019, their CEO:** Sean O'Kane, "Space Birth Startup's CEO Halts Project over 'Serious Ethical, Safety and Medical Concerns,'" The Verge, July 3, 2019, https://www.theverge.com/2019/7/3/20680006/space-birth-startup-project-ceo-serious-ethical-safety-medical-concerns-halt.

14 **In a 2022 interview:** German Lopez, "To the Moon," *New York Times,* September 1, 2022, https://www.nytimes.com/2022/09/01/briefing/nasa-moon-mars-space-travel.html.

14 **As they write:** John M. Olson, Steven J. Butow, Eric Felt, and Thomas Cooley, "State of the Space Industrial Base 2022: Winning the New Space Race for Sustainability, Prosperity and the Planet," Defense Innovation Unit, August 2022, 24, https://assets.ctfassets.net/3nanhbfkr0pc/6L5409bpVlnVyu2H5FOFnc/7595c4909616df92372a1d31be609625/State_of_the_Space_Industrial_Base_2022_Report.pdf.

16 **Consider his recent:** Elon Musk Twitter post, May 19, 2022, https://twitter.com/elonmusk/status/1527356085090545664.

1. A PREAMBLE ON SPACE MYTHS

21 **"Idyllic views of":** Andy Weir, foreword, Chris Carberry, *Alcohol in Space: Past, Present, and Future* (Jefferson, NC: McFarland and Co., 2019), 2.

23 **As Jeff Bezos says:** Noah Kulwin, "Jeff Bezos Thinks We Need to Build Industrial Zones in Space in Order to Save the Earth", CNBC, June 1, 2016, https://www.cnbc.com/2016/06/01/jeff-bezos-thinks-we-need-to-build-industrial-zones-in-space-in-order-to-save-earth.html.

26 **Although they're necessary:** World Bank, *The Changing Wealth of Nations 2021: Managing Assets for the Future* (Washington, D.C.: World Bank, 2021), 194, https://openknowledge.worldbank.org/handle/10986/36400.

28 **Dr. De Witt Kilgore:** De Witt Douglas Kilgore, *Astrofuturism: Science, Race, and Visions of Utopia in Space* (Philadelphia: University of Pennsylvania Press, 2003), 172.

28 **The prettiest version:** Carl Sagan, *Pale Blue Dot: A Vision of the Human Future in Space* (New York: Random House, 1994), xii.

30 **Yet here in 2022:** Morgan McFall-Johnsen, "The Russian Space Chief and Scott Kelly, a Former NASA Astronaut, Attack Each Other on Twitter—Calling Each Other 'Moron' and 'Child,'" *Business*

Insider, March 8, 2022, https://www.businessinsider.com/astronaut-scott-kelly-and-russia-space-chief-trade-twitter-insults-2022-3.

31 **As he says:** Frank White, *The Overview Effect: Space Exploration and Human Evolution*, 3rd ed. (Reston, VA: American Institute of Aeronautics and Astronautics, Inc., 2014), 2.

PART I: CARING FOR THE SPACEFARING

41 **Dr. Sally Ride:** Lynn Sherr, *Sally Ride: America's First Woman in Space* (New York: Simon & Schuster, 2014), 103.

2. SUFFOCATION, BONE LOSS, AND FLYING PIGS: THE SCIENCE OF SPACE PHYSIOLOGY

53 **For near-term efforts:** Lina Tran, "How NASA Will Protect Astronauts from Space Radiation at the Moon," NASA, August 7, 2019, https://www.nasa.gov/feature/goddard/2019/how-nasa-protects-astronauts-from-space-radiation-at-moon-mars-solar-cosmic-rays.

55 **In 1859, our planet:** Stuart Clark, *The Sun Kings: The Unexpected Tragedy of Richard Carrington and the Tale of How Modern Astronomy Began* (Princeton, NJ: Princeton University Press, 2007), 83–88.

56 **the words of one:** Rod Pyle, *Destination Mars: New Explorations of the Red Planet* (Amherst, NY: Prometheus, 2012), 168.

59 **you've ever wanted:** Jeffrey Chancellor et al., "Limitations in Predicting the Space Radiation Health Risk for Exploration Astronauts," *NPJ Microgravity* 4 (2018): 7, https://doi.org/10.1038/s41526-018-0043-2.

61 **Mullane recalls her:** Mike Mullane, "Backaches," in *Space Shuttle: The First 20 Years—The Astronauts' Experiences in Their Own Words*, ed. Tony Reichhardt (London: DK Publishing, 2002), 161.

61 **One study of ISS:** Scott Trappe et al., "Exercise in Space: Human Skeletal Muscle After 6 months Aboard the International Space Station," *Journal of Applied Physiology* 106 (2009): 1159–168, https://doi.org/10.1152/japplphysiol.91578.2008.

63 **In one survey:** Thomas Mader et al., "Optic Disc Edema, Globe Flattening, Choroidal Folds, and Hyperopic Shifts Observed in Astronauts after Long-Duration Space Flight," *Ophthalmology* 118 (2011): 2058–69, https://doi.org.10.1016/j.ophtha.2011.06.021.

64 **"pants that suck":** Scott Kelly, *Endurance: A Year in Space, a Lifetime of Discovery* (New York: Alfred A. Knopf, 2017), 141.

65 **For instance, according to:** S. S. Panesar and K. Ashkan, "Surgery in Space," *British Journal of Surgery* 105 (2018): 1237, https://doi.org/10.1002/bjs.10908.

66 **"Bowel floats in":** Laura Drudi et al., "Surgery in Space: Where Are We at Now?," *Acta Astronautica* 79 (2012): 63, https://doi.org/10.1016/j.actaastro.2012.04.014.

66 **"The tendency of organs":** Panesar and Ashkan, "Surgery in Space."

3. SPACE SEX AND CONSEQUENCES THEREOF

70 **As space popularizers:** James E. Oberg and Alcestis R. Oberg, *Pioneering Space: Living on the Next Frontier* (New York: McGraw-Hill, 1985), 188–89.

71 **And then, best of all:** T. A. Heppenheimer, *Colonies in Space* (New York: Warner Books, 1980), 209.

71 **Space-tourist Dennis Tito:** Mike Mullane, *Riding Rockets: The Outrageous Tales of a Space Shuttle Astronaut* (New York: Scribner, 2007), 176.

72 **Desire appears to be:** Mary Roach, *Packing for Mars: The Curious Science of Life in the Void* (New York: W. W. Norton & Co., 2010), 59.

72 **Norm Thagard remembers:** Rebecca Wright, "Edited Oral History Transcript: Norman E. Thagard," NASA Shuttle-Mir Oral History Project, September 16, 1998, https://historycollection.jsc.nasa.gov/JSCHistoryPortal/history/oral_histories/Shuttle-Mir/ThagardNE/ThagardNE_9-16-98.htm.

72 **Cosmonaut Polyakov, who you may:** Peter Pesavento, "From Aelita to the International Space Station: The Psychological and Social Effects of Isolation on Earth and in Space," *Quest* 8:2 (2000), 21.

72 **More recently, Scott Kelly:** David Plotz, "Exit Interview: Scott Kelly, an Astronaut Who Spent a Year in Space," Atlas Obscura, November 9, 2017, http://www.atlasobscura.com/articles/scott-kelly-astronaut-exit-interview-space-station-nasa.

72 **paper written by:** April Ronca et al., "Effects of Sex and Gender on Adaptations to Space: Reproductive Health," *Journal of Women's Health* 23 (2014): 969, https://doi.org/10.1089/jwh.2014.4915.

75 **"As of 1990":** G. Harry Stine, *Halfway to Anywhere: Achieving America's Destiny in Space* (M. Evans & Company, 1996), 246–47.

78 **Keeping SMACs at:** Meytar Sorek-Hamer and Marit Meyer, "Developing an Air Quality Index for Space Vehicles and Habitats," Meeting of the American Association for Aerosol Research, Portland, OR, 2019, https://ntrs.nasa.gov/citations/20190032069.

85 **Space anthropologist Dr. Cameron Smith:** Cameron Smith, *Principles of Space Anthropology: Establishing a Science of Human Space Settlement* (Switzerland: Springer, 2019): 68–69.

85 **Dr. Alexander Layendecker:** Alexander Layendecker, "Sex in Outer Space and the Advent of Astrosexology: A Philosophical Inquiry into the Implication of Human Sexuality and Reproductive Development Factors in Seeding Humanity's Future Throughout the Cosmos and the Argument for an Astrosexology Research Institute" (PhD diss., Institute for Advanced Study of Human Sexuality, 2016), 98.

86 **Perhaps most openly:** Konrad Szocik et al., "Biological and Social Challenges of Human Reproduction in a Long-Term Mars Base," *Futures* 100 (2018): 58, https://doi.org/10.1016/j.futures.2018.04.006.

4. SPACEFARER PSYCHOLOGY: IN WHICH THE ONLY THING WE'RE SURE OF IS THAT ASTRONAUTS ARE LIARS

93 **memoirs of Deke:** Deke Slayton and Michael Cassutt, *Deke! U.S. Manned Space from Mercury to the Shuttle* (New York: Forge, 1994), 62–63.

94 **more recent quote:** Mike Massimino, *Spaceman: An Astronaut's Unlikely Journey to Unlock the Secrets of the Universe* (New York: Crown Archetype, 2016), 23.

96 **Today's ISS has:** Terry Virts, *How to Astronaut: An Insider's Guide to Leaving Planet Earth* (New York: Workman Publishing Co., 2020), 176.

98 **"We had recordings":** James E. Oberg and Alcestis R. Oberg, *Pioneering Space: Living on the Next Frontier* (New York: McGraw-Hill, 1985), 10.

99 **To give one example:** AeroDork, "All the conditions necessary for murder are met if you shut two men in a cabin measuring 18 feet by 20 and leave them together for two months.—Diary of Cosmonaut Valery Ryumin, Salyut 6." Twitter, May 27, 2019. https://twitter.com/AeroDork/status/1133173589505695744.

99 **"O. Henry wrote":** Andrew Chaikin, "The Loneliness of the Long-Distance Astronaut," *Discover* 6 (1985): 20.

99 **Mission control said:** Taylor Wang, "I'm Not Going Back," in *Space Shuttle: The First 20 Years—The Astronauts' Experiences in Their Own Words*, ed. Tony Reichhardt (London: DK Publishing, Inc., 2002): 232–33.

101 **description of what:** Asif Siddiqi, "The Almaz Space Station Complex. A History, 1964–1992. Part 2: 1976–1992," *Journal of the British Interplanetary Society* 55 (2002): 35–67.

102 **He later recalled:** Robert Zimmerman, *Leaving Earth: Space Stations, Rival Superpowers, and the Quest for Interplanetary Travel* (Washington, D.C.: Joseph Henry Press, 2004), 244.

102 **He ended up:** Mary Roach, *Packing for Mars: The Curious Science of Life in the Void* (New York: W. W. Norton & Co., 2010), 56.

103 **One analysis of:** NASA Human Research Program, "Evidence Report: Risk of Adverse Cognitive or Behavioral Conditions and Psychiatric Disorders," NASA, Lyndon B. Johnson Space Center, Houston, TX, April 11, 2016, 18, https://humanresearchroadmap.nasa.gov/evidence/reports/bmed.pdf.

103 **When Eileen Collins:** Eileen Collins and Jonathan Ward, *Through the Glass Ceiling to the Stars: The Story of the First American Woman to Command a Space Mission* (New York: Arcade, 2021), 122.

104 **"the [male] cosmonauts":** Slava Gerovitch, *Voices of the Soviet Space Program: Cosmonauts, Soldiers, and Engineers Who Took the USSR into Space* (New York: Palgrave Macmillan, 2014), 220–21.

105 **Or, as veteran:** Bryan Burrough, *Dragonfly: An Epic Adventure of Survival in Outer Space* (New York: HarperCollins, 2000), 182.

105 **Mike Mullane, who removed:** Mike Mullane, *Riding Rockets: The Outrageous Tales of a Space Shuttle Astronaut* (New York: Scribner, 2007), 23.

108 **Well, we can tell you:** International Space Station Integrated Medical Group (IMG) Medical Checklist, NASA, Missions Operations Directorate Operations Division, 2001, 492–501.

NOTA BENE: ROCKETRY GOES TO THE MOVIES, OR, SPACE CAPITALISM IN DAYS OF YORE, PART 1

114 **account by Oberth's friend:** Boris Rauschenbach, *Hermann Oberth: The Father of Space Flight* (New York: West Art Publishing, 1994), 63–76.

PART II: SPOME, SPOME ON THE RANGE: WHERE WILL HUMANS LIVE OFF-WORLD?

117 **In 1835, the New York:** Meg Matthias, "The Great Moon Hoax of 1835 Was Sci-Fi Passed Off as News," Britannica, https://www.britannica.com/story/the-great-moon-hoax-of-1835-was-sci-fi-passed-off-as-news.

118 **As late as 1964:** David Portree, *Humans to Mars: Fifty Years of Mission Planning, 1950–2000* (Washington, D.C.: NASA, 2011), 17, https://www.lpi.usra.edu/lunar/strategies/HumanstoMars.pdf.

5. THE MOON: GREAT LOCATION, BIT OF A FIXER-UPPER

123 **As John Young:** "EVA-3 Closeout," NASA, Apollo 16 Lunar Surface Journal, corrected commentary and transcript by Eric M. Jones, https://www.hq.nasa.gov/alsj/a16/a16.clsout3.html, last revised June 16, 2014.

134 **As space visionary:** Krafft Ehricke, "Lunar Industrialization and Settlement—Birth of Polyglobal Civilization," in *Lunar Bases and Space Activities of the 21st Century*, ed. W. W. Mendell (Houston, TX: Lunar and Planetary Institute, 1985), 830.

135 **As NASA's international:** NASA, "Artemis Plan: NASA's Lunar Exploration Program Overview" (NP-2020-05-2853-HQ), September 2020, 9, https://www.nasa.gov/sites/default/files/atoms/files/artemis_plan-20200921.pdf.

6. MARS: LANDSCAPES OF POISON AND TOXIC SKIES, BUT WHAT AN OPPORTUNITY!

137 **"Send old men":** Walter Cunningham, *All-American Boys* (New York: iPicturebooks, 2010), 439.

145 **As late as 1968:** Arthur C. Clarke, *The Promise of Space* (New York: Harper & Row, 1968), 236.

7. GIANT ROTATING SPACE WHEELS: NOT LITERALLY THE WORST OPTION

153 **"A major shift of":** M. J. Queijo et al., "Analyses of a Rotating Advanced-Technology Space Station for the Year 2025," NASA Contractor Report 178345 (Bionetics Corporation, Hampton, VA, 1988), 107, https://ntrs.nasa.gov/api/citations/19880010196/downloads/19880010196.pdf.

NOTA BENE: SPACE IS THE PLACE FOR PRODUCT PLACEMENT, OR SPACE CAPITALISM IN DAYS OF YORE, PART 2

169 **US law allows private companies:** Title 51: National and Commercial Space Programs, USC Ch. 509: Commercial Space Launch Activities, Subtitle V: Programs Targeting Commercial Opportunities, Section 11: Space advertising.

169 **As they note in a 2002:** "Report of the Scientific and Technical Subcommittee on its 39th session, held in Vienna from 25 February to 8 March 2002," United Nations Committee on the Peaceful Uses of Outer Space, 2002, https://digitallibrary.un.org/record/465209, 20.

PART III: POCKET EDENS: HOW TO CREATE A HUMAN TERRARIUM THAT ISN'T ALL THAT TERRIBLE

9. OUTPUTS AND INPUTS: POOP, FOOD, AND "CLOSING THE LOOP"

174 **during Gemini 7:** Jeffrey Kluger, *Apollo 8: The Thrilling Story of the First Mission to the Moon* (New York: Henry Holt & Co., 2017), 80.

175 **As ISS Commander:** Hilary Brueck, "A NASA Astronaut Who Spent 665 Days Circling the Planet Reveals the Misery of Going to the Bathroom in Space," *Business Insider*, May 26, 2018, https://www.businessinsider.com/how-you-go-to-bathroom-space-nasa-astronaut-2018-5.

175 **Aboard today's ISS:** Tim Peake, *Ask an Astronaut* (London: Arrow Books, 2018), 90.
175 **details are scarce:** Scott Gleeson, "Elon Musk Says SpaceX Inspiration4 Crew Had 'Challenges' with Toilet," *USA Today*, September 23, 2021, https://www.usatoday.com/story/tech/news/2021/09/23/elon-musk-says-spacex-inspiration-4-crew-had-challenges-toilet/5825068001/.
176 **"Tang sucks.":** Mark Memmott, "Now He Tells Us: 'Tang Sucks,' Says Apollo 11's Buzz Aldrin," NPR, June 13, 2013, https://www.npr.org/sections/thetwo-way/2013/06/13/191271824/now-he-tells-us-tang-sucks-says-apollo-11s-buzz-aldrin.
177 **Pilot Sid Gutierrez:** Sid Gutierrez, "Rice Krispies and taco sauce," in *Space Shuttle: The First 20 Years—The Astronauts' Experiences in Their Own Words*, ed. Tony Reichhardt (London: DK Publishing, 2002): 219.
178 **account by food scientist:** Jennifer Ross-Nazzal, "Edited Oral History Transcript, Paul A. Lachance," NASA Johnson Space Center Oral History Project, May 4, 2006, https://historycollection.jsc.nasa.gov/JSCHistoryPortal/history/oral_histories/LachancePA/LachancePA_5-4-06.htm.
179 **answer is no:** Andy Weir, foreword, Chris Carberry, *Alcohol in Space: Past, Present, and Future* (Jefferson, NC: McFarland and Co., 2019), 1–2.
180 **As he said:** James E. Oberg and Alcestis R. Oberg, *Pioneering Space: Living on the Next Frontier* (New York: McGraw-Hill, 1985), 10.
181 **first experiment growing:** Anna-Lisa Paul, Stephen Elardo, and Robert Ferl, "Plants Grown in Apollo Lunar Regolith Present Stress-Associated Transcriptomes That Inform Prospects for Lunar Exploration," *Communications Biology* 5 (2022): 382, https://www.nature.com/articles/s42003-022-03334-8.
182 **a 2020 editorial:** Marina Carcea, "Quality and Nutritional/Textural Properties of Durum Wheat Pasta Enriched with Cricket Powder," *Foods* 9 (2020): 1298, https://doi.org/10.3390/foods9091298.
183 **a vegetarian diet:** Frank Salisbury, Josef Gitelson, and Genry Lisovsky, "Bios-3: Siberian Experiments in Bioregenerative Life Support," *BioScience* 47 (1997): 575–85.
184 **"Perhaps on Mars":** Jane Poynter, *The Human Experiment: Two Years and Twenty Minutes Inside Biosphere 2* (New York: Basic Books, 2006), 216.
185 **what bark scorpions:** Animal Fact Sheet: Bark Scorpion, Arizona-Sonora Desert Museum, https://www.desertmuseum.org/kids/oz/long-fact-sheets/Bark%20Scorp.php.

10. THERE'S NO PLACE LIKE SPOME: HOW TO BUILD OUTER-SPACE HABITATS

192 **"It is not hyperbole:"** Haym Benaroya, *Building Habitats on the Moon: Engineering Approaches to Lunar Settlements* (New York: Springer, 2018), 51.
203 **As space architect Brent Sherwood:** Brent Sherwood, "Lunar Architecture and Urbanism," in *Out of This World*, eds. Scott Howe and Brent Sherwood (Reston, VA: AIAA, 2009), 323.
208 **to create "lunarcrete":** Matthew Sparkes, "Astronauts Could Use Their Blood and Urine to Make Martian Concrete," *New Scientist*, September 14, 2021, https://www.newscientist.com/article/2290046-astronauts-could-use-their-blood-and-urine-to-make-martian-concrete/.
208 **paper in favor:** Dave Dietzler, "Making It on the Moon: Bootstrapping Lunar Industry," *NSS Space Settlement Journal* (September 2016): 15, https://space.nss.org/wp-content/uploads/NSS-JOURNAL-Bootstrapping-Lunar-Industry-2016.pdf.

NOTA BENE: THE MYSTERY OF THE TAMPON BANDOLIER

213 **It goes like this:** Lynn Sherr, *Sally Ride: America's First Woman in Space* (New York: Simon & Schuster, 2014), 144–46.
213 **"There was concern":** Jennifer Ross-Nazzal, "Edited Oral History Transcript: Margaret Rhea Seddon," NASA Johnson Space Center Oral History Project, May 21, 2010, https://historycollection.jsc.nasa.gov/JSCHistoryPortal/history/oral_histories/SeddonMR/SeddonMR_5-21-10.htm.
214 **typical NASA behavior:** NASA, "Human Integration Design Handbook: Revision 1," Report NASA/SP-2010-3407/REV1 (June 5, 2014), 604, https://www.nasa.gov/sites/default/files/atoms/files/human_integration_design_handbook_revision_1.pdf.
215 **the urination devices:** Amy Foster, *Integrating Women into the Astronaut Corps: Politics and Logistics at NASA, 1972–2004* (Baltimore: Johns Hopkins University Press, 2011), 124.
215 **As Amy Foster:** Foster, *Integrating Women into the Astronaut Corps.*
216 **According to Michael:** Michael Collins, *Carrying the Fire* (New York: Farrar, Straus and Giroux, 2019), 200.

PART IV: SPACE LAW FOR SPACE SETTLEMENTS: WEIRD, VAGUE, AND HARD TO CHANGE

217 **"No moonlit night":** Clive S. Lewis, "Onward Christian Spacemen," in *Reflections on Space: Its Implications for Domestic and International Affairs*, ed. Oscar Rechtschaffen (Colorado Springs, CO: United States Air Force Academy, 1964), 29.

11. A CYNICAL HISTORY OF SPACE

219 **"Perhaps maturity in":** Walter McDougall, . . . *The Heavens and the Earth: A Political History of the Space Age* (Baltimore: Johns Hopkins University Press, 1997), 414.
221 **the words of physicist:** William Burrows, *This New Ocean: The Story of the First Space Age* (New York: Random House, 1998), 90.
223 **describing the arrival:** Yves Beon, *Planet Dora: A Memoir of the Holocaust and the Birth of the Space Age* (New York: Basic Books, 1998), 125.
223 **In one of Beon's:** Beon, *Planet Dora*, 49.
224 **As one staff member:** McDougall, . . . *The Heavens and the Earth*, 44.
224 **As Stalin reportedly:** McDougall, . . . *The Heavens and the Earth.*
227 **they treated him:** Charles Murray and Catherine Bly Cox, *Apollo* (New York: Simon & Schuster, 1989), 46, Kindle.
227 **In an interview years later:** John Logsdon, *John F. Kennedy and the Race to the Moon* (New York: Palgrave Macmillan, 2010), 6.
227 **Kennedy assigned his:** Logsdon, *John F. Kennedy and the Race to the Moon*, 1.
227 **press conference after:** Logsdon, *John F. Kennedy and the Race to the Moon*, 71.
230 **a 1952 speech:** Fred Nadis, *Star Settlers: The Billionaires, Geniuses, and Crazed Visionaries Out to Conquer the Universe* (New York: Pegasus Books, 2020), 63.
232 **Some see the OST:** McDougall, . . . *The Heavens and the Earth*, 273.
232 **Dr. Everett Dolman:** Everett Dolman, *Astropolitik: Classical Geopolitics in the Space Age* (London: Frank Cass, 2002), 8.

12. THE OUTER SPACE TREATY: GREAT FOR REGULATING SPACE SIXTY YEARS AGO

242 **If it's not clear:** Sanjeev Miglani and Krishna Das, "Modi Hails India as Military Space Power After Anti-Satellite Missile Test," Reuters (March 27, 2019), https://www.reuters.com/article/us-india-satellite/modi-hails-india-as-military-space-power-after-anti-satellite-missile-test-idUSKCN1R80IA.
244 **As James Dunstan:** James Dunstan, "Who Wants to Step Up to a $10 Billion Risk?," SpaceNews, June 25, 2021, https://spacenews.com/op-ed-who-wants-to-step-up-to-a-10-billion-risk/.
246 **Dr. Martin Elvis:** Martin Elvis, Tony Milligan, and Alanna Krolikowski, "The Peaks of Eternal Light: A Near-Term Property Issue on the Moon," *Space Policy* 38 (2016): 31, https://doi.org/10.1016/j.spacepol.2016.05.011.
246 **As Dr. Elvis:** Elvis, Milligan, and Krolikowski, "The Peaks of Eternal Light," 30–38.

13. MURDER IN SPACE: WHO KILLED THE MOON AGREEMENT?

254 **"It is traditional":** E. B. White, "Notes and Comment," *New Yorker*, July 18, 1969, https://www.newyorker.com/magazine/1969/07/26/comment-5238.
255 **In the regrettable:** Nathan Schachner, *Space Lawyer* (New York: Gnome Press, 1953), 211.
258 **"If you touch":** Timothy Nelson, "The Moon Agreement and Private Enterprise: Lessons from Investment Law," *ILSA Journal of International & Comparative Law* 17 (2010): 399, https://nsuworks.nova.edu/ilsajournal/vol17/iss2/6.
259 **A fear among:** Nelson, "The Moon Agreement and Private Enterprise," 401.
260 **This may sound vague:** The Law Dictionary, "Sample: Definition & Legal Meaning," https://thelawdictionary.org/sample/.
261 **NASA administrator Robert Frosch:** Thomas Gangale, "Common Heritage in Magnificent Desolation," in 46th AIAA Aerospace Sciences Meeting and Exhibit, 2008, 7.

262 **Keith Henson, spouse:** Thomas Gangale, "Common Heritage in Magnificent Desolation," 5.

264 **In 2015, President:** United States Congress, "U.S. Commercial Space Law Competitiveness Act," 114th Congress, 2015, https://www.congress.gov/bill/114th-congress/house-bill/2262/text.

265 **In 2020, President:** Donald Trump, "Executive Order on Encouraging International Support for the Recovery and Use of Space Resources," April 6, 2020, https://trumpwhitehouse.archives.gov/presidential-actions/executive-order-encouraging-international-support-recovery-use-space-resources/.

270 **whose contributing authors:** John M. Olson, Steven J. Butow, Eric Felt, and Thomas Cooley, "State of the Space Industrial Base 2022: Winning the New Space Race for Sustainability, Prosperity and the Planet," Defense Innovation Unit, August 2022, 2–3, https://assets.ctfassets.net/3nanhbfkr0pc/6L5409bpVlnVyu2H5FOFnc/7595c4909616df92372a1d31be609625/State_of_the_Space_Industrial_Base_2022_Report.pdf.

NOTA BENE: SPACE CANNIBALISM FROM A LEGAL AND CULINARY PERSPECTIVE

275 **Seedhouse asks, "Imagine":** Erik Seedhouse, *Survival and Sacrifice in Mars Exploration: What We Know from Polar Expeditions* (New York: Springer, 2015), 143.

PART V: THE PATHS FORWARD: BOUND FOR MOONSYLVANIA?

14. COMMONSING THE COSMOS

280 **The language goes:** Virgiliu Pop, *Who Owns the Moon?* (New York: Springer, 2008), 96.

282 **account by geographer:** Robert Wilson, "National Interests and Claims in the Antarctic," *Arctic* 17 (March 1964): 21, https://www.jstor.org/stable/40507108.

286 **her book, *Antarctica*:** Doaa Abdel-Motaal, *Antarctica: The Battle for the Seventh Continent* (Santa Barbara, CA: Praeger: 2016), 154.

292 **as Donald Rumsfeld said:** Donald Rumsfeld, "The Law of the Sea Convention (Treaty Doc. 103-39)," in Hearings Before the Committee on Foreign Relations, United States Senate, 112th Congress, 2nd Session, 2012, https://www.govinfo.gov/content/pkg/CHRG-112shrg77375/html/CHRG-112shrg77375.htm.

293 **In his 1952 essay:** Oscar Schachter, "Who Owns the Universe?," in *Across the Space Frontier*, ed. Cornelius Ryan (New York: Viking Press, 1952), 121.

15. DIVIDING THE SKY

295 **As Locke wrote:** John Locke, *Second Treatise of Government* (Indianapolis, IN: Hackett Publishing, 1980), 19.

295 **Or, as Assistant Land Commissioner:** Virgiliu Pop, *Who Owns the Moon?* (New York: Springer, 2008), 104.

296 **Well-known space advocate:** Rand Simberg, "Multilateral Agreements for Real Property Rights in the Solar System," 70th International Astronautical Congress, Washington, D.C., 2016, 6, https://iafastro.directory/iac/paper/id/53013/summary/.

297 **Consider a proposal:** Jacob Haqq-Misra, "The Transformative Value of Liberating Mars," *New Space* 4 (2016): 64–67, https://doi.org/10.1089/space.2015.0030.

299 **Natural Law, which:** Alan Wasser, "The Space Settlement Initiative," SpaceSettlement.org, February 2016, http://www.spacesettlement.org.

301 **At the 2019:** Simberg, "Multilateral Agreements for Real Property Rights in the Solar System," 1–6.

302 **Back in 1981:** Newt Gingrich, "National Space and Aeronautics Policy Act of 1981," Pub. L. No. H.R.4286 (1981), https://www.congress.gov/bill/97th-congress/house-bill/4286.

305 **how much land:** Martin Elvis and Tony Milligan, "How Much of the Solar System Should We Leave as Wilderness?," *Acta Astronautica* 162 (2019): 574–80, https://doi.org/10.1016/j.actaastro.2019.03.014; Tony Milligan and Martin Elvis, "Mars Environmental Protection: An Application of the 1/8 Principle," in *The Human Factor in a Mission to Mars: An Interdisciplinary Approach*, ed. Konrad Szocik (Switzerland: Springer, 2019), 167–83.

307 **As Dr. Ram Jakhu:** Ram Jakhu, Joseph Pelton, and Yaw Otu Mankata Nyampong, *Space Mining and Its Regulation* (Chichester, UK: Springer, 2017), 7, https://doi.org/10.1007/978-3-319-39246-2.

16. THE BIRTH OF SPACE-STATES: LIKE THE BIRTH OF SPACE BABIES, BUT MESSIER

309 **According to Ashurbeyli:** Andrea Lo, "Asgardia, the World's First 'Space Nation,' Takes Flight," CNN, November 15, 2017, https://www.cnn.com/style/article/asgardia-satellite-launch/index.html.

310 **Dr. Robert Zubrin:** Robert Zubrin, "But the purpose of going into space is to create new nations. That can best be done on Mars," Reddit, September 25, 2021, https://www.reddit.com/r/spacex/comments/pv91cs/comment/he8xy4w/.

NOTE BENE: VIOLENCE IN ANTARCTICA, OR, HAPPY ENDINGS TO STABBY STARTS

329 **So Kelly wrote:** Theodore Shabad, "Russians Reveal Tale of Survival in the Long Polar Winter," *New York Times*, April 26, 1983.

PART VI: TO PLAN B OR NOT TO PLAN B: SPACE SOCIETY, EXPANSION, AND EXISTENTIAL RISK

17. THERE'S NO LABOR POOL ON MARS: OUTER SPACE AS A COMPANY TOWN

341 **As one author put:** Clayton Strange, *Monotown: Urban Dreams Brutal Imperatives* (San Francisco: Applied Research & Design, 2019), 56.

341 **found one account:** Eric Thompson, *On Her Majesty's Nuclear Service* (Oxford, UK: Casemate, 2018), 154–55.

342 **On September 3:** Associated Press, "AP Was There: Covering the Battle for Blair Mountain in 1921," *Seattle Times*, September 4, 2021, https://www.seattletimes.com/business/ap-was-there-covering-the-battle-for-blair-mountain-in-1921/.

18. HOW BIG IS BIG? PLAN B SETTLEMENTS WITHOUT GENETIC OR ECONOMIC CALAMITIES

354 **One recent study:** Sayaka Wakayama et al., "Evaluating the Long-Term Effect of Space Radiation on the Reproductive Normality of Mammalian Sperm Preserved on the International Space Station," *Science Advances* 7 (2021): eabg5554, https://doi.org/10.1126/sciadv.abg5554.

356 **Our source for this:** Elon Musk (@ElonMusk), "20 to 30 years from first human landing if launch rate growth is exponential. Assumes transferring ~100k each rendezvous and ~1M total people needed," Twitter, July 15, 2022, 8:43 a.m., https://twitter.com/elonmusk/status/1547924891613048833?t=d3DEDdqcKNnqpFGPiQv_Lw&s=19.

19. SPACE POLITICS BY OTHER MEANS: ON THE POSSIBILITY OF SPACE WAR

359 **"Who controls low-Earth":** Everett Dolman, *Astropolitik: Classical Geopolitics in the Space Age* (London: Frank Cass, 2002), 8.

359 **In 1999, US:** Roscoe Bartlett, "Electromagnetic Pulse," Congressional Record Online, vol. 151, no. 83, June 21, 2005, H4888—H4894, https://www.govinfo.gov/content/pkg/CREC-2005-06-21/html/CREC-2005-06-21-pt1-PgH4888-2.htm.

362 **Soon after, the Russian:** Konstantin Vorontsov, "Statement by Deputy Head of the Russian Delegation Mr. Konstantin Vorontsov at the Thematic Discussion on Outer Space (Disarmament Aspects) in the First Committee of the 77th Session of the UNGA," Permanent Mission of the Russian Federation to the United Nations, October 26, 2022, https://russiaun.ru/en/news/261022_v.

368 **"what's contested on Earth":** Neil deGrasse Tyson and Avis Lang, *Accessory to War: The Unspoken Alliance Between Astrophysics and the Military* (New York: W. W. Norton & Co., 2018), 384.

369 **Dr. Chris Blattman:** Christopher Blattman, *Why We Fight: The Roots of War and the Path to Peace* (New York: Viking, 2022), 16.

372 **James Madison, one of the framers:** James Madison, "'Helvidius' Number 4 [14 September] 1793," Founders Online, https://founders.archives.gov/documents/Madison/01-15-02-0070.

20. A BRIEF CODA ON A RARELY CONSIDERED ALTERNATIVE: WAIT-AND-GO-NOWHERE

377 **as Deudney notes:** Daniel Deudney, *Dark Skies: Space Expansionism, Planetary Geopolitics, and the Ends of Humanity* (New York: Oxford University Press, 2020), 250.

NOTA BENE: AMUSING ASTRONAUT NAMES AND THE SOVIET TENDENCY TO FUSS OVER WEIRD DETAILS

379 **"There was one":** Colin Burgess and Bert Vis, *Interkosmos: The Eastern Bloc's Early Space Program* (Chichester, UK: Springer, 2016), 146.

CONCLUSION: OF HOT TUBS AND HUMAN DESTINY

388 **our species must:** Joseph Campbell, *The Inner Reaches of Outer Space: Myth as Metaphor and as Religion* (Novato, CA: New World Library, 2018), xxi.

Partial Bibliography

Abdel-Motaal, Doaa. *Antarctica: The Battle for the Seventh Continent.* Santa Barbara, California: Praeger, 2016.

Abood, Steven. "Martian Environmental Psychology: The Choice Architecture of a Mars Mission and Colony." In *The Human Factor in a Mission to Mars: An Interdisciplinary Approach*, edited by Konrad Szocik, 3–34. Space and Society. Cham, Switzerland: Springer International Publishing, 2019. https://doi.org/10.1007/978-3-030-02059-0_1.

Adams, Gregory R., Vincent J. Caiozzo, and Kenneth M. Baldwin. "Skeletal Muscle Unweighting: Spaceflight and Ground-Based Models." *Journal of Applied Physiology* 95 (2003): 2185–201. https://doi.org/10.1152/japplphysiol.00346.2003.

Aganaba, Timiebi. "Innovative Instruments for Space Governance." Centre for International Governance Innovation, February 8, 2021. https://www.cigionline.org/articles/innovative-instruments-space-governance/.

Aganaba-Jeanty, Timiebi. "Common Benefit from a Perspective of 'Non-Traditional Partners': A Proposed Agenda to Address the Status Quo in Global Space Governance." *Acta Astronautica* 117 (2015): 172–83. https://doi.org/10.1016/j.actaastro.2015.07.014.

Ahmed, Selina. "Comparison of Soviet and U.S. Space Food and Nutrition Programs." NASA/JSC, August 22, 1988. https://ntrs.nasa.gov/citations/19890010688.

Aldrin, Buzz, and Ken Abraham. *Magnificent Desolation: The Long Journey Home from the Moon.* New York: Crown Archetype, 2009.

Aleci, Carlo. "From International Ophthalmology to Space Ophthalmology: The Threats to Vision on the Way to Moon and Mars Colonization." *International Ophthalmology* 40 (2020): 775–86. https://doi.org/10.1007/s10792-019-01212-7.

Allen, John. *Me and the Biospheres: A Memoir by the Inventor of Biosphere 2.* Santa Fe, NM: Synergetic Press, 2009.

Allen, Joseph P., and Russell Martin. *Entering Space: An Astronaut's Odyssey.* New York: Stewart Tabori & Chang, 1984.

Alshamsi, Humaid, Roy Balleste, and Michelle L. D. Hanlon. "Space Station Asgardia 2117: From Theoretical Science to a New Nation in Outer Space." In *16 Santa Clara Journal of International Law* 37 (2018): 37–60. St. Thomas University School of Law Research Paper no. 2019-02. https://papers.ssrn.com/abstract=3350627.

Andersen, Ross. "Elon Musk Puts His Case for a Multi-Planet Civilisation." Aeon, September 30, 2014. https://aeon.co/essays/elon-musk-puts-his-case-for-a-multi-planet-civilisation.

Andrews, James T. *Red Cosmos: K. E. Tsiolkovskii, Grandfather of Soviet Rocketry.* Centennial of Flight Series 18. College Station: Texas A&M University Press, 2009.

Angelis, De Giovanni, J. W. Wilson, M. S. Clowdsley, J. E. Nealy, D. H. Humes, and J. M. Clem. "Lunar Lava Tube Radiation Safety Analysis." *Journal of Radiation Research* 43 (2002): S41–45. https://doi.org/10.1269/jrr.43.S41.

Ansari, Anousheh. *My Dream of Stars.* Stanford, CT: Griffin Books, 2011.

Arendt, Josephine. "Biological Rhythms During Residence in Polar Regions." *Chronobiology International* 29 (2012): 379–94. https://doi.org/10.3109/07420528.2012.668997.

Associated Press. "Rocket to Carry Pizza Hut Logo." *New York Times*, October 1, 1999. https://www.nytimes.com/1999/10/01/business/the-media-business-rocket-to-carry-pizza-hut-logo.html.

Atal, Maha Rafi. "When Companies Rule: Corporate Political Authority in India, Kenya and South Africa." Thesis, University of Cambridge, 2019. https://doi.org/10.17863/CAM.37015.

Badescu, Viorel, ed. *Asteroids: Prospective Energy and Material Resources.* Heidelberg; New York: Springer, 2013.

———, ed. *Mars: Prospective Energy and Material Resources*. New York: Springer, 2009. https://www.springer.com/gp/book/9783642036286.

———, ed. *Moon: Prospective Energy and Material Resources*. Berlin: Springer, 2012.

Badescu, Viorel, and Kris Zacny, eds. *Outer Solar System: Prospective Energy and Material Resources*. New York: Springer, 2018.

Baker, Philip. *The Story of Manned Space Stations: An Introduction*. Chichester, UK: Springer / Praxis, 2007.

Balmforth, Tom. "First Woman in Space Brought Down to Earth by Anger over Bid to Prolong Putin Rule." Reuters, March 13, 2020. https://www.reuters.com/article/us-russia-putin-lawmaker-idUSKBN21021H.

Banet, Catherine. *The Law of the Seabed Access, Uses, and Protection of Seabed Resources*. Leiden and Boston: Brill, 2020.

Bank of America/Merrill Lynch. "To Infinity and Beyond—Global Space Primer." Thematic Investing. Bank of America/Merrill Lynch, October 30, 2017. https://newspaceglobal.com/wp-content/uploads/imce/u3479/MerrillLynchSpace-Oct2017.pdf.

Barger, Laura K., Erin E. Flynn-Evans, Alan Kubey, Lorcan Walsh, Joseph M Ronda, Wei Wang, Kenneth P. Wright, and Charles A Czeisler. "Prevalence of Sleep Deficiency and Use of Hypnotic Drugs in Astronauts Before, During, and After Spaceflight: An Observational Study." *Lancet Neurology* 13 (2014): 904–12. https://doi.org/10.1016/S1474-4422(14)70122-X.

Barrett, Emma, and Paul Martin. *Extreme: Why Some People Thrive at the Limits*. Oxford: Oxford University Press, 2016.

Barth, Bob. *Sea Dwellers: The Humor, Drama, and Tragedy of the U.S. Navy Sealab Programs*. Houston, TX: Doyle Publishing Co., 2000.

Barthel, Joseph, and Nesrin Sarigul-Klijn. "A Review of Radiation Shielding Needs and Concepts for Space Voyages Beyond Earth's Magnetic Influence." *Progress in Aerospace Sciences* 110 (2019): 100553. https://doi.org/10.1016/j.paerosci.2019.100553.

Basner, Mathias, David F. Dinges, Daniel Mollicone, Adrian Ecker, Christopher W. Jones, Eric C. Hyder, Adrian Di Antonio, et al. "Mars 520-d Mission Simulation Reveals Protracted Crew Hypokinesis and Alterations of Sleep Duration and Timing." *Proceedings of the National Academy of Sciences* 110 (2013): 2635–40. https://doi.org/10.1073/pnas.1212646110.

Bazzi, Samuel, Martin Fiszbein, and Mesay Gebresilasse. "Frontier Culture: The Roots and Persistence of 'Rugged Individualism' in the United States." *Econometrica* 88 (2020): 2329–68. https://doi.org/10.3982/ECTA16484.

———. "'Rugged Individualism' and Collective (in)Action during the COVID-19 Pandemic." *Journal of Public Economics* 195 (2021): 104357. https://doi.org/10.1016/j.jpubeco.2020.104357.

BBC. "Kosovo Profile." BBC News, April 12, 2021. https://www.bbc.com/news/world-europe-18328859.

Beardslee, Luke A., Ben D. Lawson, and David P. Regis. "An Overview of the Unique Field of Submarine Medicine." Naval Submarine Medical Research Laboratory, Groton, CT, March 10, 2019. https://apps.dtic.mil/sti/pdfs/AD1082304.pdf.

Belanger, Dian Olson. *Deep Freeze: The United States, the International Geophysical Year, and the Origins of Antarctica's Age of Science*. Boulder: University Press of Colorado, 2006.

Bell, Suzanne T., Shanique G. Brown, and Tyree Mitchell. "What We Know About Team Dynamics for Long-Distance Space Missions: A Systematic Review of Analog Research." *Frontiers in Psychology* 10 (2019): 811. https://doi.org/10.3389/fpsyg.2019.00811.

Benaroya, Haym. *Building Habitats on the Moon: Engineering Approaches to Lunar Settlements*. New York: Springer, 2018.

Bencke, Matthew Von. *The Politics of Space: A History of U.S.-Soviet/Russian Competition and Cooperation in Space*. Boulder, CO: Westview Press, 1996.

Bennett, John. "How Antarctic Isolation Affects the Mind." *Canadian Geographic*, September 15, 2016. https://canadiangeographic.ca/articles/how-antarctic-isolation-affects-the-mind/.

Beon, Yves. *Planet Dora: A Memoir of the Holocaust and the Birth of the Space Age*. New York: Basic Books, 1998.

Bernal, J. D., and McKenzie Wark. *The World, the Flesh and the Devil: An Enquiry into the Future of the Three Enemies of the Rational Soul*. London: Verso, 2017.

Bhatt, Mukesh Chiman. "Space for Dissent: Disobedience on Artificial Habitats and Planetary Settlements." In *Dissent, Revolution and Liberty Beyond Earth*, edited by Charles S. Cockell, 71–92. Space and Society. Cham, Switzerland: Springer International Publishing, 2016. https://doi.org/10.1007/978-3-319-29349-3_6.

Bignami, Giovanni, and Andrea Sommariva. *The Future of Human Space Exploration*. London: Palgrave Macmillan, 2016.

Bishop, Sheryl L. "From Earth Analogues to Space: Learning How to Boldly Go." In *On Orbit and Beyond: Psychological Perspectives on Human Spaceflight*, edited by Douglas A. Vakoch, 25–50. Space Technology Library. Berlin, Heidelberg: Springer, 2013. https://doi.org/10.1007/978-3-642-30583-2_2.

PARTIAL BIBLIOGRAPHY

Blair, David M., Loic Chappaz, Rohan Sood, Colleen Milbury, Antonio Bobet, H. Jay Melosh, Kathleen C. Howell, and Andrew M. Freed. "The Structural Stability of Lunar Lava Tubes." *Icarus* 282 (2017): 47–55. https://doi.org/10.1016/j.icarus.2016.10.008.

Blair, Roger D., and Jeffrey L. Harrison. *Monopsony in Law and Economics*. New York: Cambridge University Press, 2010.

Blattman, Christopher. *Why We Fight: The Roots of War and the Paths to Peace*. New York: Viking, 2022.

Blonsky, George B., and Charlotte E. Blonsky. Apparatus for Facilitating the Birth of a Child by Centrifugal Force. Patent application United States US3216423A, filed January 15, 1963, and issued November 9, 1965. https://patents.google.com/patent/US3216423/en.

Boal, William M. "Testing for Employer Monopsony in Turn-of-the-Century Coal Mining." *RAND Journal of Economics* 26 (1995): 519–36. https://www.jstor.org/stable/2556001.

Bolonkin, Alexander A. "Economic Development of Mercury: A Comparison with Mars Colonization." In *Inner Solar System: Prospective Energy and Material Resources*, edited by Viorel Badescu and Kris Zacny, 407–19. Cham, Switzerland: Springer International Publishing, 2015. https://doi.org/10.1007/978-3-319-19569-8_19.

Borman, Frank, and Robert J. Serling. *Countdown: An Autobiography*. New York: Silver Arrow, 1988.

Botella, Cristina, Rosa M. Baños, Ernestina Etchemendy, Azucena García-Palacios, and Mariano Alcañiz. "Psychological Countermeasures in Manned Space Missions: 'EARTH' System for the Mars-500 Project." *Computers in Human Behavior* 55 (2016): 898–908. https://doi.org/10.1016/j.chb.2015.10.010.

Bourland, Charles T., and Gregory L. Vogt. *The Astronaut's Cookbook: Tales, Recipes, and More*. New York and London: Springer, 2010.

Bowen, Bleddyn E. *War in Space: Strategy, Spacepower, Geopolitics*. Edinburgh: Edinburgh University Press, 2020.

Bown, Stephen R. *Scurvy: How a Surgeon, a Mariner, and a Gentlemen Solved the Greatest Medical Mystery of the Age of Sail*. New York: St. Martin's / Griffin, 2005.

Boyd, Jennifer E., Nick Kanas, Vadim I. Gushin, and Stephanie Saylor. "Cultural Differences in Patterns of Mood States on Board the International Space Station." *Acta Astronautica* 61 (2007): 668–71. https://doi.org/10.1016/j.actaastro.2006.12.002.

Boyle, Alan. "NBC Launching 2001 Space Odyssey." NBC News, October 26, 2006. https://www.nbcnews.com/id/wbna15424344.

Brand, Stewart, ed. *Space Colonies*. Harmondsworth, UK: Penguin, 1977.

Braun, Wernher von. "Prelude to Space Travel." In *Across the Space Frontier*, edited by Cornelius Ryan, 12–70. New York: Viking Press, 1952.

Brown, H. M., A. K. Boyd, B. W. Denevi, M. R. Henriksen, M. R. Manheim, M. S. Robinson, E. J. Speyerer, and R. V. Wagner. "Resource Potential of Lunar Permanently Shadowed Regions." *Icarus* 377 (2022): 114874. https://doi.org/10.1016/j.icarus.2021.114874.

Bruhaug, Gerrit, and William Phillips. "Nuclear Fuel Resources of the Moon: A Broad Analysis of Future Lunar Nuclear Fuel Utilization." *NSS Space Settlement Journal*, no. 5 (2021): 11. https://space.nss.org/wp-content/uploads/NSS-JOURNAL-Nuclear-Fuel-Resources-of-the-Moon-2021-June.pdf.

Bruhns, Sara, and Jacob Haqq-Misra. "A Pragmatic Approach to Sovereignty on Mars." *Space Policy* 38 (2016): 57–63. https://doi.org/10.1016/j.spacepol.2016.05.008.

Brünner, Christian, and Alexander Soucek, eds. *Outer Space in Society, Politics and Law*. Studies in Space Policy 8. Vienna: Springer, 2011. https://link.springer.com/book/10.1007/978-3-7091-0664-8.

Bruno, Claudio, ed. *Nuclear Space Power and Propulsion Systems*. Reston, VA: AIAA, 2008.

Buckley, James Michael. "A Factory Without a Roof: The Company Town in the Redwood Lumber Industry." *Perspectives in Vernacular Architecture* 7 (1997): 75–92. https://doi.org/10.2307/3514386.

Buden, David. *Space Nuclear Fission Electric Power Systems*. Lakewood, CO: Polaris Books, 2011.

Burgess, Colin, and Chris Dubbs. *Animals in Space: From Research Rockets to the Space Shuttle*. Berlin; New York: Springer, 2007.

Burgess, Colin, and Bert Vis. *Interkosmos: The Eastern Bloc's Early Space Program*. Chichester, UK: Springer / Praxis, 2016.

Burns, Robin. *Just Tell Them I Survived!: Women in Antarctica*. Crows Nest, NSW: Allen & Unwin, 2001.

Burrough, Bryan. *Dragonfly: An Epic Adventure of Survival in Outer Space*. New York: HarperCollins, 2000.

Burrows, William E. *This New Ocean : The Story of the First Space Age*. New York: Random House, 1998.

Butler, Carol. "Robert E. Stevenson Oral History." NASA Johnson Space Center Oral History Project, Houston, TX, 1999. https://historycollection.jsc.nasa.gov/JSCHistoryPortal/history/oral_histories/StevensonRE/StevensonRE_5-13-99.htm.

Cain, John R. "Space Terrorism—A New Environment; New Causes." In *Dissent, Revolution and Liberty Beyond Earth*, edited by Charles S. Cockell, 93–109. Space and Society. Cham, Switzerland: Springer International Publishing, 2016. https://doi.org/10.1007/978-3-319-29349-3_7.

PARTIAL BIBLIOGRAPHY

Cannon, Kevin M., and Daniel T. Britt. "Feeding One Million People on Mars." *New Space* 7 (2019): 245–54. https://doi.org/10.1089/space.2019.0018.

Cannon, Kevin M., Ariel N. Deutsch, James W. Head, and Daniel T. Britt. "Stratigraphy of Ice and Ejecta Deposits at the Lunar Poles." *Geophysical Research Letters* 47 (2020): e2020GL088920. https://doi.org/10.1029/2020GL088920.

Cantor, Bruce A., Nicholas B. Pickett, Michael C. Malin, Steven W. Lee, Michael J. Wolff, and Michael A. Caplinger. "Martian Dust Storm Activity near the Mars 2020 Candidate Landing Sites: MRO-MARCI Observations from Mars Years 28–34." *Icarus* 321 (2019): 161–70. https://doi.org/10.1016/j.icarus.2018.10.005.

Capova, Klara Anna. "Human Extremophiles: Mars as a Camera Obscura of the Extraterrestrial Scientific Culture." In *The Human Factor in a Mission to Mars: An Interdisciplinary Approach*, edited by Konrad Szocik, 115–32. Space and Society. Cham: Springer International Publishing, 2019. https://doi.org/10.1007/978-3-030-02059-0_6.

Carberry, Chris. *Alcohol in Space: Past, Present and Future.* Illustrated edition. Jefferson, North Carolina: McFarland, 2019.

Cassini, John. "Space Fission Power: NASA's Best Bet to Continue to Explore the Outer Solar System." In *Nuclear and Emerging Technologies for Space*, American Nuclear Society Topical Meeting, Richland, Washington, February 25–28, 2019, 5. Pasadena, CA: Jet Propulsion Laboratory, National Aeronautics and Space Administration, 2019. https://trs.jpl.nasa.gov/handle/2014/50405.

CBS. "Cold-Blooded: Scientist in Antarctica Accused of Stabbing Colleague for Spoiling the Endings of Books." CBS News Los Angeles, October 30, 2018. https://www.cbsnews.com/losangeles/news/antarctica-stabbing-sergey-savitsky-oleg-beloguzov-bellinghausen/.

Chaikin, Andrew. "The Loneliness of the Long-Distance Astronaut." *Discover*, February 1985, 20–31.

Chancellor, Jeffery C., Rebecca S. Blue, Keith A. Cengel, Serena M. Auñón-Chancellor, Kathleen H. Rubins, Helmut G. Katzgraber, and Ann R. Kennedy. "Limitations in Predicting the Space Radiation Health Risk for Exploration Astronauts." *NPJ Microgravity* 4 (2018): 1–11. https://doi.org/10.1038/s41526-018-0043-2.

Chancellor, Jeffery C., Graham B. I. Scott, and Jeffrey P. Sutton. "Space Radiation: The Number One Risk to Astronaut Health beyond Low Earth Orbit." *Life* 4 (2014): 491–510. https://doi.org/10.3390/life4030491.

Chancellor, Jeffery, Craig Nowadly, Jacqueline Williams, Serena Aunon-Chancellor, Megan Chesal, Jayme Looper, and Wayne Newhauser. "Everything You Wanted to Know about Space Radiation but Were Afraid to Ask." *Journal of Environmental Science and Health, Part C* 39 (2021): 113–28. https://doi.org/10.1080/26896583.2021.1897273.

Chen, Stephen. "Chinese Scientists Plan Monkey Reproduction Experiment in Space Station." *South China Morning Post*, November 3, 2022. https://www.scmp.com/news/china/science/article/3198222/chinese-scientists-plan-monkey-reproduction-experiment-space-station.

Cherry-Garrard, Apsley. *The Worst Journey in the World: Antarctic, 1910–1913.* New York: Penguin Classics, 2006.

Clarke, Arthur C. *The Promise of Space.* New York: Harper & Row, 1968.

Clarke, Jonathan D. A., ed. *Mars Analog Research*, vol. 111. Science and Technology Series. San Diego, CA: Univelt, 2006.

Cockell, Charles. "Mars Is an Awful Place to Live." *Interdisciplinary Science Reviews* 27 (2002): 32–38. https://doi.org/10.1179/030801802225002881.

Cockell, Charles S. "A Simple Land Use Policy for Mars." In *Mars Analog Research*, edited by Jonathan D. A. Clarke, 111: 301–11. Science and Technology Series. San Diego, CA: American Astronautical Society, 2006.

———. "Extraterrestrial Liberty: Can It Be Planned?" In *Human Governance Beyond Earth: Implications for Freedom*, edited by Charles S. Cockell, 23–42. Space and Society. Cham, Switzerland: Springer International Publishing, 2015. https://doi.org/10.1007/978-3-319-18063-2_3.

———. "Freedom in a Box: Paradoxes in the Structure of Extraterrestrial Liberty." In *The Meaning of Liberty Beyond Earth*, edited by Charles S. Cockell, 47–68. Space and Society. Cham: Springer International Publishing, 2015. https://doi.org/10.1007/978-3-319-09567-7_4.

———. "Introduction: Human Governance and Liberty Beyond Earth." In *Human Governance Beyond Earth: Implications for Freedom*, edited by Charles S. Cockell, 1–8. Space and Society. Cham, Switzerland: Springer International Publishing, 2015. https://doi.org/10.1007/978-3-319-18063-2_1.

———. "Introduction: The Meaning of Liberty Beyond the Earth." In *The Meaning of Liberty Beyond Earth*, edited by Charles S. Cockell, 1–9. Space and Society. Cham, Switzerland: Springer International Publishing, 2015. https://doi.org/10.1007/978-3-319-09567-7_1.

———, ed. *Martian Expedition Planning*, vol. 107. Science and Technology Series. San Diego, CA: American Astronautical Society, 2004.

PARTIAL BIBLIOGRAPHY

Cockell, Charles S, and Gerda Horneck. "Planetary Parks—Formulating a Wilderness Policy for Planetary Bodies." *Space Policy* 22 (2006): 256–61. https://doi.org/10.1016/j.spacepol.2006.08.006.

Colaprete, Anthony, Peter Schultz, Jennifer Heldmann, Diane Wooden, Mark Shirley, Kimberly Ennico, Brendan Hermalyn, et al. "Detection of Water in the LCROSS Ejecta Plume." *Science* 330 (2010): 463–68. https://doi.org/10.1126/science.1186986.

Cole, Dandridge M., and Donald W. Cox. *Islands in Space: The Challenge of the Planetoids*. Philadelphia: Chilton Books, 1964.

Collins, David A. "Efficient Allocation of Property Rights on the Planet Mars." *Boston University Journal of Law and Technology* 14 (2008): 201–20. https://papers.ssrn.com/abstract=1556665.

Collins, Eileen, and Jonathan H. Ward. *Through the Glass Ceiling to the Stars: The Story of the First American Woman to Command a Space Mission*. New York: Arcade, 2021.

Collins, Michael. *Carrying the Fire*. New York: Farrar, Straus and Giroux, 2019.

Commander, Simon. "One-Company Towns: Scale and Consequences." *IZA World of Labor* 433 (2018): 1–9. https://doi.org/10.15185/izawol.433.

Compton, W. David, and Charles D. Benson. *Living and Working in Space: A NASA History of Skylab*. US: CreateSpace Independent Publishing Platform, 2014.

Condon, Stephanie. "Romney Tells Gingrich: I'd Fire You for Your Moon Proposal." CBS News, January 27, 2012. https://www.cbsnews.com/news/romney-tells-gingrich-id-fire-you-for-your-moon-proposal/.

Conrad, Nancy, and Howard A. Klausner. *Rocketman: Astronaut Pete Conrad's Incredible Ride to the Moon and Beyond*. New York: NAL Hardcover, 2005.

Crawford, Ian A. "Interplanetary Federalism: Maximising the Chances of Extraterrestrial Peace, Diversity and Liberty." In *The Meaning of Liberty Beyond Earth*, edited by Charles S. Cockell, 199–218. Space and Society. Cham, Switzerland: Springer International Publishing, 2015. https://doi.org/10.1007/978-3-319-09567-7_13.

———. "Lunar Resources: A Review." *Progress in Physical Geography* 39 (2015): 137–67. https://doi.org/10.1177/0309133314567585.

Crawford, James R. *The Creation of States in International Law*. Oxford: Oxford University Press, 2011.

Cristoforetti, Samantha. *Diary of an Apprentice Astronaut*. London: Allen Lane, 2020.

Cronon, William. "Revisiting the Vanishing Frontier: The Legacy of Frederick Jackson Turner." *Western Historical Quarterly* 18 (1987): 157–76. https://doi.org/10.2307/969581.

Crossman, Frank, ed. *MARS COLONIES: Plans for Settling the Red Planet*. Lakewood, CO: Polaris Books, 2019.

Cucinotta, Francis A. "Space Radiation Risks for Astronauts on Multiple International Space Station Missions." *PLOS ONE* 9 (2014): e96099. https://doi.org/10.1371/journal.pone.0096099.

Cucinotta, Francis A., and Eliedonna Cacao. "Predictions of Cognitive Detriments from Galactic Cosmic Ray Exposures to Astronauts on Exploration Missions." *Life Sciences in Space Research* 25 (2020): 129–35. https://doi.org/10.1016/j.lssr.2019.10.004.

Cucinotta, Francis A., Eliedonna Cacao, Myung-Hee Y. Kim, and Premkumar B. Saganti. "Benchmarking Risk Predictions and Uncertainties in the NSCR Model of GCR Cancer Risks with Revised Low Let Risk Coefficients." *Life Sciences in Space Research* 27 (2020): 64–73. https://doi.org/10.1016/j.lssr.2020.07.008.

———. "Cancer and Circulatory Disease Risks for a Human Mission to Mars: Private Mission Considerations." *Acta Astronautica* 166 (2020): 529–36. https://doi.org/10.1016/j.actaastro.2018.08.022.

Cucinotta, Francis A., Nobuyuki Hamada, and Mark P. Little. "No Evidence for an Increase in Circulatory Disease Mortality in Astronauts Following Space Radiation Exposures." *Life Sciences in Space Research* 10 (2016): 53–56. https://doi.org/10.1016/j.lssr.2016.08.002.

Cunningham, Walter. *All-American Boys*. New York: iPicturebooks, 2010.

Curley, Su, Imelda Stambaugh, Michael Swickrath, Molly S. Anderson, and Henry Rotter. "Deep Space Habitat ECLSS Design Concept," 15. American Institute of Aeronautics and Astronautics, Reston, VA, 2012. https://arc.aiaa.org/doi/abs/10.2514/6.2012-3417.

Dapremont, Angela M. "Mars Land Use Policy Implementation: Approaches and Best Methods." *Space Policy* 57 (2021): 101442. https://doi.org/10.1016/j.spacepol.2021.101442.

Davenport, Christian. *The Space Barons: Elon Musk, Jeff Bezos, and the Quest to Colonize the Cosmos*. New York: PublicAffairs, 2018.

David, Leonard. *Moon Rush: The New Space Race*. Washington, DC: National Geographic, 2019.

Davila, Alfonso F., David Willson, John D. Coates, and Christopher P. McKay. "Perchlorate on Mars: A Chemical Hazard and a Resource for Humans." *International Journal of Astrobiology* 12 (2013): 321–25. https://doi.org/10.1017/S1473550413000189.

Dawson, Linda. *War in Space: The Science and Technology Behind Our Next Theater of Conflict*. New York: Springer, 2019.

PARTIAL BIBLIOGRAPHY

Day, Dwayne A. "The Reality Space Race." Space Review, June 21, 2004. https://www.thespacereview.com/article/165/1.

Deb, Dorothy. "Language, Culture and the Creation of Bangladesh." *Journal of Defense Studies* 15 (2021): 59–76. https://www.idsa.in/jds/language-culture-and-the-creation-of-bangladesh-dorothy-deb.

Del Rey, Lester. *Space Flight: The Coming Exploration of the Universe*. New York: Golden Press, 1959.

Delledonne, Giacomo, and Giuseppe Martinico, eds. *The Canadian Contribution to a Comparative Law of Secession: Legacies of the Quebec Secession Reference*. London: Palgrave Macmillan, 2019.

Demontis, Gian C., Marco M. Germani, Enrico G. Caiani, Ivana Barravecchia, Claudio Passino, and Debora Angeloni. "Human Pathophysiological Adaptations to the Space Environment." *Frontiers in Physiology* 8 (2017): 547. https://doi.org/10.3389/fphys.2017.00547.

Dempster, W. F., M. Van Thillo, A. Alling, J. P. Allen, S. Silverstone, and M. Nelson. "Technical Review of the Laboratory Biosphere Closed Ecological System Facility." *Advances in Space Research: Space Life Sciences: Life Support Systems and Biological Systems under Influence of Physical Factor*s, 34 (2004): 1477–82. https://doi.org/10.1016/j.asr.2003.10.034.

Dempster, William F. "Biosphere 2 Engineering Design." *Ecological Engineering* 13 (1999): 31–42. https://doi.org/10.1016/S0925-8574(98)00090-1.

Deudney, Daniel. *Dark Skies: Space Expansionism, Planetary Geopolitics, and the Ends of Humanity*. New York: Oxford University Press, 2020.

Dinius, Oliver J., and Angela Vergara, eds. *Company Towns in the Americas: Landscape, Power, and Working-Class Communities*. Athens: University of Georgia Press, 2011.

Dismukes, Kim. "History of Shuttle-Mir Home Page." NASA, June 14, 1999. https://spaceflight.nasa.gov/history/shuttle-mir/.

Dolman, Everett C. *Astropolitik: Classical Geopolitics in the Space Age*. London: Frank Cass Publishers, 2002.

Drudi, Laura, Chad G. Ball, Andrew W. Kirkpatrick, Joan Saary, and S. Marlene Grenon. "Surgery in Space: Where Are We at Now?," *Acta Astronautica* 79 (2012): 61–66. https://doi.org/10.1016/j.actaastro.2012.04.014.

Dunbar, Brian. "Food for Space Flight." NASA. November 25, 2019. http://www.nasa.gov/audience/forstudents/postsecondary/features/F_Food_for_Space_Flight.html.

Dunbar, Robin I. M., and Richard Sosis. "Optimising Human Community Sizes." *Evolution and Human Behavior* 39 (2018): 106–11. https://doi.org/10.1016/j.evolhumbehav.2017.11.001.

Dunk, Frans von der, and Fabio Tronchetti, eds. *Handbook of Space Law*. Cheltenham, UK: Edward Elgar Publishing, 2017.

Ehrenfreund, Pascale, Margaret Race, and David Labdon. "Responsible Space Exploration and Use: Balancing Stakeholder Interests." *New Space* 1 (2013): 60–72. https://doi.org/10.1089/space.2013.0007.

Ehrenfried, Manfred "Dutch" von. *From Cave Man to Cave Martian: Living in Caves on the Earth, Moon and Mars*. New York: Springer, 2019.

Elvis, Martin. *Asteroids: How Love, Fear, and Greed Will Determine Our Future in Space*. New Haven, CT: Yale University Press, 2021.

Elvis, Martin, Alanna Krolikowski, and Tony Milligan. "Concentrated Lunar Resources: Imminent Implications for Governance and Justice." *Philosophical Transactions of the Royal Society A: Mathematical, Physical and Engineering Sciences* 379 (2021): 20190563. https://doi.org/10.1098/rsta.2019.0563.

Elvis, Martin, Tony Milligan, and Alanna Krolikowski. "The Peaks of Eternal Light: A Near-Term Property Issue on the Moon." *Space Policy* 38 (2016): 30–38. https://doi.org/10.1016/j.spacepol.2016.05.011.

Escobar, Christine M., and James A. Nabity. "Past, Present, and Future of Closed Human Life Support Ecosystems—A Review," 18. 47th International Conference on Environmental Systems, July 16–20, 2017, Charleston, SC. https://ttu-ir.tdl.org/bitstream/handle/2346/73083/ICES_2017_311.pdf?sequence=1&isAllowed=y.

Evans, Ben. *At Home in Space: The Late Seventies into the Eighties*. New York: Springer, 2011.

Fearon, James D. "Rationalist Explanations for War." *International Organization* 49 (1995): 379–414. https://www.jstor.org/stable/2706903.

Fernholz, Tim. *Rocket Billionaires: Elon Musk, Jeff Bezos, and the New Space Race*. Boston: Mariner Books, 2018.

Ferrero-Turrión, Ruth. "The Consequences of State Non-Recognition: The Cases of Spain and Kosovo." *European Politics and Society* 22 (2021): 347–58. https://doi.org/10.1080/23745118.2020.1762958.

Finney, Ben R., and Eric M. Jones, eds. *Interstellar Migration and the Human Experience*. Berkeley: University of California Press, 1985.

Fishback, Price. "The Economics of Company Housing: Historical Perspectives from the Coal Fields." *Journal of Law, Economics and Organization* 8 (1992): 346–65. https://doi.org/10.1093/oxfordjournals.jleo.a037042.

Fishback, Price V. "Did Coal Miners 'Owe Their Souls to the Company Store'? Theory and Evidence from the Early 1900s." *Journal of Economic History* 46 (1986): 1011–29. https://doi.org/10.1017/S0022050700050695.

———. *Soft Coal, Hard Choices: The Economic Welfare of Bituminous Coal Miners, 1890–1930*. New York: Oxford University Press, 1992.

PARTIAL BIBLIOGRAPHY

Flynn-Evans, Erin E., Laura K. Barger, Alan A. Kubey, Jason P. Sullivan, and Charles A. Czeisler. "Circadian Misalignment Affects Sleep and Medication Use Before and During Spaceflight." *NPJ Microgravity* 2 (2016): 15019. https://doi.org/10.1038/npjmgrav.2015.19.

Foale, Colin. *Waystation to the Stars: The Story of Mir, Michael and Me*. London: Headline Books, 1999.

Fong, Kevin. *Extreme Medicine: How Exploration Transformed Medicine in the Twentieth Century*. New York: Penguin Press, 2014.

Foss, Richard. *Food in the Air and Space: The Surprising History of Food and Drink in the Skies*. New York: Rowman & Littlefield Publishers, 2016.

Foster, Amy E. *Integrating Women into the Astronaut Corps: Politics and Logistics at NASA, 1972–2004*. Baltimore, MD: Johns Hopkins University Press, 2011.

Foust, Jeff. "Pepsi Drops Plans to Use Orbital Billboard." SpaceNews, April 16, 2019. https://spacenews.com/pepsi-drops-plans-to-use-orbital-billboard/.

Freeman, Marsha. *How We Got to the Moon: The Story of the German Space Pioneers*. Washington, DC: 21st Century Science Associates, 1993.

Freitas, Jr., Robert A. "Survival Homicide in Space." Nanomedicine.com, November 20, 1978. http://www.nanomedicine.com/Papers/SurvivalHomicideInSpace1978.pdf.

Friedman, Louis. *Human Spaceflight: From Mars to the Stars*. Tucson: University of Arizona Press, 2015.

Fu, Yuming, Rong Guo, and Hong Liu. "An Optimized 4-day Diet Meal Plan for 'Lunar Palace 1.'" *Journal of the Science of Food and Agriculture* 99 (2019): 696–702. https://doi.org/10.1002/jsfa.9234.

Fu, Yuming, Leyuan Li, Beizhen Xie, Chen Dong, Mingjuan Wang, Boyang Jia, Lingzhi Shao, et al. "How to Establish a Bioregenerative Life Support System for Long-Term Crewed Missions to the Moon or Mars." *Astrobiology* 16 (2016): 925–36. https://doi.org/10.1089/ast.2016.1477.

Gagarin, Yuri, and Vladimir Lebedev. *Psychology and Space*. Honolulu, HI: University Press of the Pacific, 2003.

Gangale, Thomas. "Common Heritage in Magnificent Desolation." In *46th AIAA Aerospace Sciences Meeting and Exhibit*, 10. Reno, NV: American Institute of Aeronautics and Astronautics, Inc., 2008.

Gangale, Thomas, and Marilyn Dudley-Rowley. "To Build Bifrost: Developing Space Property Rights and Infrastructure," 1–11. Long Beach, CA: American Institute of Aeronautics and Astronautics, 2005. https://doi.org/10.2514/6.2005-6762.

Garcia, Mark. "NASA Station Astronaut Record Holders." NASA, April 20, 2022. http://www.nasa.gov/feature/nasa-station-astronaut-record-holders.

Gardner, Amy. "Gingrich Pledges Moon Colony during Presidency." *Washington Post*, January 25, 2012. https://www.washingtonpost.com/blogs/election-2012/post/gingrich-pledges-moon-colony-during-presidency/2012/01/25/gIQAmQxiRQ_blog.html.

Garshnek, Victoria. "Soviet Space Flight: The Human Element." *Aviation, Space, and Environmental Medicine* 60 (1989): 695–705.

Gemignani, Angelo, Andrea Piarulli, Danilo Menicucci, Marco Laurino, Giuseppina Rota, Francesca Mastorci, Vadim Gushin, et al. "How Stressful Are 105 Days of Isolation? Sleep EEG Patterns and Tonic Cortisol in Healthy Volunteers Simulating Manned Flight to Mars." *International Journal of Psychophysiology* 93 (2014): 211–19. https://doi.org/10.1016/j.ijpsycho.2014.04.008.

Genta, Giancarlo. *Next Stop Mars: The Why, How, and When of Human Missions*. New York: Springer, 2017.

Gibson, Marc A., Steven R. Oleson, David I. Poston, and Patrick McClure. "NASA's Kilopower Reactor Development and the Path to Higher Power Missions." In *2017 IEEE Aerospace Conference*, 1–14. Big Sky, MT: IEEE, 2017. https://doi.org/10.1109/AERO.2017.7943946.

Gibson, Marc A., David I. Poston, Patrick McClure, Thomas Godfroy, James Sanzi, and Maxwell H. Briggs. "The Kilopower Reactor Using Stirling TechnologY (KRUSTY) Nuclear Ground Test Results and Lessons Learned." In *2018 International Energy Conversion Engineering Conference*. Cincinnati, OH: American Institute of Aeronautics and Astronautics, 2018. https://doi.org/10.2514/6.2018-4973.

Gingrich, Newt. National Space and Aeronautics Policy Act of 1981, Pub. L. No. H.R.4286 (1981). https://www.congress.gov/bill/97th-congress/house-bill/4286.

Gläser, P., F. Scholten, D. De Rosa, R. Marco Figuera, J. Oberst, E. Mazarico, G. A. Neumann, and M. S. Robinson. "Illumination Conditions at the Lunar South Pole Using High Resolution Digital Terrain Models from LOLA." *Icarus* 243 (2014): 78–90. https://doi.org/10.1016/j.icarus.2014.08.013.

Glenn, John, and Nick Taylor. *John Glenn: A Memoir*. New York: Bantam, 2000.

Globus, Al. "Countering Objections to Space Settlement." Space.AlGlobus.net, May 2021. http://space.alglobus.net/papers/Countering-Objections-to-Space-Settlement.pdf.

Globus, Al, Stephen Covey, and Daniel Faber. "Space Settlement: An Easier Way." *NSS Space Settlement Journal*, July 2017, 1–46. http://space.alglobus.net/papers/Easy.pdf.

Globus, Al, and Theodore Hall. "Space Settlement Population Rotation Tolerance." *NSS Space Settlement Journal*, June 2017, 1–25. http://space.alglobus.net/papers/RotationPaper.pdf.

PARTIAL BIBLIOGRAPHY

Globus, Al, and Joe Strout. "Orbital Space Settlement Radiation Shielding." *NSS Space Settlement Journal*, April 2017, 1–38. https://space.nss.org/wp-content/uploads/Orbital-Space-Settlement-Radiation-Shielding-2015.pdf.

Goemaere, Sophie, Thomas Van Caelenberg, Wim Beyers, Kim Binsted, and Maarten Vansteenkiste. "Life on Mars from a Self-Determination Theory Perspective: How Astronauts' Needs for Autonomy, Competence and Relatedness Go Hand in Hand with Crew Health and Mission Success—Results from HI-SEAS IV." *Acta Astronautica* 159 (2019): 273–85. https://doi.org/10.1016/j.actaastro.2019.03.059.

Goldwater, Barry. "Space Shuttle Processing Worth $2 Billion Annually." Congressional Record, Washington, D.C., April 3, 1974. https://www.govinfo.gov/content/pkg/GPO-CRECB-1974-pt7/pdf/GPO-CRECB-1974-pt7-6-1.pdf.

Goswami, Namrata, and Peter A. Garretson. *Scramble for the Skies: The Great Power Competition to Control the Resources of Outer Space*. Lanham, MD: Lexington Books, 2020.

Grandin, Greg. *Fordlandia: The Rise and Fall of Henry Ford's Forgotten Jungle City*. New York: Picador, 2010.

Green, Hardy. *The Company Town*. New York: Basic Books, 2012. https://www.basicbooks.com/titles/hardy-green/the-company-town/9780465022649/.

Green, James. *The Devil Is Here in These Hills: West Virginia's Coal Miners and Their Battle for Freedom*. New York: Grove Press, 2016.

Guinness World Records. "First Commercial Filmed in Space." August 22, 1997. https://www.guinnessworldrecords.com/world-records/first-commercial-filmed-in-space.

Guo, Jin-Hu, Wei-Min Qu, Shan-Guang Chen, Xiao-Ping Chen, Ke Lv, Zhi-Li Huang, and Yi-Lan Wu. "Keeping the Right Time in Space: Importance of Circadian Clock and Sleep for Physiology and Performance of Astronauts." *Military Medical Research* 1 (2014): 23. https://doi.org/10.1186/2054-9369-1-23.

Gupta, Anika. "The Cola Wars. Smear Campaigns in Space?" *Smithsonian*, August 1, 2008. https://www.smithsonianmag.com/smithsonian-institution/the-cola-wars-smear-campaigns-in-space-27520139/.

Haber, Heinz. "Can We Survive in Space?" In *Across the Space Frontier*, edited by Cornelius Ryan, 71–97. New York: Viking Press, 1952.

Hall, Rex D., Shayler David, and Bert Vis. *Russia's Cosmonauts: Inside the Yuri Gagarin Training Center*. Berlin; Heidelberg: Praxis, 2005.

Handmer, Casey. *How to Industrialize Mars: A Strategy for Self-Sufficiency*. EBook., 2018. https://docs.google.com/document/d/1pxQg51rGP6JtdD4Eix1xpVek1xX05eQVgc7Jbm6VtDw/edit; https://www.amazon.com/How-Industrialize-Mars-Strategy-Self-Sufficiency-ebook/dp/B07GN3BJX3/

Hapgood, Mike. "The Impact of Space Weather on Human Missions to Mars: The Need for Good Engineering and Good Forecasts." In *The Human Factor in a Mission to Mars: An Interdisciplinary Approach*, edited by Konrad Szocik, 69–91. Space and Society. Cham, Switzerland: Springer International Publishing, 2019. https://doi.org/10.1007/978-3-030-02059-0_4.

Haqq-Misra, Jacob. "Toward a Sustainable Land Use Policy for Mars." In *Human Governance Beyond Earth: Implications for Freedom*, edited by Charles S. Cockell, 43–49. Space and Society. Cham, Switzerland: Springer International Publishing, 2015. https://doi.org/10.1007/978-3-319-18063-2_4.

Harbaugh, Jennifer. "Fission Surface Power." NASA, May 6, 2021. http://www.nasa.gov/mission_pages/tdm/fission-surface-power/index.html.

Harland, David M. *The Story of Space Station Mir*. New York: Praxis, 2005. https://doi.org/10.1007/978-0-387-73977-9.

Harrison, Albert A. *Spacefaring: The Human Dimension*. Berkeley: University of California Press, 2001.

Harrison, Albert A., Yvonne A. Clearwater, and Christopher P. McKay, eds. *From Antarctica to Outer Space: Life in Isolation and Confinement*. New York: Springer, 1990.

Harrison, Jean-Pierre. *The Edge of Time: The Authoritative Biography of Kalpana Chawla*. Saratoga, CA: Harrison Publishing, 2011.

Harrison, Todd, Andrew Hunter, Kaitlyn Johnson, and Thomas Roberts. *Implications of Ultra-Low-Cost Access to Space*. Washington, D.C.: Center for Strategic & International Studies, 2017.

Harrison, Todd, Kaitlyn Johnson, and Thomas G. Roberts. "Space Threat Assessment 2019." Aerospace Security Project. Washington, D.C.: Center for Strategic and International Studies, April 2019. https://csis-website-prod.s3.amazonaws.com/s3fs-public/publication/190404_SpaceThreatAssessment_interior.pdf.

Hartmann, William K., Ron Miller, and Pamela Lee. *Out of the Cradle: Exploring the Frontiers Beyond Earth*. New York: Workman Publishing, 1985.

Harvey, Brian. *China in Space: The Great Leap Forward*. Chichester, UK: Springer, 2013.

Hassler, Donald M., Cary Zeitlin, Robert F. Wimmer-Schweingruber, Bent Ehresmann, Scot Rafkin, Jennifer L. Eigenbrode, David E. Brinza, et al. "Mars' Surface Radiation Environment Measured with the Mars Science Laboratory's Curiosity Rover." *Science* 343 (2014): 1244797. https://doi.org/10.1126/science.1244797.

Heaps, Leo. *Operation Morning Light: Terror in Our Skies: The True Story of Cosmos 954*. New York: Paddington Press , 1978.

Hearon, Christopher M., Jr., Katrin A. Dias, Gautam Babu, John E. T. Marshall, James Leidner, Kirsten Peters, Erika Silva, James P. MacNamara, Joseph Campain, and Benjamin D. Levine. "Effect of Nightly Lower Body Negative Pressure on Choroid Engorgement in a Model of Spaceflight-Associated Neuro-Ocular Syndrome: A Randomized Crossover Trial." *JAMA Ophthalmology* 140 (2022): 59–65. https://doi.org/10.1001/jamaophthalmol.2021.5200.

Heavens, Nicholas G., Armin Kleinböhl, Michael S. Chaffin, Jasper S. Halekas, David M. Kass, Paul O. Hayne, Daniel J. McCleese, Sylvain Piqueux, James H. Shirley, and John T. Schofield. "Hydrogen Escape from Mars Enhanced by Deep Convection in Dust Storms." *Nature Astronomy* 2 (2018): 126–32. https://doi.org/10.1038/s41550-017-0353-4.

Hendrickx, Bart. "Illness in Orbit." *Spaceflight* 53 (2011): 104–9.

Heppenheimer, T. A. *Colonies in Space*. New York: Warner Books, 1980.

Hernández, José M., and Monica Rojas Rubin. *Reaching for the Stars: The Inspiring Story of a Migrant Farmworker Turned Astronaut*. New York: Center Street, 2012.

Highfield, Roger. "Colonies in Space May Be Only Hope, Says Hawking." *Telegraph* (London), October 16, 2001. https://www.telegraph.co.uk/news/uknews/1359562/Colonies-in-space-may-be-only-hope-says-Hawking.html.

Hilbig, Reinhard, Albert Gollhofer, Otmar Bock, and Dietrich Manzey. *Sensory Motor and Behavioral Research in Space*. New York: Springer, 2017.

Hobbs, Harry, and George Williams. "Micronations: A Lacuna in the Law." *International Journal of Constitutional Law* 19 (2021): 71–97. https://doi.org/10.1093/icon/moab020.

Hong, Xiaoman, Anamika Ratri, Sungshin Y. Choi, Joseph S. Tash, April E. Ronca, Joshua S. Alwood, and Lane K. Christenson. "Effects of Spaceflight Aboard the International Space Station on Mouse Estrous Cycle and Ovarian Gene Expression." *NPJ Microgravity* 7 (2021): 1–8. https://doi.org/10.1038/s41526-021-00139-7.

Hood, Frank, and Charles Hood. *Poopie Suits & Cowboy Boots: Tales of a Submarine Officer During the Height of the Cold War*. Independently published, 2021.

Horst, Felix, Daria Boscolo, Marco Durante, Francesca Luoni, Christoph Schuy, and Uli Weber. "Thick Shielding Against Galactic Cosmic Radiation: A Monte Carlo Study with Focus on the Role of Secondary Neutrons." *Life Sciences in Space Research* 33 (2022): 58–68. https://doi.org/10.1016/j.lssr.2022.03.003.

Horvath, Tyler, Paul O. Hayne, and David A. Paige. "Thermal and Illumination Environments of Lunar Pits and Caves: Models and Observations from the Diviner Lunar Radiometer Experiment." *Geophysical Research Letters* 49 (2022): e2022GL099710. https://doi.org/10.1029/2022GL099710.

Howe, A. Scott, and Brent Sherwood, eds. *Out of This World*. Reston, VA: AIAA, 2009.

"Human Exploration of Mars Design Reference Architecture 5.0." NASA Johnson Space Center, Houston, TX, 2014. http://ston.jsc.nasa.gov/collections/TRS/.

Humphreys, Richard. *Under Pressure: Living Life and Avoiding Death on a Nuclear Submarine*. New York: Hanover Square Press, 2020.

Huntford, Roland. *Race for the South Pole: The Expedition Diaries of Scott and Amundsen*. London; New York: Continuum, 2011.

Ihle, Eva C., Jennifer B. Ritsher, and Nick Kanas. "Positive Psychological Outcomes of Spaceflight: An Empirical Study." *Aviation, Space, and Environmental Medicine* 77, no. 2 (2006): 93–101. https://www.ingentaconnect.com/content/asma/asem/2006/00000077/00000002/art00001.

Ijiri, K. "Mating Behavior of the Fish (Medaka) and Development of Their Eggs in Space." Second International Microgravity Laboratory (IML-2) Final Report. NASA Reference Publication 1405, July 1997. https://ntrs.nasa.gov/api/citations/19970035095/downloads/19970035095.pdf.

Inoue, Natsuhiko, Ichiyo Matsuzaki, and Hiroshi Ohshima. "Group Interactions in SFINCSS-99: Lessons for Improving Behavioral Support Programs." *Aviation, Space, and Environmental Medicine* 75 (2004): C28–35.

Ivanovich, Grujica S. *Salyut—The First Space Station: Triumph and Tragedy*. Berlin: Praxis, 2008.

Jakhu, Ram S., and Steven Freeland. "McGill Manual on International Law Applicable to Military Uses of Outer Space: Volume I—Rules." Montreal: Centre for Research in Air and Space Law, 2022. https://www.mcgill.ca/milamos/files/milamos/mcgill_manual_volume_i_-_rules_final.pdf.

Jakhu, Ram S., Joseph N. Pelton, and Yaw Otu Mankata Nyampong. *Space Mining and Its Regulation*. Cham, Switzerland: Springer International Publishing, 2017. https://doi.org/10.1007/978-3-319-39246-2.

James, Tom, ed. *Deep Space Commodities: Exploration, Production and Trading*. London: Palgrave Macmillan, 2018.

Jansen, Jan C., and Jürgen Osterhammel. *Decolonization*. Princeton, NJ: Princeton University Press, 2017. https://press.princeton.edu/books/hardcover/9780691165219/decolonization.

Jemison, Dr. Mae. *Find Where the Wind Goes: Moments from My Life*. New York: Scholastic Press, 2001.

Jerusalem Post Staff. "Tnuva and the Silicon Affair." *Jerusalem Post*, October 14, 2008. https://www.jpost.com/opinion/editorials/tnuva-and-the-silicon-affair.

PARTIAL BIBLIOGRAPHY

Johnson, Nicholas. *Big Dead Place: Inside the Strange and Menacing World of Antarctica*. Los Angeles, CA: Feral House, 2005.

Johnson-Freese, Joan. *Heavenly Ambitions: America's Quest to Dominate Space*. Philadelphia: University of Pennsylvania Press, 2009.

———. *Space Warfare in the 21st Century: Arming the Heavens*. London; New York: Routledge, 2016.

Johnston, Smith L., Mark R. Campbell, Roger D. Billica, and Stevan M. Gilmore. "Cardiopulmonary Resuscitation in Microgravity: Efficacy in the Swine During Parabolic Flight." *Aviation, Space, and Environmental Medicine* 75 (2004): 546–50.

Jones, Eric M. "Apollo 16 Lunar Surface Journal: EVA-3 Closeout." NASA, June 16, 2014. https://www.hq.nasa.gov/alsj/a16/a16.clsout3.html.

Kamin, Debra. "The Future of Space Tourism Is Now. Well, Not Quite." *New York Times*, May 7, 2022. https://www.nytimes.com/2022/05/07/travel/space-travel-tourism.html.

Kanas, N., G. Sandal, J. E. Boyd, V. I. Gushin, D. Manzey, R. North, G. R. Leon, et al. "Psychology and Culture during Long-Duration Space Missions." *Acta Astronautica* 64 (2009): 659–77. https://doi.org/10.1016/j.actaastro.2008.12.005.

Kanas, Nick. *Humans in Space: The Psychological Hurdles*. Chichester, UK: Springer / Praxis, 2015.

Kanas, Nick, Matthew Harris, Thomas Neylan, Jennifer Boyd, Daniel S. Weiss, Colleen Cook, and Stephanie Saylor. "High versus Low Crewmember Autonomy during a 105-Day Mars Simulation Mission." *Acta Astronautica* 69 (2011): 240–44. https://doi.org/10.1016/j.actaastro.2011.04.014.

Kanas, Nick, and Dietrich Manzey. *Space Psychology and Psychiatry*. El Segundo, CA; Dordrecht: Springer, 2008.

Karacalıoğlu, Göktuğ. "Energy Resources for Space Missions." *Space Safety*, January 16, 2014. http://www.spacesafetymagazine.com/aerospace-engineering/nuclear-propulsion/energy-resources-space-missions/.

Kaszeta, Dan. *Toxic: A History of Nerve Agents, from Nazi Germany to Putin's Russia*. New York: Oxford University Press, 2021.

Keener, Chris and Ben Youngerman, dirs. *Spacedrop*. Documentary, 2020.

Kelly, Scott. *Endurance: My Year in Space, A Lifetime of Discovery*. New York: Vintage, 2018.

Kilgore, De Witt Douglas. *Astrofuturism: Science, Race, and Visions of Utopia in Space*. Philadelphia: University of Pennsylvania Press, 2003.

Kimhi, Shaul. "Understanding Good Coping: A Submarine Crew Coping with Extreme Environmental Conditions." *Psychology* 2 (2011): 961–67. https://doi.org/10.4236/psych.2011.29145.

Kirkpatrick, Andrew W., Chad G. Ball, Mark Campbell, David R. Williams, Scott E. Parazynski, Kenneth L. Mattox, and Timothy J. Broderick. "Severe Traumatic Injury during Long Duration Spaceflight: Light Years beyond ATLS." *Journal of Trauma Management & Outcomes* 3 (2009): 4. https://doi.org/10.1186/1752-2897-3-4.

Kirton, Allan G., and Stephen C. Vasciannie. "Deep Seabed Mining Under the Law of the Sea Convention and the Implementation Agreement: Developing Country Perspectives." *Social and Economic Studies* 51 (2002): 63–115. http://www.jstor.org/stable/27865277.

Kleiman, Matthew J. *The Little Book of Space Law*. Chicago, IL: American Bar Association, 2014.

Kluger, Jeffrey. *Apollo 8: The Thrilling Story of the First Mission to the Moon*. New York: Henry Holt and Co., 2017.

Koerth, Maggie. "Space Sex Is Serious Business." *FiveThirtyEight*, March 14, 2017. https://fivethirtyeight.com/features/space-sex-is-serious-business/.

Koren, Marina. "Everything You Never Thought to Ask About Astronaut Food." *The Atlantic*, December 15, 2017. https://www.theatlantic.com/science/archive/2017/12/astronaut-food-international-space-station/548255/.

Kramer, Larry A., Khader M. Hasan, Michael B. Stenger, Ashot Sargsyan, Steven S. Laurie, Christian Otto, Robert J. Ploutz-Snyder, Karina Marshall-Goebel, Roy F. Riascos, and Brandon R. Macias. "Intracranial Effects of Microgravity: A Prospective Longitudinal MRI Study." *Radiology* 295 (2020): 640–48. https://doi.org/10.1148/radiol.2020191413.

Landon, Lauren Blackwell, Christina Rokholt, Kelley J. Slack, and Yvonne Pecena. "Selecting Astronauts for Long-Duration Exploration Missions: Considerations for Team Performance and Functioning." *REACH* 5 (2017): 33–56. https://doi.org/10.1016/j.reach.2017.03.002.

Lang, Thomas, Adrian LeBlanc, Harlan Evans, Ying Lu, Harry Genant, and Alice Yu. "Cortical and Trabecular Bone Mineral Loss from the Spine and Hip in Long-Duration Spaceflight." *Journal of Bone and Mineral Research* 19 (2004): 1006–12. https://doi.org/10.1359/JBMR.040307.

Lang, Thomas, Jack J. W. A. Van Loon, Susan Bloomfield, Laurence Vico, Angele Chopard, Joern Rittweger, Antonios Kyparos, et al. "Towards Human Exploration of Space: The THESEUS Review Series on Muscle and Bone Research Priorities." *NPJ Microgravity* 3 (2017): 1–10. https://doi.org/10.1038/s41526-017-0013-0.

PARTIAL BIBLIOGRAPHY

Lansing, Alfred. *Endurance*. New York: Basic Books, 2015.

Larsson, Magnus, and Alex Kaiser. "Cloud Ten." In *Inner Solar System: Prospective Energy and Material Resources*, edited by Viorel Badescu and Kris Zacny, 451–98. Cham, Switzerland: Springer International Publishing, 2015. https://doi.org/10.1007/978-3-319-19569-8_22.

Lasser, David. *The Conquest of Space*. Apogee Books Space Series 27. Ontario, Canada: Collector's Guide Publishing, Inc., 2002.

Layendecker, Alexander. "Sex in Outer Space and the Advent of Astrosexology: A Philosophical Inquiry into the Implication of Human Sexuality and Reproductive Development Factors in Seeding Humanity's Future Throughout the Cosmos and the Argument for an Astrosexology Research Institute." Dissertation, Institute for Advanced Study of Human Sexuality, 2016.

Layendecker, Alexander B., and Shawna Pandya. "Logistics of Reproduction in Space." In *Handbook of Life Support Systems for Spacecraft and Extraterrestrial Habitats*, edited by Erik Seedhouse and David J Shayler, 1–16. Cham, Switzerland: Springer International Publishing, 2018. https://doi.org/10.1007/978-3-319-09575-2_211-1.

Lebedev, Valentine. *Diary of a Cosmonaut: 211 Days in Space*. New York: Bantam Books, 1990.

Lei, Xiaohua, Yujing Cao, Ying Zhang, and Enkui Duan. "Advances of Mammalian Reproduction and Embryonic Development Under Microgravity." In *Life Science in Space: Experiments on Board the SJ-10 Recoverable Satellite*, edited by Enkui Duan and Mian Long, 281–315. Research for Development. Singapore: Springer, 2019. https://doi.org/10.1007/978-981-13-6325-2_11.

Leib, Karl. "State Sovereignty in Space: Current Models and Possible Futures." *Astropolitics* 13 (2015): 1–24. https://doi.org/10.1080/14777622.2015.1015112.

Lele, Ajey. *50 Years of the Outer Space Treaty: Tracing the Journey*. New Delhi: Pentagon Press, 2017.

Leonov, Alexei, David Scott, and Christine Toomey. *Two Sides of the Moon: Our Story of the Cold War Space Race*. New York: St. Martin's / Griffin, 2006.

Levesque, Emily. *The Last Stargazers: The Enduring Story of Astronomy's Vanishing Explorers*. London: Oneworld Publications, 2020.

Levy, Buddy. *Labyrinth of Ice: The Triumphant and Tragic Greely Polar Expedition*. St. Martin's Griffin, 2021.

Ley, Willy. "A Station in Space." In *Across the Space Frontier*, edited by Cornelius Ryan, 98–117. New York: Viking Press, 1952.

———. *Satellites, Rockets and Outer Space*. New York: Signet Key Book, 1958.

Limerick, Patricia Nelson. "Imagined Frontiers: Westward Expansion and the Future of the Space Program." In Radford Byerly Jr., *Space Policy Alternatives*, 249–61. New York: Routledge, 1992.

———. *The Legacy of Conquest: The Unbroken Past of the American West*. New York: W. W. Norton & Co., 2006.

———. *Something in the Soil: Legacies and Reckonings in the New West*. New York: W. W. Norton & Co., 2001.

Linenger, Jerry. *Off the Planet: Surviving Five Perilous Months Aboard the Space Station Mir*. New York: McGraw-Hill, 2000.

Linklater, Andro. *Owning the Earth: The Transforming History of Land Ownership*. London: Bloomsbury Publishing, 2015.

Livingston, David M. "Broadcast 470 (Special Edition). Sunday 19 Mar 2006. Guest: Dr. Vadim Y. Rygalov." *The Space Show*. Accessed October 3, 2022. https://www.thespaceshow.com/show/19-mar-2006/broadcast-470-special-edition.

Logsdon, J. *John F. Kennedy and the Race to the Moon*. New York: Palgrave Macmillan, 2011.

Lovell, Jim, and Jeffrey Kluger. *Apollo 13: Lost Moon*. New York: Pocket Books, 1995.

Lucas, Rex A. *Minetown, Milltown, Railtown: Life in Canadian Communities of Single Industry*. Ontario: Oxford University Press, 2008.

Lucid, Shannon. *Tumbleweed: Six Months Living on Mir*. Middleton, DE: MkEk Publishing, 2020.

Lyall, Francis, and Paul B. Larsen. *Space Law: A Treatise*. New York: Routledge, 2017.

MacDonald, Alexander. *The Long Space Age: The Economic Origins of Space Exploration from Colonial America to the Cold War*. New Haven, CT: Yale University Press, 2017.

Mader, Thomas H., C. Robert Gibson, Anastas F. Pass, Larry A. Kramer, Andrew G. Lee, Jennifer Fogarty, William J. Tarver, et al. "Optic Disc Edema, Globe Flattening, Choroidal Folds, and Hyperopic Shifts Observed in Astronauts after Long-Duration Space Flight." *Ophthalmology* 118 (2011): 2058–69. https://doi.org/10.1016/j.ophtha.2011.06.021.

Mankins, John. *The Case for Space Solar Power*. Houston, TX: Virginia Edition, 2014.

Mann, Michael E., and Lee R. Kump. *Dire Predictions: The Visual Guide to the Findings of the IPCC*. New York: DK Publishing, 2015.

Manning, Alan. "Monopsony in Labor Markets: A Review." *ILR Review* 74 (2021): 3–26. https://doi.org/10.1177/0019793920922499.

Mars Architecture Steering Group. "Human Exploration of Mars Design Reference Architecture 5.0." NASA Johnson Space Center, Houston, TX, 2009. http://ston.jsc.nasa.gov/collections/TRS/.

PARTIAL BIBLIOGRAPHY

Masunaga, Samantha. "Elon Musk Wants to Create His Own City. Here's How That Could Work." *Los Angeles Times*, April 2, 2021. https://www.latimes.com/business/story/2021-04-02/tech-governments-starbase-elon-musk.

Matsumura, Takafumi, Taichi Noda, Masafumi Muratani, Risa Okada, Mutsumi Yamane, Ayako Isotani, Takashi Kudo, Satoru Takahashi, and Masahito Ikawa. "Male Mice, Caged in the International Space Station for 35 Days, Sire Healthy Offspring." *Scientific Reports* 9 (2019): 13733. https://doi.org/10.1038/s41598-019-50128-w.

May, Kate Torgovnick. "What Will We Eat on Mars?" Ideas.Ted.Com, November 21, 2014. https://ideas.ted.com/comfort-food-in-space-the-final-frontier/.

McCall, Robert, and Isaac Asimov. *Our World in Space*. Greenwich, CT: New York Graphic Society, 1974.

McClure, Patrick R., David I. Poston, Marc A. Gibson, Lee S. Mason, and R. Chris Robinson. "Kilopower Project: The KRUSTY Fission Power Experiment and Potential Missions." *Nuclear Technology* 206 (2020): S1–12. https://doi.org/10.1080/00295450.2020.1722554.

McCurdy, Howard E. *Space and the American Imagination*. Baltimore, MD: Johns Hopkins University Press, 2011.

McDonald, F. B. "Review of Galactic and Solar Cosmic Rays." In *Second Symposium on Protection Against Radiation in Space*, edited by Arthur Reetz, 19–29. Gatlinburg, TN: Scientific and Technical Information Division, National Aeronautics and Space Administration, 1964.

McDougall, Walter A. . . . *The Heavens and the Earth: A Political History of the Space Age*. Baltimore, MD: Johns Hopkins University Press, 1997.

McDowell, Jonathan. "Jonathan's Space Pages: Starlink Statistics." Jonathan's Space Report, October 12, 2022. https://planet4589.org/space/stats/star/starstats.html.

Meierding, Emily. *The Oil Wars Myth: Petroleum and the Causes of International Conflict*. Ithaca, NY: Cornell University Press, 2020.

Melvin, Leland. *Chasing Space: An Astronaut's Story of Grit, Grace, and Second Chances*. New York: Amistad, 2017.

Mersmann, Kathryn. "The Fact and Fiction of Martian Dust Storms." NASA, September 18, 2015. http://www.nasa.gov/feature/goddard/the-fact-and-fiction-of-martian-dust-storms.

Milligan, Tony. *Nobody Owns the Moon: The Ethics of Space Exploitation*. Jefferson, NC: McFarland & Company, 2015.

Milligan, Tony, and Martin Elvis. "Mars Environmental Protection: An Application of the 1/8 Principle." In *The Human Factor in a Mission to Mars: An Interdisciplinary Approach*, edited by Konrad Szocik, 167–83. Space and Society. Cham, Switzerland: Springer International Publishing, 2019. https://doi.org/10.1007/978-3-030-02059-0_10.

Ming, D. W., and R. V. Morris. "Chemical, Mineralogical, and Physical Properties of Martian Dust and Soil," 5. NASA Johnson Space Center, Houston, TX, 2017. https://ntrs.nasa.gov/citations/20170005414.

Mishra, Birendra, and Ulrike Luderer. "Reproductive Hazards of Space Travel in Women and Men." *Nature Reviews Endocrinology* 15 (2019): 713–30. https://doi.org/10.1038/s41574-019-0267-6.

Mitchell, Edgar, and Ellen Mahoney. *Earthrise: My Adventures as an Apollo 14 Astronaut*. Chicago, IL: Chicago Review Press, 2014.

Mohanty, Susmita, Sue Fairburn, Barbara Imhof, Stephen Ransom, and Andreas Vogler. "Survey of Past, Present and Planned Human Space Mission Simulators." SAE Technical Papers 2008-01-2020, 24. https://doi.org/10.4271/2008-01-2020.

Moltz, James Clay. *The Politics of Space Security: Strategic Restraint and the Pursuit of National Interests, Third Edition*. Stanford, CA: Stanford University Press, 2019.

———. "Toward Cooperation or Conflict on the Moon?: Considering Lunar Governance in Historical Perspective." *Strategic Studies Quarterly* 3 (2009): 82–103. http://www.jstor.org/stable/26268666.

Moore, Kimberly C. *Star Crossed: The Story of Astronaut Lisa Nowak*. Gainesville, FL: University Press of Florida, 2020.

Moran, Norah. "Ep 107: The Overview Effect." NASA, August 29, 2019. http://www.nasa.gov/johnson/HWHAP/the-overview-effect.

Morgan Stanley. "Space: Investing in the Final Frontier." Morgan Stanley, July 24, 2020. https://www.morganstanley.com/ideas/investing-in-space.

Morozov, Sergei. "Asgardia's Calendar and Its Role in Space Industrialisation Strategy," Room: The Space Journal of Asgardia, 2019. https://room.eu.com/article/asgardias-calendar-and-its-role-in-space-industrialisation-strategy.

Moses, Robert, and Dennis M. Bushnell. "Frontier In-Situ Resource Utilization for Enabling Sustained Human Presence on Mars." NASA, Langley Research Center, Hampton, VA, April 1, 2016. https://ntrs.nasa.gov/search.jsp?R=20160005963.

Mountfield, David. *A History of Polar Exploration*. New York: Dial Press, 1974.

Muir-Harmony, Teasel. *Operation Moonglow: A Political History of Project Apollo*. New York: Basic Books, 2020.

Mullane, Mike. *Riding Rockets: The Outrageous Tales of a Space Shuttle Astronaut*. New York: Scribner, 2007.

Myers, Steven Lee. "The Moon, Mars and Beyond: China's Ambitious Plans in Space." *New York Times*, October 15, 2021. https://www.nytimes.com/article/china-mars-space.html.

Nadis, Fred. "Star Children: Can Humans Be Fruitful and Multiply Off-Planet?" Space Review, September 14, 2020. https://www.thespacereview.com/article/4024/1.

———. *Star Settlers: The Billionaires, Geniuses, and Crazed Visionaries Out to Conquer the Universe*. New York: Pegasus Books, 2020.

Nansen, Fridtjof. *Farthest North*. Macmillan's Colonial Library, vols. 1 and 2. London: Macmillan and Co., Ltd., 1897.

NASA, "Departing Space Station Commander Provides Tour of Orbital Laboratory." International Space Station, 2012. https://www.youtube.com/watch?v=doN4t5NKW-k.

———. "Heat Probe | Instruments." NASA's InSight Mars Lander. Accessed October 4, 2022. https://mars.nasa.gov/insight/spacecraft/instruments/hp3.

———. "Human Integration Design Handbook." Washington, D.C.: NASA, June 5, 2014. https://www.nasa.gov/sites/default/files/atoms/files/human_integration_design_handbook_revision_1.pdf.

———. "Human Research Program: Integrated Research Plan (Revision N)." NASA, Johnson Space Center, Houston, TX, July 2022. https://humanresearchroadmap.nasa.gov/Documents/IRP_Rev-Current.pdf.

———. "NASA InSight's 'Mole' Ends Its Journey on Mars." NASA Mars InSight Mission, February 21, 2020. https://mars.nasa.gov/news/8836/nasa-insights-mole-ends-its-journey-on-mars?site=insight.

———. "NASA Space Flight Human-System Standard: Volume 1: Crew Health." NASA, January 5, 2022. https://www.nasa.gov/sites/default/files/atoms/files/2022-01-05_nasa-std-3001_vol.1_rev._b_final_draft_with_signature_010522.pdf.

NASA Human Research Program. "Evidence Report: Risk of Adverse Cognitive or Behavioral Conditions and Psychiatric Disorders." NASA, Johnson Space Center, Houston, TX, April 11, 2016. https://humanresearchroadmap.nasa.gov/evidence/reports/bmed.pdf.

National Academies of Sciences, Engineering, and Medicine. *Space Radiation and Astronaut Health: Managing and Communicating Cancer Risks*. Washington, D.C.: National Academies Press, 2021. https://doi.org/10.17226/26155.

National Council on Radiation Protection and Measurements. "Radiation Protection for Space Activities: Supplement to Previous Recommendations." Commentary No. 23, 2014. https://ncrponline.org/shop/commentaries/commentary-no-23-radiation-protection-for-space-activities-supplement-to-previous-recommendations-2014/?attribute_book=PDF.

Nelson, Peter Lothian, and Walter E. Block. *Space Capitalism: How Humans Will Colonize Planets, Moons, and Asteroids*. New York: Palgrave Macmillan, 2018.

Nelson, Timothy G. "The Moon Agreement and Private Enterprise: Lessons from Investment Law." *ILSA Journal of International & Comparative Law* 17 (2010): 391–416. https://nsuworks.nova.edu/ilsajournal/vol17/iss2/6.

Neufeld, Michael. *Von Braun: Dreamer of Space, Engineer of War*. New York: Vintage Books, 2008.

Nicogossian, Arnauld E., Richard S. Williams, Carolyn L. Huntoon, Charles R. Doarn, James D. Polk, and Victor S. Schneider, eds. *Space Physiology and Medicine: From Evidence to Practice*, 4th ed. New York: Springer, 2016.

Niederwieser, Tobias, Robert Aaron, Stefanie Countryman, Shankini Doraisingam, Howard Fultz, Ryan Griffith, Alexander Hoehn, Mark Rupert, Jim Wright, and Louis Stodieck. "FRIDGE—The Next Generation Freezer/Refrigerator/Incubator for Food and Experiment Conditioning Onboard the ISS." *Journal of Space Safety Engineering* 9 (2022): 291–97. https://doi.org/10.1016/j.jsse.2022.06.001.

Niederwieser, Tobias, Patrick Kociolek, and David Klaus. "Spacecraft Cabin Environment Effects on the Growth and Behavior of Chlorella Vulgaris for Life Support Applications." *Life Sciences in Space Research* 16 (2018): 8–17. https://doi.org/10.1016/j.lssr.2017.10.002.

Nielsen, Dr. Jerri, and Maryanne Vollers. *Ice Bound: A Doctor's Incredible Battle for Survival at the South Pole*. New York: Miramax, 2001.

Oberg, James E. *Uncovering Soviet Disasters: Exploring the Limits of Glasnost*. New York: Random House, 1988.

Oberg, James E., and Alcestis R. Oberg. *Pioneering Space: Living on the Next Frontier*. New York: McGraw-Hill, 1985.

Office of the Historian. "Milestones: 1989–1992; The Breakup of Yugoslavia, 1990–1992." Foreign Service Institute, United States Department of State. Accessed October 20, 2022. https://history.state.gov/milestones/1989-1992/breakup-yugoslavia.

Ogneva, Irina V., Maria A. Usik, Sergey S. Loktev, Yuliya S. Zhdankina, Nikolay S. Biryukov, Oleg I. Orlov, and Vladimir N. Sychev. "Testes and Duct Deferens of Mice during Space Flight: Cytoskeleton Structure, Sperm-Specific Proteins and Epigenetic Events." *Scientific Reports* 9 (2019): 9730. https://doi.org/10.1038/s41598-019-46324-3.

O'Kane, Sean. "Space Birth Startup's CEO Halts Project over 'Serious Ethical, Safety and Medical Concerns.'" The Verge, July 3, 2019. https://www.theverge.com/2019/7/3/20680006/space-birth-startup-project-ceo-serious-ethical-safety-medical-concerns-halt.

O'Leary, Brian. *The Making of an Ex-Astronaut.* Boston: Houghton Mifflin Co., 1970.

Oliver, Dave. *Against the Tide: Rickover's Leadership and the Rise of the Nuclear Navy.* Annapolis, MD: Naval Institute Press, 2018.

Oluwafemi, Funmilola A., Rayan Abdelbaki, James C.-Y. Lai, Jose G. Mora-Almanza, and Esther M. Afolayan. "A Review of Astronaut Mental Health in Manned Missions: Potential Interventions for Cognitive and Mental Health Challenges." *Life Sciences in Space Research* 28 (2021): 26–31. https://doi.org/10.1016/j.lssr.2020.12.002.

O'Neill, Gerard K. *The High Frontier: Human Colonies in Space.* Princeton, NJ: Space Studies Institute Press, 1989. https://www.thriftbooks.com/w/the-high-frontier-human-colonies-in-space_gerard-k-oneill/311174/.

———. *2081: A Hopeful View of the Human Future.* New York: Simon & Schuster, 1981.

Ostrom, Elinor, and Dan H. Cole, eds. *Property in Land and Other Resources.* Cambridge, MA: Lincoln Institute of Land Policy, 2011.

Pagel, J. I., and A. Choukèr. "Effects of Isolation and Confinement on Humans-Implications for Manned Space Explorations." *Journal of Applied Physiology* 120 (2016): 1449–57. https://doi.org/10.1152/japplphysiol.00928.2015.

Palmer, Suzanne. "The Turkish Republic of Northern Cyprus: Should the United States Recognize It as an Independent State." *Boston University International Law Journal* 4 (1986): 423. https://heinonline.org/HOL/LandingPage?handle=hein.journals/builj4&div=29&id=&page=.

Panesar, S. S., and K. Ashkan. "Surgery in Space." *British Journal of Surgery* 105 (2018): 1234–43. https://doi.org/10.1002/bjs.10908.

Patel, Zarana S., Tyson J. Brunstetter, William J. Tarver, Alexandra M. Whitmire, Sara R. Zwart, Scott M. Smith, and Janice L. Huff. "Red Risks for a Journey to the Red Planet: The Highest Priority Human Health Risks for a Mission to Mars." *NPJ Microgravity* 6 (2020): 1–13. https://doi.org/10.1038/s41526-020-00124-6.

Peake, Tim. *Ask an Astronaut.* London: Arrow Books, 2018.

Pearlman, Robert Z. "As Seen on TV: These Commercials Were Filmed in Space!," Space.com, June 21, 2017. https://www.space.com/37263-commercials-in-space.html.

Pelt, Michel van. *Dream Missions: Space Colonies, Nuclear Spacecraft and Other Possibilities.* New York: Springer, 2017.

Pershing, Abigail D. "Interpreting the Outer Space Treaty's Non-Appropriation Principle: Customary International Law from 1967 to Today." *Yale Journal of International Law* 44 (2019): 30. https://digitalcommons.law.yale.edu/yjil/vol44/iss1/5.

Persson, Erik. "Citizens of Mars Ltd." In *Human Governance Beyond Earth: Implications for Freedom*, edited by Charles S. Cockell, 121–37. Space and Society. Cham, Switzerland: Springer International Publishing, 2015. https://doi.org/10.1007/978-3-319-18063-2_9.

Pinault, Lewis. "Towards a World Space Agency." In *Human Governance Beyond Earth: Implications for Freedom*, edited by Charles S. Cockell, 173–96. Space and Society. Cham, Switzerland: Springer International Publishing, 2015. https://doi.org/10.1007/978-3-319-18063-2_12.

Pitjeva, E. V., and N. P. Pitjev. "Masses of the Main Asteroid Belt and the Kuiper Belt from the Motions of Planets and Spacecraft." *Astronomy Letters* 44 (2018): 554–66. https://doi.org/10.1134/S1063773718090050.

Pizza Hut. "Hut Life—Official Pizza Hut Blog." 2022. http://blog.pizzahut.com/.

Pletser, Vladimir. *On to Mars! Chronicles of Martian Simulations.* Singapore: Springer, 2018.

Pleus, Richard, and Corey, Lisa. "Environmental Exposure to Perchlorate: A Review of Toxicology and Human Health." *Toxicology and Applied Pharmacology* 358 (2018), 102–9. 10.1016/j.taap.2018.09.001.

Pogue, William R. *But for the Grace of God: An Autobiography of an Aviator and Astronaut.* Golden, CO: CreateSpace Independent Publishing Platform, 2016.

Poláčková Šolcová, Iva, Iva Šolcová, Iva Stuchlíková, and Yvona Mazehóová. "The Story of 520 Days on a Simulated Flight to Mars." *Acta Astronautica*, Space Flight Safety, 126 (2016): 178–89. https://doi.org/10.1016/j.actaastro.2016.04.026.

Pop, Virgiliu. *Who Owns the Moon?* New York: Springer, 2008.

Portree, David S. F. "Humans to Mars: Fifty Years of Mission Planning, 1950–2000." Monographs in Aerospace History #21. Washington, D.C.: NASA, February 2001. https://www.lpi.usra.edu/lunar/strategies/HumanstoMars.pdf.

Poston, David I., Marc A. Gibson, Thomas Godfroy, and Patrick R. McClure. "KRUSTY Reactor Design." *Nuclear Technology* 206 (2020): S13–30. https://doi.org/10.1080/00295450.2020.1725382.

PARTIAL BIBLIOGRAPHY

Potter, Sean. "Demonstration Proves Nuclear Fission Can Provide Exploration Power." NASA, May 2, 2018. http://www.nasa.gov/press-release/demonstration-proves-nuclear-fission-system-can-provide-space-exploration-power.

Poynter, Jane. *The Human Experiment: Two Years and Twenty Minutes Inside Biosphere 2*. New York: Basic Books, 2006.

Preston, Robert, Dana J. Johnson, Sean J. A. Edwards, Michael D. Miller, and Calvin Shipbaugh. "Space Weapons Earth Wars." RAND Corporation, January 1, 2002. https://www.rand.org/pubs/monograph_reports/MR1209.html.

Proshchina, Alexandra, Victoria Gulimova, Anastasia Kharlamova, Yuliya Krivova, Nadezhda Besova, Rustam Berdiev, and Sergey Saveliev. "Reproduction and the Early Development of Vertebrates in Space: Problems, Results, Opportunities." *Life* 11 (2021): 109. https://doi.org/10.3390/life11020109.

Pyle, Rod. *Destination Mars: New Explorations of the Red Planet*. Amherst, NY: Prometheus, 2012.

Rader, Andrew. *Beyond the Known: How Exploration Created the Modern World and Will Take Us to the Stars*. New York: Scribner, 2019.

Rajendran, Anushri, Franz K. Fuss, and Yehuda Weizman. "Designing a Technology Ecosystem for the Integration of Environmental Analysis and Health Diagnostics to Assist Humans in the Colonisation of Mars." Paper presented at the 70th International Astronautical Congress (IAC), Washington, D.C., October 20–26, 2019. https://iafastro.directory/iac/paper/id/53619/summary/.

Rapp, Donald. *Use of Extraterrestrial Resources for Human Space Missions to Moon or Mars*. New York: Springer, 2018.

Rappaport, Margaret Boone, and Christopher Corbally. "Program Planning for a Mars Hardship Post: Social, Psychological, and Spiritual Services." In *The Human Factor in a Mission to Mars: An Interdisciplinary Approach*, edited by Konrad Szocik, 35–58. Space and Society. Cham, Switzerland: Springer International Publishing, 2019. https://doi.org/10.1007/978-3-030-02059-0_2.

Rauschenbach, Boris V. *Hermann Oberth: The Father of Space Flight 1894–1989*. New York: West Art Publishing, 1994.

Red Heaven. Documentary. Raised by Wolves Production, LLC, 2021. https://www.redheavenfilm.com/.

Reichhardt, Tony, ed. *Space Shuttle: The First 20 Years—The Astronauts' Experiences in Their Own Words*. New York: DK Publishing, Inc, 2002.

Reid, Helen. "Pacific Island of Nauru Sets Two-Year Deadline for U.N. Deep-Sea Mining Rules." Reuters, June 29, 2021. https://www.reuters.com/business/environment/pacific-island-nauru-sets-two-year-deadline-deep-sea-mining-rules-2021-06-29/.

RFE/RL. "Russia's Constitutional Court Approves Amendments Allowing Putin to Rule Until 2036." Radio Free Europe/Radio Liberty, March 16, 2020. https://www.rferl.org/a/russia-constitutional-court-approves-constitutional-amendments/30490913.html.

RIA Novosti. "The Court in St. Petersburg Dismissed the Case of a Polar Explorer Who Hit a Colleague with a Knife." RIA Novosti, February 8, 2019. https://ria.ru/20190208/1550592240.html.

Ride, Sally, and Susan Okie. *To Space and Back*. New York: HarperCollins, 1986.

Rivolier, J., R. Goldsmith, D. J. Lugg, and A. J. W. Taylor, eds. *Man in the Antarctic: The Scientific Work of the International Biomedical Expedition to the Antarctic (IBEA)*. London: Taylor & Francis, 1988.

Roach, Mary. *Packing for Mars: The Curious Science of Life in the Void*. New York: W. W. Norton & Co., 2010.

Roberts, Thomas G. "Space Launch to Low Earth Orbit: How Much Does It Cost?" Aerospace Security, September 1, 2022. https://aerospace.csis.org/data/space-launch-to-low-earth-orbit-how-much-does-it-cost/.

Rogers, Susan Fox, ed. *Antarctica: Life on the Ice*. Palo Alto, CA: Travelers' Tales, 2007.

Ronca, April E., Ellen S. Baker, Tamara G. Bavendam, Kevin D. Beck, Virginia M. Miller, Joseph S. Tash, and Marjorie Jenkins. "Effects of Sex and Gender on Adaptations to Space: Reproductive Health." *Journal of Women's Health* 23 (2014): 967–74. https://doi.org/10.1089/jwh.2014.4915.

Ross, Amia, Sephora Ruppert, Philipp Gläser, and Martin Elvis. "Towers on the Peaks of Eternal Light: Quantifying the Available Solar Power." arXiv, February 23, 2021. https://doi.org/10.48550/arXiv.2102.11766.

Ross-Nazzal, Jennifer. "Edited Oral History Transcript—Paul A. Lachance." NASA Johnson Space Center Oral History Project, Houston, TX, May 4, 2006. https://historycollection.jsc.nasa.gov/JSCHistoryPortal/history/oral_histories/LachancePA/LachancePA_5-4-06.htm.

Ryder, Valerie. "Spacecraft Maximum Allowable Concentrations for Airborne Contaminants: Revision A." NASA Johnson Space Center, Houston, TX, April 2020. https://www.nasa.gov/sites/default/files/atoms/files/jsc_20584_signed.pdf.

Sagan, Carl, and Steven J. Ostro. "Dangers of Asteroid Deflection." *Nature* 368 (1994): 501. https://doi.org/10.1038/368501a0.

Salmeri, Antonino. "Developing and Managing Moon and Mars Settlements in Accordance with International Space Law," 1–9. International Astronautical Federation, 2020. https://orbilu.uni.lu/handle/10993/44630.
———. "No, Mars Is Not a Free Planet, No Matter What SpaceX Says." SpaceNews, December 5, 2020. https://spacenews.com/op-ed-no-mars-is-not-a-free-planet-no-matter-what-spacex-says/.
Salotti, Jean-Marc. "Minimum Number of Settlers for Survival on Another Planet." *Scientific Reports* 10 (2020): 9700. https://doi.org/10.1038/s41598-020-66740-0.
Salter, Alexander William. "Settling the Final Frontier: The ORBIS Lease and the Possibilities of Proprietary Communities in Space." *Journal of Air Law and Commerce* 84 (2019): 85–114. http://dx.doi.org/10.2139/ssrn.3331215.
Salter, Alexander William, and Peter T. Leeson. "Celestial Anarchy." *Cato Journal* 34 (2014): 581–96. https://papers.ssrn.com/abstract=2379599.
Sandal, Gro M. "Culture and Tension During an International Space Station Simulation: Results from SFINCSS'99." *Aviation, Space, and Environmental Medicine* 75 (2004): C44–51.
Sandoval, Luis, Kathryn Keeton, Camille Shea, Christian Otto, Holly Patterson, and Lauren Leveton. "Perspectives on Asthenia in Astronauts and Cosmonauts: Review of the International Research Literature." Technical Report, NASA Johnson Space Center, Houston, TX, January 1, 2012. https://ntrs.nasa.gov/search.jsp?R=20110023297.
Santy, Patricia A. *Choosing the Right Stuff: The Psychological Selection of Astronauts and Cosmonauts*. Westport, CT: Praeger, 1994.
Sauro, Francesco, Riccardo Pozzobon, Matteo Massironi, Pierluigi De Berardinis, Tommaso Santagata, and Jo De Waele. "Lava Tubes on Earth, Moon and Mars: A Review on Their Size and Morphology Revealed by Comparative Planetology." *Earth-Science Reviews* 209 (2020): 103288. https://doi.org/10.1016/j.earscirev.2020.103288.
Schachter, Oscar. "Who Owns the Universe?" In *Across the Space Frontier*, edited by Cornelius Ryan, 118–31. New York: Viking Press, 1952.
Scharmen, Fred. *Space Forces: A Critical History of Life in Outer Space*. Brooklyn, NY: Verso, 2021.
Schirra, Wally, and Richard Billings. *Schirra's Space*. Boston: Quinlan Press, 1988.
Schmidt, Stanley, and Robert Zubrin, eds. *Islands in the Sky: Bold New Ideas for Colonizing Space*. New York: Wiley, 1996.
Schulte-Ladbeck, Dr. Regina E. *Basics of Spaceflight for Space Exploration, Space Commercialization, and Space Colonization*. Np: RESLscience, 2016.
Schuster, Haley, and Steven L. Peck. "Mars Ain't the Kind of Place to Raise Your Kid: Ethical Implications of Pregnancy on Missions to Colonize Other Planets." *Life Sciences, Society and Policy* 12 (2016): 10. https://doi.org/10.1186/s40504-016-0043-5.
Schwartz, James S. J. "Myth-Free Space Advocacy Part I—The Myth of Innate Exploratory and Migratory Urges." *Acta Astronautica* 137 (2017): 450–60. https://doi.org/10.1016/j.actaastro.2017.05.002.
———. "Myth-Free Space Advocacy Part II: The Myth of the Space Frontier." *Astropolitics* 15 (2017): 167–84. https://doi.org/10.1080/14777622.2017.1339255.
———. *The Value of Science in Space Exploration*. New York: Oxford University Press, 2020.
Scoles, Sarah. "The Good Kind of Crazy: The Quest for Exotic Propulsion." *Scientific American*, August 1, 2019. https://doi.org/10.1038/scientificamerican0819-58.
Scott, Captain Winston E. *Reflections from Earth Orbit*. Ontario: Collector's Guide Publishing, Inc., 2005.
Seedhouse, Erik. *Lunar Outpost: The Challenges of Establishing a Human Settlement on the Moon*. Berlin; New York: Praxis, 2008.
———. *Survival and Sacrifice in Mars Exploration: What We Know from Polar Expeditions*. New York: Springer, 2015.
Serra, João Falcão. "The Legal Framework for Conventional Military Activities in Outer Space: Past, Present and Future." Thesis, Universidade NOVA de Lisboa, 2020. http://www.proquest.com/openview/5bd1a4f3e43e02d0ffa76b19eaffaa89/1.pdf?pq-origsite=gscholar&cbl=2026366&diss=y.
Sgobba, Tommaso, Barbara G. Kanki, Jean-Francois Clervoy, and Gro Sandal, eds. *Space Safety and Human Performance*. Oxford: Butterworth-Heinemann, 2017.
Sharman, Helen, and Christopher Priest. *Seize the Moment: The Autobiography of Britain's First Astronaut*. London: Orion Publishing, 1993.
Sheehan, Michael. *The International Politics of Space*. New York: Routledge, 2007.
Sheetz, Michael. "The Space Industry Will Be Worth Nearly $3 Trillion in 30 Years, Bank of America Predicts." CNBC, October 31, 2017. https://www.cnbc.com/2017/10/31/the-space-industry-will-be-worth-nearly-3-trillion-in-30-years-bank-of-america-predicts.html.
———. "Space Tourism Pioneer Dennis Tito Books Private Moon Trip on SpaceX's Starship." CNBC, October 12, 2022. https://www.cnbc.com/2022/10/12/spacex-starship-seats-space-tourism-pioneer-dennis-tito-books-private-moon-trip.html.

Shelhamer, Mark. "Enabling and Enhancing Human Health and Performance for Mars Colonies: Smart Spacecraft and Smart Habitats." In *The Human Factor in a Mission to Mars: An Interdisciplinary Approach*, edited by Konrad Szocik, 59–67. Space and Society. Cham, Switzerland: Springer International Publishing, 2019. https://doi.org/10.1007/978-3-030-02059-0_3.

Sherr, Lynn. *Sally Ride: America's First Woman in Space*. New York: Simon & Schuster, 2014.

Sherwood, Brent. "Lunar Architecture and Urbanism." In *Out of This World: The New Field of Space Architecture*, 317–30. Library of Flight. Reston, VA: American Institute of Aeronautics and Astronautics, 2009. https://doi.org/10.2514/5.9781563479878.0317.0330.

Shiba, Dai, Hiroyasu Mizuno, Akane Yumoto, Michihiko Shimomura, Hiroe Kobayashi, Hironobu Morita, Miki Shimbo, et al. "Development of New Experimental Platform 'MARS'—Multiple Artificial-Gravity Research System—to Elucidate the Impacts of Micro/Partial Gravity on Mice." *Scientific Reports* 7 (2017): 10837. https://doi.org/10.1038/s41598-017-10998-4.

Shipman, Harry L. *Humans in Space: 21st Century Frontiers*. New York: Plenum, 1989.

Siddiqi, Asif A. *Challenge to Apollo: The Soviet Union and the Space Race*. Washington, D.C.: National Aeronautics and Space Administration, 2000.

———. *The Red Rockets' Glare: Spaceflight and the Russian Imagination, 1857–1957*. Cambridge: Cambridge University Press, 2014.

Sides, Hampton. *In the Kingdom of Ice: The Grand and Terrible Polar Voyage of the USS Jeannette*. New York: Anchor Books, 2015.

Sidorchik, Andrey. "Stabbing in Antarctica. What Happened at Bellingshausen Station?" AiF, October 24, 2018. https://aif.ru/incidents/ponozhovshchina_v_antarktide_chto_proizoshlo_na_stancii_bellinsgauzen.

Silverstone, Sally. *Eating In: From the Field to the Kitchen in Biosphere 2*. Oracle, AZ: Biosphere Press, 1993.

Simonsen, Lisa C., John W. Wilson, Myung H. Kim, and Francis A. Cucinotta. "Radiation Exposure for Human Mars Exploration." *Health Physics* 79 (2000): 515–25. https://journals.lww.com/health-physics/Abstract/2000/11000/Radiation_Exposure_For_Human_Mars_Exploration.8.aspx.

Slayton, Deke, and Michael Cassutt. *Deke! U.S. Manned Space from Mercury to the Shuttle*. New York: Forge Books, 1994.

Smith, Cameron M. *Principles of Space Anthropology: Establishing a Science of Human Space Settlement*. Edited by Douglas A. Vakoch. Cham, Switzerland: Springer, 2019.

Smith, Kelly C. "Cultural Evolution and the Colonial Imperative." In *Dissent, Revolution and Liberty Beyond Earth*, edited by Charles S. Cockell, 169–87. Space and Society. Cham, Switzerland: Springer International Publishing, 2016. https://doi.org/10.1007/978-3-319-29349-3_13.

Smithsonian. "What Really Is Astronaut Food?" National Air and Space Museum, November 8, 2021. https://airandspace.si.edu/stories/editorial/what-really-astronaut-food.

Solomon, Scott. "Evolutionary Lessons for an Interplanetary Future." Room—The Space Journal, 2020. https://room.eu.com/article/evolutionary-lessons-for-an-interplanetary-future.

———. *Future Humans: Inside the Science of Our Continuing Evolution*. New Haven, CT: Yale University Press, 2016.

"The Space Race Is Dominated by New Contenders." *The Economist*, October 18, 2019. https://www.economist.com/graphic-detail/2018/10/18/the-space-race-is-dominated-by-new-contenders.

Spielmann, Peter J. "FBI Agents to Visit Antarctica in Rare Investigation of Assault." *Spokesman-Review*, October 14, 1996. https://www.spokesman.com/stories/1996/oct/14/fbi-agents-to-visit-antarctica-in-rare/.

Spudis, Paul D. *The Value of the Moon: How to Explore, Live, and Prosper in Space Using the Moon's Resources*. Washington, D.C.: Smithsonian Books, 2016.

Staff. "Russian Space Station Module Has Extra Topping." *Guardian*, July 12, 2000. https://www.theguardian.com/science/2000/jul/12/spaceexploration.

Statista. "Number of Active Satellites by Year 1957–2021." Accessed October 12, 2022. https://www.statista.com/statistics/897719/number-of-active-satellites-by-year/.

Steger, Manfred B. *Globalization: A Very Short Introduction*. New York: Oxford University Press, 2020.

Steller, Jon G., Jeffrey R. Alberts, and April E. Ronca. "Oxidative Stress as Cause, Consequence, or Biomarker of Altered Female Reproduction and Development in the Space Environment." *International Journal of Molecular Sciences* 19 (2018): 3729. https://doi.org/10.3390/ijms19123729.

Stepanova, Anastasia. "How Russian Space Food Has Evolved over the Years." Russia Beyond, June 24, 2018. https://www.rbth.com/russian-kitchen/328572-russian-space-food-evolution.

Stewart, Will. "Antarctic Scientist 'Stabs Colleague Who Kept Telling Him the Endings of Books He Was Reading on Remote Research Station.'" *The Sun*, October 30, 2018. https://www.thesun.co.uk/news/7615571/antarctic-scientist-stabs-colleague-who-kept-telling-him-endings-of-books-he-was-reading/.

Stine, G. Harry. *The Third Industrial Revolution*. New York: Ace Books, 1979.

Stine, Harry G. *Halfway to Anywhere: Achieving America's Destiny in Space*. New York: M. Evans & Company, 1996.

PARTIAL BIBLIOGRAPHY

Stone, Brad. *Amazon Unbound: Jeff Bezos and the Invention of a Global Empire*. New York: Simon & Schuster, 2021.

Stott, Nicole. *Back to Earth: What Life in Space Taught Me About Our Home Planet—And Our Mission to Protect It*. New York: Seal Press, 2021.

Strange, Clayton. *Monotown: Urban Dreams Brutal Imperatives*. San Francisco: Applied Research & Design, 2019.

Stross, Charlie. "Insufficient Data." *Antipope* (blog). July 23, 2010. http://www.antipope.org/charlie/blog-static/2010/07/insufficient-data.html.

Stuster, Jack. *Bold Endeavors: Lessons from Polar and Space Exploration*. Annapolis, MD: Naval Institute Press, 2011.

Suedfeld, Peter, Jelena Brcic, Phyllis J. Johnson, and Vadim Gushin. "Personal Growth Following Long-Duration Spaceflight." *Acta Astronautica* 79 (2012): 118–23. https://doi.org/10.1016/j.actaastro.2012.04.039.

Szocik, Konrad, ed. *Human Enhancements for Space Missions: Lunar, Martian, and Future Missions to the Outer Planets*. Space and Society. Cham, Switzerland: Springer, 2020.

———. "Human Place in the Outer Space: Skeptical Remarks." In *The Human Factor in a Mission to Mars: An Interdisciplinary Approach*, edited by Konrad Szocik, 233–52. Space and Society. Cham, Switzerland: Springer International Publishing, 2019. https://doi.org/10.1007/978-3-030-02059-0_14.

Szocik, Konrad, Steven Abood, Chris Impey, Mark Shelhamer, Jacob Haqq-Misra, Erik Persson, Lluis Oviedo, et al. "Visions of a Martian Future." *Futures* 117 (2020): 102514. https://doi.org/10.1016/j.futures.2020.102514.

Taco Bell. "Taco Bell, Our History," 2022. https://www.tacobell.com/history.

Tako, Y., S. Tsuga, T. Tani, R. Arai, O. Komatsubara, and M. Shinohara. "One-Week Habitation of Two Humans in an Airtight Facility with Two Goats and 23 Crops—Analysis of Carbon, Oxygen, and Water Circulation." *Advances in Space Research* 41 (2008): 714–24. https://doi.org/10.1016/j.asr.2007.09.023.

Tako, Yasuhiro, Ryuji Arai, Sho-ichi Tsuga, Osamu Komatsubara, Tsuyoshi Masuda, Susumu Nozoe, and Keiji Nitta. "CEEF: Closed Ecology Experiment Facilities." *Gravitational and Space Research* 23 (2010): 13–24. http://gravitationalandspaceresearch.org/index.php/journal/article/view/489.

Taylor, Alan. "The Secret City." *The Atlantic*, June 25, 2012. https://www.theatlantic.com/photo/2012/06/the-secret-city/100326/.

Taylor, Andrew J., Jonathan D. Beauchamp, Loïc Briand, Martina Heer, Thomas Hummel, Christian Margot, Scott McGrane, Serge Pieters, Paola Pittia, and Charles Spence. "Factors Affecting Flavor Perception in Space: Does the Spacecraft Environment Influence Food Intake by Astronauts?" *Comprehensive Reviews in Food Science and Food Safety* 19 (2020): 3439–75. https://doi.org/10.1111/1541-4337.12633.

Terdiman, Daniel. "How NASA Reinvented the Tortilla, and Other Tales of Food in Space." CNET, July 7, 2014. https://www.cnet.com/science/houston-we-have-a-tortilla-problem/.

Thompson, Eric. *On Her Majesty's Nuclear Service*. Oxford: Casemate, 2018.

Tomlinson, Don E., and Rob L. Wiley. "People Do Read Large Ads: The Law of Advertising from Outer Space." *Federal Communications Law Journal* 47 (1995): 535–69. https://www.repository.law.indiana.edu/fclj/vol47/iss3/3/.

Torchinsky, Rina. "Elon Musk Hints at a Crewed Mission to Mars in 2029." NPR, March 17, 2022. https://www.npr.org/2022/03/17/1087167893/elon-musk-mars-2029.

Trappe, Scott, David Costill, Philip Gallagher, Andrew Creer, Jim R. Peters, Harlan Evans, Danny A. Riley, and Robert H. Fitts. "Exercise in Space: Human Skeletal Muscle After 6 Months Aboard the International Space Station." *Journal of Applied Physiology* 106 (2009): 1159–68. https://doi.org/10.1152/japplphysiol.91578.2008.

Trousselard, Marion, Damien Leger, Pascal van Beers, Olivier Coste, Arnaud Vicard, Julien Pontis, Sylvain-Nicolas Crosnier, and Mounir Chennaoui. "Sleeping Under the Ocean: Despite Total Isolation, Nuclear Submariners Maintain Their Sleep and Wake Patterns Throughout Their Under Sea Mission." *PLOS ONE* 10 (2015): e0126721. https://doi.org/10.1371/journal.pone.0126721.

Turkina, Olessya. *Soviet Space Dogs*. Translated by Inna Cannon and Lisa Wasserman. London FUEL Design & Publishing, 2014.

Turner, Frederick Jackson, and John Mack Faragher. *Rereading Frederick Jackson Turner: The Significance of the Frontier in American History and Other Essays*. New York: Henry Holt & Co, 1994.

Turrini, Paolo. "The Sky's Not the Limit: Legal Bonds and Boundaries in Claiming Sovereignty over Celestial Bodies." In *Borders, Legal Spaces and Territories in Contemporary International Law: Within and Beyond*, edited by Tommaso Natoli and Alice Riccardi, 173–209. Cham, Switzerland: Springer International Publishing, 2019. https://doi.org/10.1007/978-3-030-20929-2_7.

United Press International. "Soyuz to Lift Off After Word from Sponsor." *Los Angeles Times*, November 30, 1990s. https://www.latimes.com/archives/la-xpm-1990-11-30-mn-5835-story.html.

PARTIAL BIBLIOGRAPHY

United States of America, Appellee, v. Mario Jaime Escamilla, Appellant, 467 F.2d 341 (4th Cir. 1972), No. 71-1575 (US Court of Appeals for the Fourth Circuit August 17, 1972).

United States Environmental Protection Agency. "What Is the Average Level of Carbon Monoxide in Homes?" Overviews and Factsheets. EPA, December 30, 2021. https://www.epa.gov/indoor-air-quality-iaq/what-average-level-carbon-monoxide-homes.

Usachev, Yuri. *Cosmonaut's Diary. Three Lives in Space.* Russia: Geleos, 2004. https://libking.ru/books/sci-/sci-history/147744-yuriy-usachev-dnevnik-kosmonavta-tri-zhizni-v-kosmose.html.

Vakoch, Douglas A., ed. *On Orbit and Beyond: Psychological Perspectives on Human Spaceflight.* New York: Springer, 2012.

———, ed. *Psychology of Space Exploration: Contemporary Research in Historical Perspective.* NASA History Series. Middletown, DE: U.S. Government Printing Office, 2020.

Van Ombergen, Angelique, Andrea Rossiter, and Thu Jennifer Ngo-Anh. "'White Mars'—Nearly Two Decades of Biomedical Research at the Antarctic Concordia Station." *Experimental Physiology* 106 (2021): 6–17. https://doi.org/10.1113/EP088352.

Vance, Ashlee. *Elon Musk: Tesla, SpaceX, and the Quest for a Fantastic Future.* New York: Avon Books, 2016.

Vasquez, John A. *The War Puzzle Revisited.* Cambridge Studies in International Relations. Cambridge: Cambridge University Press, 2009. https://doi.org/10.1017/CBO9780511627224.

Vida, István Kornél. "The 'Great Moon Hoax' of 1835." *Hungarian Journal of English and American Studies* 18 (2012): 431–41. http://www.jstor.org/stable/43488485.

Vigo, Daniel E., Francis Tuerlinckx, Barbara Ogrinz, Li Wan, Guido Simonelli, Evgeny Bersenev, Omer Van den Bergh, and André E. Aubert. "Circadian Rhythm of Autonomic Cardiovascular Control During Mars500 Simulated Mission to Mars." *Aviation, Space, and Environmental Medicine* 84 (2013): 1023–28. https://doi.org/10.3357/ASEM.3612.2013.

Virts, Terry. *How to Astronaut: An Insider's Guide to Leaving Planet Earth.* New York: Workman Publishing Co., 2020.

Vyborny, Lee, and Don Davis. *Dark Waters: An Insider's Account of the NR-1, the Cold War's Undercover Nuclear Sub.* New York: New American Library, 2004.

Wagner, R. V., and M. S. Robinson. "Lunar Pit Morphology: Implications for Exploration." *Journal of Geophysical Research: Planets* 127 (2022): e2022JE007328. https://doi.org/10.1029/2022JE007328.

Wang, Yue, Xiaolu Jing, Ke Lv, Bin Wu, Yanqiang Bai, Yuejia Luo, Shanguang Chen, and Yinghui Li. "During the Long Way to Mars: Effects of 520 Days of Confinement (Mars500) on the Assessment of Affective Stimuli and Stage Alteration in Mood and Plasma Hormone Levels." *PLoS ONE* 9 (2014). https://doi.org/10.1371/journal.pone.0087087.

Wanjek, Christopher. *Spacefarers: How Humans Will Settle the Moon, Mars, and Beyond.* Cambridge, MA: Harvard University Press, 2020.

Wattles, Jackie. "Colonizing Mars Could Be Dangerous and Ridiculously Expensive. Elon Musk Wants to Do It Anyway." CNN, September 8, 2020. https://www.cnn.com/2020/09/08/tech/spacex-mars-profit-scn/index.html.

Weeden, Brian, and Victoria Samson, eds. *Global Counterspace Capabilities: An Open Source Assessment.* Broomfield, CO: Secure World Foundation, 2022. https://swfound.org/media/207344/swf_global_counterspace_capabilities_2022.pdf.

Weisberger, Mandy. "No Evidence Russian Engineer Stabbed Antarctica Colleague for Spoiling Book Endings." livescience.com, November 5, 2018. https://www.livescience.com/64012-antarctica-stabbing.html.

White, Frank. *The Overview Effect: Space Exploration and Human Evolution.* Reston, VA: American Institute of Aeronautics and Astronautics, Inc., 2014.

White, Neil. *Company Towns: Corporate Order and Community.* Toronto: University of Toronto Press, 2012.

White, Richard. *"It's Your Misfortune and None of My Own": A New History of the American West.* Norman: University of Oklahoma Press, 1993.

Wilford, John Noble. "Skylab Astronauts Are Reprimanded in 1st Day Aboard." *New York Times*, November 18, 1973. https://www.nytimes.com/1973/11/18/archives/skylab-astronauts-are-reprimanded-in-1st-day-aboard-learned-from.html.

Williams, Dave. "Mercury Fact Sheet." NASA, December 23, 2021. https://nssdc.gsfc.nasa.gov/planetary/factsheet/mercuryfact.html.

Williams, Jeffrey N. *The Work of His Hands.* St. Louis, MO: Concordia Publishing, 2010.

Winchester, Simon. *Land: How the Hunger for Ownership Shaped the Modern World.* New York: Harper, 2021.

Wohlforth, Charles, and Amanda R. Hendrix. *Beyond Earth: Our Path to a New Home in the Planets.* New York: Pantheon Books, 2016.

Wolfe, Tom. *The Right Stuff.* New York: Picador, 2008.

PARTIAL BIBLIOGRAPHY

Wolverton, Mark. *Burning the Sky: Operation Argus and the Untold Story of the Cold War Nuclear Tests in Outer Space.* New York: Overlook Press / Harry N. Abrams, 2018.

Woodmansee, Laura S. *Sex in Space.* Ontario: Collector's Guide Publishing, Inc., 2006.

World Nuclear Association. "Nuclear Reactors and Radioisotopes for Space." World Nuclear Association, May 2021. https://world-nuclear.org/information-library/non-power-nuclear-applications/transport/nuclear-reactors-for-space.aspx#:~:text=Radioisotope%20power%20sources%20have%20been,source%20for%20deep%20space%20missions.

———. "Plutonium." World Nuclear Association, April 2021. https://world-nuclear.org/information-library/nuclear-fuel-cycle/fuel-recycling/plutonium.aspx.

Wotring, Virginia E. "Medication Use by U.S. Crewmembers on the International Space Station." *FASEB Journal* 29 (2015): 4417–23. https://doi.org/10.1096/fj.14-264838.

Wright, Rebecca. "Edited Oral History Transcript: Norman E. Thagard." NASA Shuttle-Mir Oral History Project, September 16, 1998. https://historycollection.jsc.nasa.gov/JSCHistoryPortal/history/oral_histories/Shuttle-Mir/ThagardNE/ThagardNE_9-16-98.htm.

Yazawa, Yuuki, Akira Yamaguchi, and Hiroshi Takeda. "Lunar Minerals and Their Resource Utilization with Particular Reference to Solar Power Satellites and Potential Roles for Humic Substances for Lunar Agriculture." In *Moon: Prospective Energy and Material Resources,* edited by Viorel Badescu, 105–38. Berlin, Heidelberg: Springer, 2012. https://doi.org/10.1007/978-3-642-27969-0_5.

Young, John W., and James R. Hansen. *Forever Young: A Life of Adventure in Air and Space.* Gainesville: University Press of Florida, 2013.

Zak, Anatoly. "A Rare Look at the Russian Side of the Space Station." *Air & Space Magazine,* September 2015. https://www.airspacemag.com/space/rare-look-russian-side-space-station-180956244/.

———. "Here Is the Soviet Union's Secret Space Cannon." *Popular Mechanics,* November 16, 2015. https://www.popularmechanics.com/military/weapons/a18187/here-is-the-soviet-unions-secret-space-cannon/.

Zeitlin, C., L. Narici, R. R. Rios, A. Rizzo, N. Stoffle, D. M. Hassler, B. Ehresmann, et al. "Comparisons of High-Linear Energy Transfer Spectra on the ISS and in Deep Space." *Space Weather* 17 (2019): 396–418. https://doi.org/10.1029/2018SW002103.

Zell, Holly. "Halloween Storms of 2003 Still the Scariest." NASA. May 12, 2015. http://www.nasa.gov/topics/solarsystem/features/halloween_storms.html.

Zhang, Li-Fan, and Alan R. Hargens. "Spaceflight-Induced Intracranial Hypertension and Visual Impairment: Pathophysiology and Countermeasures." *Physiological Reviews* 98 (2017): 59–87. https://doi.org/10.1152/physrev.00017.2016.

Zimmerman, Robert. *Leaving Earth: Space Stations, Rival Superpowers, and the Quest for Interplanetary Travel.* Washington, D.C.: Joseph Henry Press, 2004.

Zubrin, Robert. *The Case for Space: How the Revolution in Spaceflight Opens Up a Future of Limitless Possibility.* Amherst, NY: Prometheus, 2019.

———."The Economic Viability of Mars Colonization." In *Deep Space Commodities: Exploration, Production and Trading,* edited by Tom James, 159–80. Cham, Switzerland: Springer International Publishing, 2018. https://doi.org/10.1007/978-3-319-90303-3_12.

———. "Mars Direct 2.0: How to Send Humans to Mars Using Starships." Presentation at the 70th International Astronautical Congress (IAC), Washington, D.C., October 20–26, 2019.

Zubrin, Robert, and Richard Wagner. *The Case for Mars: The Plan to Settle the Red Planet and Why We Must.* New York: Free Press, 2021.

Zupagrafika. *Monotowns.* Poznan, Poland: Zupagrafika, 2021.

Index

Italicized page numbers indicate material in tables or illustrations.

INDEX